AF806509

Atlas of Terrestrial Mammal Limbs

Atlas of Terrestrial Mammal Limbs

Christine Böhmer
Postdoctoral Researcher
Museum National d'Histoire Naturelle
Paris, France

Jean-Christophe Theil
Osteological Preparation Technician in Charge of Dissections
Museum National d'Histoire Naturelle
Paris, France

Anne-Claire Fabre
Researcher Co-Investigator, Life Sciences
The Natural History Museum
London, United Kingdom

Anthony Herrel
Director of Research at the French National Centre for Scientific Research
Museum National d'Histoire Naturelle
Paris, France

CRC Press
Taylor & Francis Group
Boca Raton London New York

CRC Press is an imprint of the
Taylor & Francis Group, an **informa** business

Illustrator of original cover drawing: Hermann Dittrich, 1905
Permalink: https://digital.library.wisc.edu/1711.dl/FZB3CO27UDVU58J

CRC Press
Taylor & Francis Group
6000 Broken Sound Parkway NW, Suite 300
Boca Raton, FL 33487-2742

© 2020 by Taylor & Francis Group, LLC
CRC Press is an imprint of Taylor & Francis Group, an Informa business

No claim to original U.S. Government works

Printed on acid-free paper

International Standard Book Number-13: 978-1-138-70590-6 (Hardback)

This book contains information obtained from authentic and highly regarded sources. Reasonable efforts have been made to publish reliable data and information, but the author and publisher cannot assume responsibility for the validity of all materials or the consequences of their use. The authors and publishers have attempted to trace the copyright holders of all material reproduced in this publication and apologize to copyright holders if permission to publish in this form has not been obtained. If any copyright material has not been acknowledged, please write and let us know so we may rectify in any future reprint.

Except as permitted under U.S. Copyright Law, no part of this book may be reprinted, reproduced, transmitted, or utilized in any form by any electronic, mechanical, or other means, now known or hereafter invented, including photocopying, microfilming, and recording, or in any information storage or retrieval system, without written permission from the publishers.

For permission to photocopy or use material electronically from this work, please access www.copyright.com (http://www.copyright.com/) or contact the Copyright Clearance Center, Inc. (CCC), 222 Rosewood Drive, Danvers, MA 01923, 978-750-8400. CCC is a not-for-profit organization that provides licenses and registration for a variety of users. For organizations that have been granted a photocopy license by the CCC, a separate system of payment has been arranged.

Trademark Notice: Product or corporate names may be trademarks or registered trademarks, and are used only for identification and explanation without intent to infringe.

Visit the Taylor & Francis Web site at
http://www.taylorandfrancis.com

and the CRC Press Web site at
http://www.crcpress.com

In memory of our colleague Stéphane Peigné.

Photo and Illustration Credits

All illustrations by C. Böhmer.

LIST OF PHOTO CONTRIBUTORS:

Photos *Acinonyx jubatus* (Cheetah): C. Böhmer.
Photos *Bradypus tridactylus* (Pale-throated three-toed sloth): C. Böhmer.
Photos *Capreolus capreolus* (European roe deer): C. Böhmer.
Photos *Chiropotes satanas* (Brown-bearded saki): C. Böhmer.
Photos *Cryptoprocta ferox* (Fossa): C. Böhmer.
Photos *Cuon alpinus* (Dhole): C. Böhmer.
Photos *Dasyurus viverrinus* (Eastern quoll): J-C. Theil
Photos *Elephantulus brachyrhynchus* (Short-snouted elephant shrew): J-C. Theil.
Photos *Erinaceus europaeus* (European hedgehog): C. Böhmer.
Photos *Herpestes auropunctatus* (Small Indian mongoose): C. Böhmer.
Photos *Hyaena hyaena* (Striped hyena): C. Böhmer.
Photos *Laonastes aenigmamus* (Laotian rock rat): C. Böhmer.
Photos *Manis tricuspis* (Tree pangolin): J-C. Theil.
Photos *Martes foina* (European stone marten): C. Böhmer.
Photos *Meles meles* (Eurasian badger): C. Böhmer.
Photos *Meriones unguiculatus* (Mongolian gerbil): C. Böhmer.
Photos *Mesocricetus auratus* (Golden hamster): C. Böhmer.
Photos *Nasua nasua* (South American coati): C. Böhmer.
Photos *Octodon degus* (Degu): C. Böhmer.
Photos *Oryctolagus cuniculus* (European rabbit): C. Böhmer.
Photos *Panthera leo* (Lion): C. Böhmer.
Photos *Pedetes capensis* (Springhare): J-C. Theil.
Photos *Philander opossum* (Gray four-eyed opossum): A-C. Fabre/A. Herrel.
Photos *Potos flavus* (Kinkajou): C. Böhmer.
Photos *Rattus norvegicus* (Norway rat): C. Böhmer.
Photos *Sciurus vulgaris* (Eurasian red squirrel): C. Böhmer.
Photos *Sus scrofa* (Wild boar): J-C. Theil.
Photos *Vulpes vulpes* (Red fox): C. Böhmer.

Contents

PART III

Preface

Specialized anatomical works for individual species and veterinary anatomy textbooks for domestic animals exist, but to date we lack a clear and visual reference book of the muscular anatomy in a diversity of terrestrial mammals. In this atlas, the musculature of 28 different mammalian species is presented in a series of standardized, black-and-white and full-color photographs of detailed dissections. The aim of these dissections and photographs is to reveal the topographical anatomy of the animals allowing for comparison across phylogenetically and ecologically diverse taxa. We targeted terrestrial mammals with different locomotor ecologies highlighting features which are of particular interest in the study of the limbs. Along with feeding and reproduction, locomotion is a key functional capacity of animals. Movement is powered by muscles, and consequently, differences in muscular anatomy in the limbs may reflect differences in locomotor mode, such as running, climbing, digging or hopping.

This book is intended to be of interest to students, lecturers and researchers studying comparative anatomy, functional morphology and veterinary medicine. Due to my paleontological background, the intention of this project was also to create an atlas that would aid paleontologists in their study of anatomy and when inferring muscles in extinct animals only known from fossil bones.

I would like to thank my co-authors for this pleasant collaboration, especially, for their remarkable patience when I was trying to take the perfect picture. It paid off! In particular, I owe a great debt of thanks to Anthony for the opportunity to publish this atlas. I deeply acknowledge his trust in me to handle such a huge project!

Preparing this book has been an extraordinary joy because it has allowed me to start a journey of learning and progression. I remember well a photograph of my younger sister and me dissecting our first animals (a rabbit and some pig eyes) in our family's kitchen at the age of 5 (under the careful supervision of our mother – a veterinarian). Several years later, it has been a special fortune for me to discover the anatomy of a diversity of mammals as a professional researcher and to even publish an anatomy atlas.

Composing a book is also hard work, and doubts may arise about the success. I wish to thank my family who patiently supported me in many ways during the project, including theoretical and practical help! They patiently accepted my devotion to anatomy, and over time, they got used to receiving my most recent dissection pictures.

Christine Böhmer
Paris, France/Munich, Germany

Acknowledgments

We acknowledge our dear colleague Stéphane Peigné for his unbridled enthusiasm. He sadly passed away at the beginning of this project. We thank Aurélien Lowie, Fanny Pagès, Maxime Taverne, Camille Lacroux and Kévin Le Verger for their precious assistance and inestimable help in preparing the largest specimens for dissection.

Financial support was provided by a grant from Agence Nationale de la Recherche (LabEx ANR-10-LABX-0003-BCDiv) and a Marie-Skłodowska Curie fellowship (H2020-EU.1.3.2. EU project 655694 – GETAGRIP to A-C. F).

All animals used for this work were generous donations from professional institutions or were available through the fluid collections of the Natural History Museum (MNHN) in Paris (France). We are especially grateful to Marc Herbin and Vilaine Nicolas for their support and access to specimens from the collections of the MNHN. We would like to thank Grégory Breton and Jérôme Catinaud from the Parc des Félins and the animal park of la Haute-Touche for providing specimens for the study, and Eric Pellé, Zoé Thalaud and Christophe Voisin from the taxidermy facilities for the preparation of these specimens. We also thank Thomas Bauer, Sven Reese and Estella Böhmer (Faculty of Veterinary Medicine of the Ludwig-Maximilians-Universität in Munich); Opale Robin and Benoît Clavel (INRAP Centre de Recherches Archologiques de l'Oise in Compiegne); Christophe Gottini (Taxidermy facility of the MNHN in Paris); Géraldine Veron; and Jérôme Fuchs; Natalie Dogna (Parc Zoologique de Paris) for providing us with specimens for dissection. Finally, we would like to thank Benoit de Thoisy and the association Kwata for access to specimens for dissection.

The authors are also indebted to Chuck Crumly for editorial advice and who kindly encouraged us to complete this book.

Authors

Christine Böhmer is a vertebrate paleobiologist with expertise in anatomy, evolutionary developmental biology (EvoDevo) and functional morphology. She graduated from the Technische Universität in Munich (BSc) and Ludwig-Maximilians-Universität in Munich (MSc) (Germany). After a visiting fellowship at the University of Chicago (USA), she went back to Germany and earned her doctoral degree from the Ludwig-Maximilians-Universität (PhD) in 2013 with *summa cum laude*. Subsequently, Christine spent one year as postdoctoral researcher at the RIKEN Institute in Kobe (Japan). Since 2015, she works as Postdoc at the Muséum National d'Histoire Naturelle in Paris (France). In 2019, she was a visiting professor for vertebrate paleobiology at the University of Vienna (Austria). Recently, Christine was awarded a Marie-Skłodowska Curie fellowship from the European Commission.

Anne-Claire Fabre is an evolutionary biologist and functional morphologist principally interested in shape evolution. Her research on macroevolution is highly integrative, linking different research areas in biology in order to understand the evolution of the shape of a structure in relation to development, function and behavior. She earned her PhD from University College London and did postdocs as a Fondation Fyssen fellow at Duke University (USA) and as a Marie-Skłodowska Curie fellow at the Museum National d'Histoire Naturelle (France). She is currently working as a research co-investigator at the Natural History Museum (UK).

Jean-Christophe Theil. Interested in animals, anatomy and osteology since his childhood he started an osteological collection leading him to learn more about animals. Re-assembling skulls and skeletons of all kinds of vertebrates is the crossing of a scientific and artistic process that he has come to love. During his studies, he met Anthony Herrel who introduced him to the world of muscles. After earning a Master's degree in Systematics, Evolution and Paleontology at the Muséum National d'Histoire Naturelle in Paris, he joined the FUNEVOL team as temporary dissection worker and bone preparator. He is very happy to combine his understanding of both muscles and bones which are very complementary.

Anthony Herrel is a research director of the CNRS working at the Muséum National d'Histoire Naturelle. His main interests are the evolution of the vertebrate feeding and locomotor systems. He is by training a comparative anatomist with a keen interest in functional morphology and biomechanics. He earned his PhD from the University of Antwerp in Belgium and did postdocs at the University of Antwerp in Belgium, Northern Arizona University, Tulane University and Harvard University in the US before landing a job at the CNRS. He now runs the Function and Evolution team of the UMR7179 at the Muséum in Paris.

Introduction

Studies of anatomy and more specifically of muscle have a long history. The learning of human anatomy by means of the dissection of cadavers was already common practice in Greece during the 3rd century BC and culminated with the works of Herphilus and Erasistratus (Von Staden, 1992; Ghosh, 2015). The ancient Greeks were also interested in the structure and behavior of animals, and Aristotle is generally considered as the first well-known comparative anatomist (Russell, 1982). Galen, famous for his work on human anatomy, in fact used mostly animal cadavers to describe anatomy, resulting in many errors in his books. Yet, he was the first to explicitly state that muscles were responsible for movement in both humans and animals (Russell, 1982). After this period, dissections became less and less common, and eventually, human dissection was completely banned following the introduction of Christianity in Europe. Only in the 14th century, Italy did dissection-based human anatomy come back into practice culminating in the work of Vesalius who stressed the importance of empirical observation and dissection, and pointing out the errors in the work of Galen (Joutsivuo, 1997; Ghosh, 2015). Although cadaveric dissections were mostly destined to teach medical students, artists and even the general public showed great interest in human anatomy (Mavrodi et al., 2013). Following the renaissance of anatomy, more scholars began to study the anatomy of different animals in a systematic way. Belon, a French naturalist from the 16th century, is considered by many as the forefather of modern comparative anatomy. Edward Tyson pushed research in comparative anatomy one step further and began systematic comparative studies of the anatomy of different animals, humans included. Later comparative anatomy flourished and became the foundation of the work of many great anatomists such as Cuvier, Owen, Huxley or Romer.

Over the past decade, there has been a true revival of comparative anatomy (Ashley-Ross and Gillis, 2002), partly due to the availability of novel methods and techniques including digital dissections based on contrast-enhanced μCT data or synchrotron μCT data (e.g., Cox and Jeffrey, 2011; Cox and Faulkes, 2014; Descamps et al., 2014; Lautenschlager et al., 2014; Gignac et al., 2016; Porro and Richards, 2017; Brocklehurst et al., 2019). Moreover, advances in embryology and the establishment of gene ontologies have prompted the need for better anatomical descriptions in model organisms and the establishment of muscle ontologies (DeLaurier et al., 2008; Cox and Jeffrey, 2011; Druzinsky et al., 2016; Porro and Richards, 2017; Constantinescu, 2018). Better phylogenies and a better understanding of the relationships between organisms have generated also a renewed interest in questions of homology and lead to a re-evaluation of the nomenclature used for muscles in different taxa (Diogo and Abdala, 2010; Zierman et al., 2019). Despite the recent boom of papers using digital dissection, for many animals, especially terrestrial mammals, this is unfeasible due to the size of the specimens. Moreover, the cost associated with scanning many specimens for a comparative study is prohibitive for many researchers, in addition to the difficulties of having access to CT scanners. However, carefully conducted classic dissections provide high-quality quantitative data at little or no cost, and are available to all researchers that have access to cadavers or museum collections. Moreover, the quality of the collected data is on-par with that collected by means of digital dissection, including measures of, for example, fiber lengths (Cox et al., 2012).

A comprehensive knowledge of anatomy in general, and muscular anatomy more specifically, is a necessary prerequisite for understanding of muscle function (McMahon, 1984; Basmajian and De Luca, 1985). Moreover, understanding the position, architecture and line of action of a muscle is essential for creating accurate biomechanical models (e.g., Gröning et al., 2013). Understanding and quantifying muscles, their architecture, and their insertions, allows for better and more accurate reconstructions of the function, ecology and general paleobiology of both extant and extinct animals (Argot, 2001, 2002; Lowie et al., 2018; Taverne et al., 2018, Böhmer et al., 2019). Indeed, the only remains that are available for extinct species are often bones. Many researchers in paleontology have tried to infer function from the bony material alone (e.g., Fabre et al., 2015), yet this can be difficult, especially if the relationships between muscle size, architecture and orientation are not known for closely related extant animals. As such, quantitative studies linking muscle size and architecture to bone shape (e.g., Fabre et al., 2014, 2018) are particularly promising for making better inferences in paleontology. Finally, a deep understanding of anatomy and more specifically muscle anatomy is important for creating solid data matrices allowing phylogenetic reconstructions based on anatomical traits (e.g., Diogo and Wood, 2012; Blanke and Wesener, 2014).

Surprisingly, and despite the wealth of studies devoted to mammalian limb muscle anatomy and function (e.g., Howard, 1973, 1975; Argot, 2001, 2002; Ercoli et al., 2013, 2015; Cuff, 2016a,b; Böhmer et al., 2018), no single comparative work exists illustrating the diversity in form, function and phylogeny across terrestrial mammals. By creating this atlas, we hope to fill this void and to stimulate further research on limb muscle anatomy in terrestrial mammals. To do so, we have created an atlas that provides easy and visual access to the anatomy of the musculature of the forelimb and hind limb in terrestrial mammals. Moreover, by providing tables of nomenclature used in different studies and schematic drawings on the origins and insertions of the muscles onto bones, we hope to make the comparative study of muscle anatomy more accessible to a wider range of researchers.

OVERVIEW OF TAXA

The aim of this atlas is to illustrate the diversity in muscle size and organization across terrestrial mammals encompassing both ecological (locomotor ecology and diet) and phylogenetic diversity. We provide illustrations of forelimb and hind limb muscle anatomy for 28 species of terrestrial mammals belonging to eight of the major clades of mammals and ranging from burrowers (e.g., the badger) over cursorial species (canids) to arboreal taxa. For example, we included arboreal taxa from six major groups including the gray four-eyed opossum, the three-toed sloth, the tree pangolin, the fossa, the kinkajou, the black-bearded saki and the red squirrel (**Figure 1**). For each species, we provide a general overview as well as more detailed anatomical plates of both the forelimb and hind limb. We decided not to focus on primates as several excellent atlases have been published over the past ten years (Diogo et al., 2012; 2013a,b; 2014). We also specifically did not include domestic animals as their morphology has been shaped by human selection, largely for functional or aesthetic reasons. Moreover, excellent reference works and atlases on veterinary anatomy are available in the literature (Barone, 1968; Popesko, 1972; Hudson and Hamilton, 2010).

The present atlas is organized into three main sections. Part I comprises a general introduction to limbs and related aspects such as locomotor ecology, anatomical terms of location and position, osteology (terminology of forelimb and hind limb bones to understand the muscle origins and insertions) and myology (muscle terminology; origins and insertions, musculoskeletal system-figures, muscle maps). Part II provides the species-specific and comparative dissection guide that shows and compares the muscular anatomy of the limbs in a diverse array of terrestrial mammals. Finally, Part III provides a list of synonyms and references for further reading.

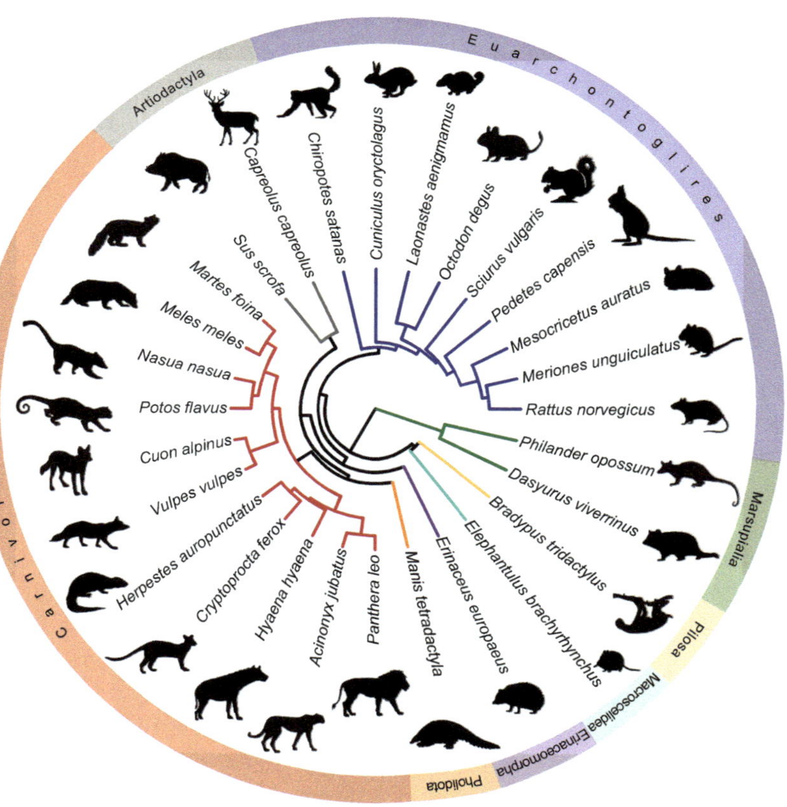

FIGURE 1 Phylogenetic relationships of the mammals dissected in this atlas. Different colors depict different clades. Phylogeny derived from the TimeTree project (Hedges et al., 2006)

REFERENCES

Argot C. 2001. Functional-adaptive anatomy of the forelimb in the Didelphidae, and the paleobiology of the Paleocene marsupials *Mayulestes ferox* and *Pucadelphys andinus*. *J. Morphol.* **247**: 51–79.

Argot C. 2002. Functional-adaptative analysis of the hind limb anatomy of extant marsupials and the paleobiology of the Paleocene marsupials *Mayulestes ferox* and *Pucadelphis andinus*. *J. Morphol.* **253**: 76–108.

Ashley-Ross M, Gillis GB. 2002. A brief history of vertebrate functional morphology. *Integr. Comp. Biol.* **42**: 183–189.

Basmajian JV, De Luca CJ. 1985. Muscles Alive: Their Functions Revealed by Electromyography. Williams and Wilkins: Baltimore, MD.

Barone R. 1968. *Anatomie compare des mammifères domestiques*. Arthrologie et myologie. Ecole Nationale Vétérinaire de Lyon: Tome Second.

Blanke A, Wesener T. 2014. Revival of forgotten characters and modern imaging techniques help to produce a robust phylogeny of the Diplopoda (Arthropoda, Myriapoda). *Arthr. Struct. Dev.* **43**: 63–75.

Böhmer C, Fabre A-C, Herbin M, Peigné S, Herrel A. 2018. Anatomical basis of differences in locomotor behavior in martens: A comparison of the forelimb musculature between two sympatric species of *Martes*. *Anat. Rec.* **301**: 449–472.

Böhmer C, Fabre A-C, Taverne M, Herbin M, Peigné S, Herrel A. 2019. Functional relationship between myology and ecology in carnivores: do forelimb muscles reflect adaptations to prehension? *Biol. J. Linn. Soc.* **127**: 661–680.

Brocklehurst R, Porro L, Herrel A, Adriaens D, Rayfield E. 2019. A digital dissection of two teleost fishes: Comparative functional anatomy of the cranial musculoskeletal system in pike (*Esox lucius*) and eel (*Anguilla anguilla*). *J. Anat.* **235**: 189–204.

Constantinescu GM. 2018. Comparative Anatomy of the Mouse and the Rat: A Color Atlas and Text. CRC Press: Boca Raton, FL.

Cox PG, Jeffery N. 2011. Reviewing the morphology of the jaw-closing musculature in squirrels, rats, and guinea pigs with contrast-enhanced microCT. *Anat. Rec.* **294**: 915–928.

Cox PG, Rayfield EJ, Fagan MJ, Herrel A, Pataky TC, Jeffery N. 2012. Functional evolution of the feeding system in rodents. *PloS ONE* **7**: e36299.

Cox PG, Faulkes CG. 2014. Digital dissection of the masticatory muscles of the naked mole-rat, *Heterocephalus glaber* (Mammalia, Rodentia). *Peer J.* **2**: e448.

Cuff AR, Sparkes EL, Randau M, Pierce SE, Kitchener AC, Goswami A, Hutchinson JR. 2016a. The scaling of postcranial muscles in cats (Felidae) I: Forelimb, cervical, and thoracic muscles. *J. Anat.* **229**: 128–141.

Cuff AR, Sparkes EL, Randau M, Pierce SE, Kitchener AC, Goswami A, Hutchinson JR. 2016b. The scaling of postcranial muscles in cats (Felidae) II: Hindlimb and lumbosacral muscles. *J. Anat.* **229**: 142–152.

DeLaurier A, Burton N, Bennett M, Baldock R, Davidson D, Mohun TJ, Logan MPO. 2008. The mouse limb anatomy atlas: An interactive 3D tool for studying embryonic limb patterning. *BMC Devel. Biol.* **8**: 83. Doi: 0.1186/1471-213X-8-83.

Descamps E, Sochacka A, De Kegel B, Van Loo D, Van Hoorebeke L, Adriaens D. 2014. Soft tissue discrimination with contrast agents using micro-CT scanning. *Belg. J. Zool.* **144**: 20–40.

Diogo R, Abdala V. 2010. Muscles of Vertebrates: Comparative Anatomy, Evolution, Homologies and Development. CRC Press: Boca Raton, FL.

Diogo R, Wood B. 2012. Comparative Anatomy and Phylogeny of Primate Muscles and Human Evolution. CRC Press: Boca Raton, FL.

Diogo R, Potau JM, Pastor JF, de Paz FJ, Ferrero EM, Bello G, Barbosa M, Ashraf Aziz M, Burrows AM, Arias-Martorell J, Wood B. 2012. Photographic and Descriptive Musculoskeletal Atlas of Gibbons and Siamangs (*Hylobates*): With Notes on the Attachments, Variations, Innervation, Synonymy and Weight of the Muscles. CRC Press: Boca Raton, FL.

Diogo R, Potau JM, Pastor JF. 2013a. Photographic and Descriptive Musculoskeletal Atlas of Chimpanzees: With Notes on the Attachments, Variations, Innervation, Function and Synonymy and Weight of the Muscles. CRC Press: Boca Raton, FL.

Diogo R, Potau JM, Pastor JF, de Paz FJ, Barbosa M, Ferrero EM, Bello G, Ashraf Aziz M, Arias-Martorell J, Wood B. 2013b. Photographic and Descriptive Musculoskeletal Atlas of Orangutans: With Notes on the Attachments, Variations, Innervations, Function and Synonymy and Weight of the Muscles. CRC Press: Boca Raton, FL.

Diogo R, Pastor JF, Hartstone-Rose A, Muchlinski MN. 2014. Baby Gorilla: Photographic and Descriptive Atlas of Skeleton, Muscles and Internal Organs. CRC Press: Boca Raton, FL.

Druzinsky RE, Balhoff JP, Crompton AW, Done J, German RZ, Haendel MA, Herrel A, Herring SW, Lapp H, Mabee PM, Muller H, Mungall CJ, Sternberg PW, Van Auken K, Vinyard CJ, Williams SH, Wall CE. 2016. Muscle logic: New knowledge resource for anatomy enables comprehensive searches of the literature on the feeding muscles of mammals. *PLoS ONE* 11(2): e0149102.

Ercoli MD, Echarri S, Busker F, Alvarez A, Morales MM, Turazzini GF. 2013. The functional and phylogenetic implications of the myology of the lumbar region, tail, hind limbs of the lesser grison (*Galictis cuja*). *J. Mamm. Evol.* 20: 309–336.

Ercoli MD, Alvarez A, Stefanini MI, Busker F, Morales MM. 2015. Muscular anatomy of the forelimbs of the lesser grison (*Galictis cuja*), and a functional and phylogenetic overview of Mustelidae and other Caniformia. *J. Mamm. Evol.* 22: 57–91.

Fabre A-C, Andrade DV, Huyghe K, Cornette R, Herrel A. 2014. Interrelationships between bones, muscles, and performance: Biting in the lizard *Tupinambis merianae*. *Evol. Biol.* **41**: 518–527.

Fabre A-C, Salesa MJ, Cornette R, Antón M, Morales J, Peigné S. 2015. Quantitative inferences on the locomotor behaviour of extinct species applied to *Simocyon batalleri* (Ailuridae, Late Miocene, Spain). *Sci. Nat.* 102: 30.

Fabre A-C, Perry JMG, Hartstone-Rose A, Lowie A, Boens A, Dumont M. 2018. Do muscles constrain skull shape evolution in Strepsirrhines? *Anat. Rec.* 301: 291–310.

Gignac PM, Kley NJ, Clarke JA, Colbert MW, Morhardt AC, Cerio D, Cost IN, Cox PG, Daza JD, Early CM, Echols MS, Henkelman RM, Herdina AN, Holliday CM, Li Z, Mahlow K, Merchant S, Müller J, Orsbon CP, Paluh DJ, Thies ML, Tsai HP, Witmer LM. 2016. Diffusible iodine-based contrast-enhanced computed tomography (diceCT): An emerging tool for rapid, high-resolution, 3-D imaging of metazoan soft tissues. *J. Anat.* 228: 889–909.

Gosh SK. 2015. Human cadaveric dissection: A historical account from ancient Greece to the modern era. Anat. *Cell Biol.* 48: 153–169.

Gröning F, Jones MEH, Curtis N, Herrel A, O'Higgins P, Evans SE, Fagan MJ. 2013. The importance of accurate muscle modelling for biomechanical analyses: A case study with a lizard skull. *J. R. Soc. Interface* 10: 20130216.

Hedges SB, Dudley J, & Kumar S (2006). TimeTree: A public knowledge-base of divergence times among organisms. *Bioinformatics* 22: 2971–2972.

Howard LD. 1973. Muscular anatomy of the forelimb of the sea otter (*Enhydra lutris*). *Proc. Calif. Acad. Sci.* 39: 411–500.

Howard LD. 1975. Muscular anatomy of the hind limb of the sea otter (*Enhydra lutris*). *Proc. Calif. Acad. Sci.* 40: 335–416.

Hudson L, Hamilton W. 2010. Atlas of Feline Anatomy for Veterinarians. CRC Press: Boca Raton, FL.

Joutsivuo T. 1997. Vesalius and De humani corporis fabrica: Galen's errors and the change of anatomy in the sixteenth century. *Hippokrates* 1997: 98–112.

Lautenschlager S, Bright JA, Rayfield EJ. 2014. Digital dissection–using contrast-enhanced computed tomography scanning to elucidate hard- and soft-tissue anatomy in the common buzzard *Buteo buteo*. *J. Anat.* 224: 412–431.

Lowie A, Herrel A, Abdala V, Manzano AS, Fabre A-C. 2018. Does the morphology of the forelimb flexor muscles differ between lizards using different habitats? *Anat. Rec.* 301: 424–433.

Mavrodi A, Paraskevasa G, Kitsoulis P. 2013. The history and the art of anatomy: a source of inspiration even nowadays. *Ital. J. Anat. Embryol.* 118: 267–276.

McMahon TA. 1984. Muscles, Reflexes, and Locomotion. Princeton University Press: Princeton.

Popesko P. 1972. Atlas of Topographic Anatomy of Domestic Animals. Priroda: Bratislava.

Porro LB, Richards CT. 2017. Digital dissection of the model organism *Xenopus laevis* using contrast-enhanced computed tomography. *J. Anat.* 231: 169–191.

Russell ES. 1982. Form and Function: A Contribution to the History of Animal Morphology. 2nd ed. University of Chicago Press: Chicago, IL.

Taverne M, Fabre A-C, Herbin M, Herrel A, Lacroux C, Lowie A, Pagès F, Peigné S, Theil J-C, Böhmer C. 2018. Convergence in the functional properties of forelimb muscles in carnivorans: Adaptations to an arboreal lifestyle? *Biol. J. Linn. Soc.* 125: 250–263.

von Staden H. 1992. The discovery of the body: Human dissection and its cultural contexts in ancient Greece. *Yale J. Biol. Med.* 65: 223–241.

Zierman JM, Diaz Jr RE, Diogo R. 2019. Heads, Jaws, and Muscles. Anatomical, Functional and Developmental Diversity in Chordate Evolution. Springer Verlag: Berlin.

Part I

Limbs and Locomotion

The muscles of the appendicular apparatus function to move the limbs and serve the biological role of locomotion and body support. Locomotion has a crucial impact on the survival of animals because many relevant activities, such as foraging, escaping from predators and searching for reproductive partners, depend on how animals move. Locomotor performance is strongly affected by the physical structure of the habitat in which an animal lives. Mammals are globally distributed and can move in many different ways including terrestrial, fossorial, arboreal, aerial and aquatic locomotor modes. Acknowledging some limitations in establishing clear limits, eight general types of locomotor categories can be recognized in mammals (Table I.1.1). Primates are all very good climbers. The huge amount of literature on primate anatomy and locomotion offers a number of additional locomotor categories that highlight differences in the degree to which the forelimbs and hind limbs are used (Table I.1.2).

The assignment of animals to one of these categories is obviously not mutually exclusive. Generalists may be able to move in several different environments, and there may be intermediate forms that are adapted to move in two environments with radically divergent physical properties, such as semiaquatic mammals. Furthermore, there are also scaling effects. For small mammals, such as many rodents, the terrestrial substrate involves many vertical or steeply inclined surfaces and, consequently, they are very good at climbing although they do not necessarily live in an arboreal habitat. Nevertheless, the locomotor categories allow to appreciate the anatomy of the limbs in the light of the prevalent mode of locomotion of the animal.

TABLE I.1.1

General Locomotor Categories in Mammals

Category	Definition	Examples
Ambulatorial	Moving by walking (generalist)	Hyaena, hedgehog, racoon
Cursorial	Moving by running (specialist)	Cheetah, fox, deer
Fossorial	Moving by tunneling; adapted to burrowing and digging into the ground	Badger, mole
Natatorial	Moving by swimming	Sea otter, seal
Saltatorial	Moving by jumping (quadrupedal) or hopping (bipedal)	Rabbit, kangaroo
Scansorial	Moving by climbing (locomotion over vertical or steeply inclined surfaces)	Coati, squirrel
Suspensorial	Moving by upside-down body posture	Sloth
Volant	Moving by flying or gliding	Bat

Mammals can move in different ways in terrestrial, arboreal, aerial and aquatic environments.

TABLE I.1.2

Specialized Locomotor Categories in Primates

Category	Definition	Examples
Quadrupedal	Moving by using all four limbs (hind limb-dominated or forelimb-hind limb-equivalence)	Capuchin monkey, baboon, slender loris
Vertical clinging and leaping	Moving by jumping off from one vertical support and landing on another vertical support using the hind limbs (hind limb-dominated locomotion)	Galago, sifaka
Brachiation	Moving by swinging from branch to branch using the forelimbs (forelimb-dominated locomotion)	Gibbon
Bipedal	Moving by exclusively using the hind limbs (hind limb-dominated locomotion)	Homo

Primate locomotion can further be classified into four major modes. The differences lie principally in the degree to which the forelimbs and hind limbs are used.

Comparative Anatomy

ANATOMICAL TERMS OF LOCATION

The anatomical terms of location used in this book (Figure I.2.1) follow those of Nickel et al. (2003) and the *Nomina Anatomica Veterinaria* (NAV, 2017) prepared by the International Committee on Veterinary Anatomical Nomenclature – the standard reference for anatomical terminology in the field of veterinary science. The term **dorsal** refers to the back (dorsum) of the trunk or the corresponding surface of another body part. The opposing term **ventral** refers to the belly (venter) or the corresponding surface of another body part. **Cranial** structures lie toward the head (cranium) and **caudal** structures toward the tail (cauda). The terms cranial and caudal apply to the neck and trunk, and to the limbs proximal to the carpus and tarsus. Within the hand and foot, the term dorsal refers to the back. The opposing terms are **palmar** in the forelimb (palm of the hand) and **plantar** in the hind limb (sole of the foot).

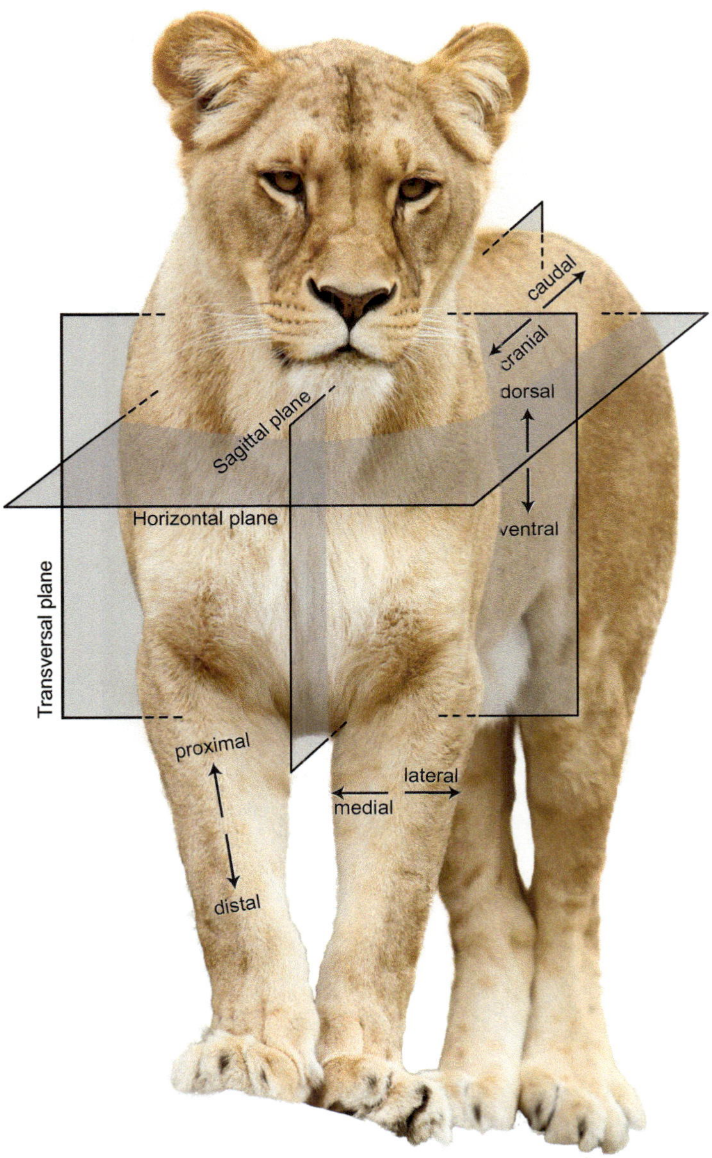

FIGURE I.2.1 **Terms.** Directional terms and anatomical planes used to describe the relative position of structures in the body.

Proximal structures are closer to a point of attachment, and **distal** structures are at a greater distance from the point of attachment. The term **lateral** refers to a structure being away from the middle of the body (toward the side of the animal). The opposing term **medial** refers to a structure lying toward the center line. The terms lateral and medial are used on the whole body. The **sagittal** plane bisects the body from cranial to caudal, dividing it into left and right parts. The **horizontal** plane divides the body from dorsal to ventral, and the **transverse** plane bisects the body laterally from side to side, dividing it into cranial and caudal parts.

LIMB SKELETON

From proximal to distal, the skeleton of the forelimb is composed of the scapula, the humerus, the radius, the ulna, and the carpal and metacarpal bones as well as the phalanges (Figures I.2.2 and I.2.3). In some mammals (e.g., rodents and primates), the clavicle is present connecting

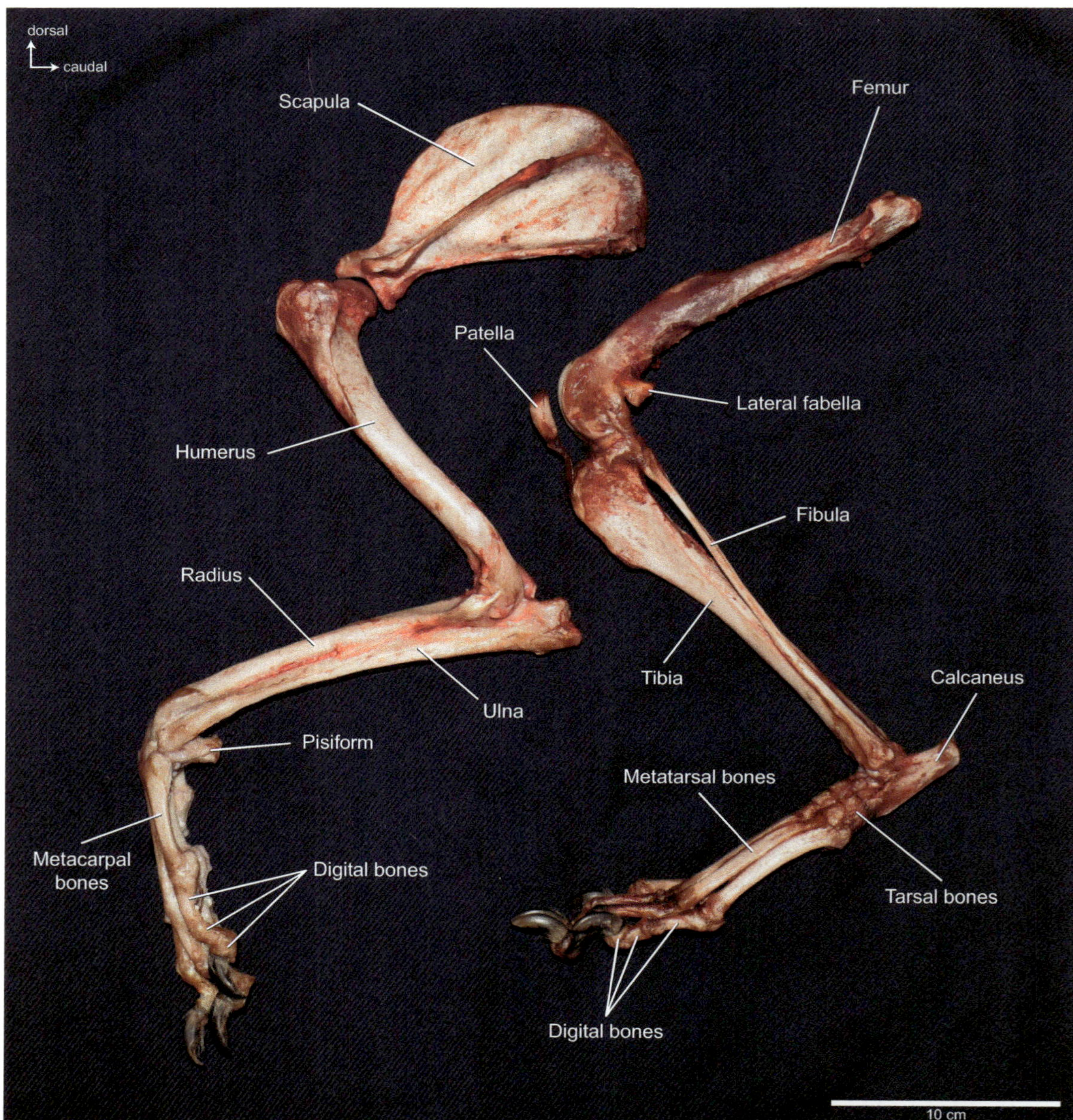

FIGURE I.2.2 **Articulated forelimb and hind limb of the dhole (*Cuon alpinus*).** Forelimb (left) and hind limb (right) in lateral view. The pisiform is a sesamoid bone found in the wrist. The lateral fabella is a sesamoid bone found in the knee of some mammals. It is embedded in a tendon caudal to the lateral condyle of the femur.

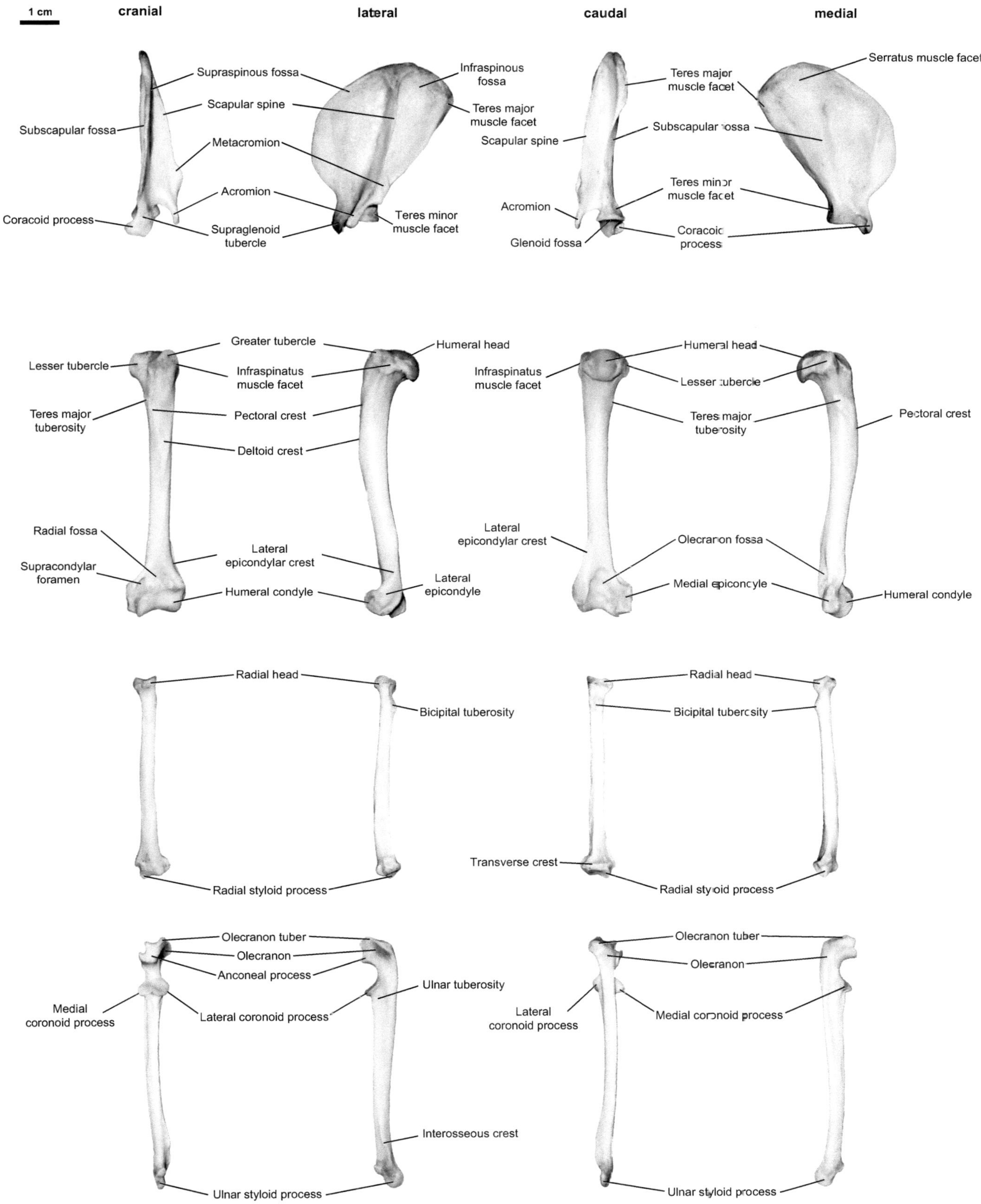

FIGURE I.2.3 Shoulder and forelimb bones of the European pine marten (*Martes martes*). From top to bottom: scapula, humerus, radius and ulna in cranial, lateral, caudal and medial view.

the sternum with the scapula. In most carnivores, the clavicle is absent, or it is a rudimentary, small bone that has lost its attachment to the sternum and/or scapula. From proximal to distal, the skeleton of the hind limb consists of the femur, the tibia, the fibula, and the tarsal and metatarsal bones as well as the phalanges (Figures I.2.2 and I.2.4). The interosseous membrane is a more or less dense fibrous sheet of connective tissue that spans the space between radius and ulna in the forelimb and between tibia and fibula in the hind limb (Figure I.2.5).

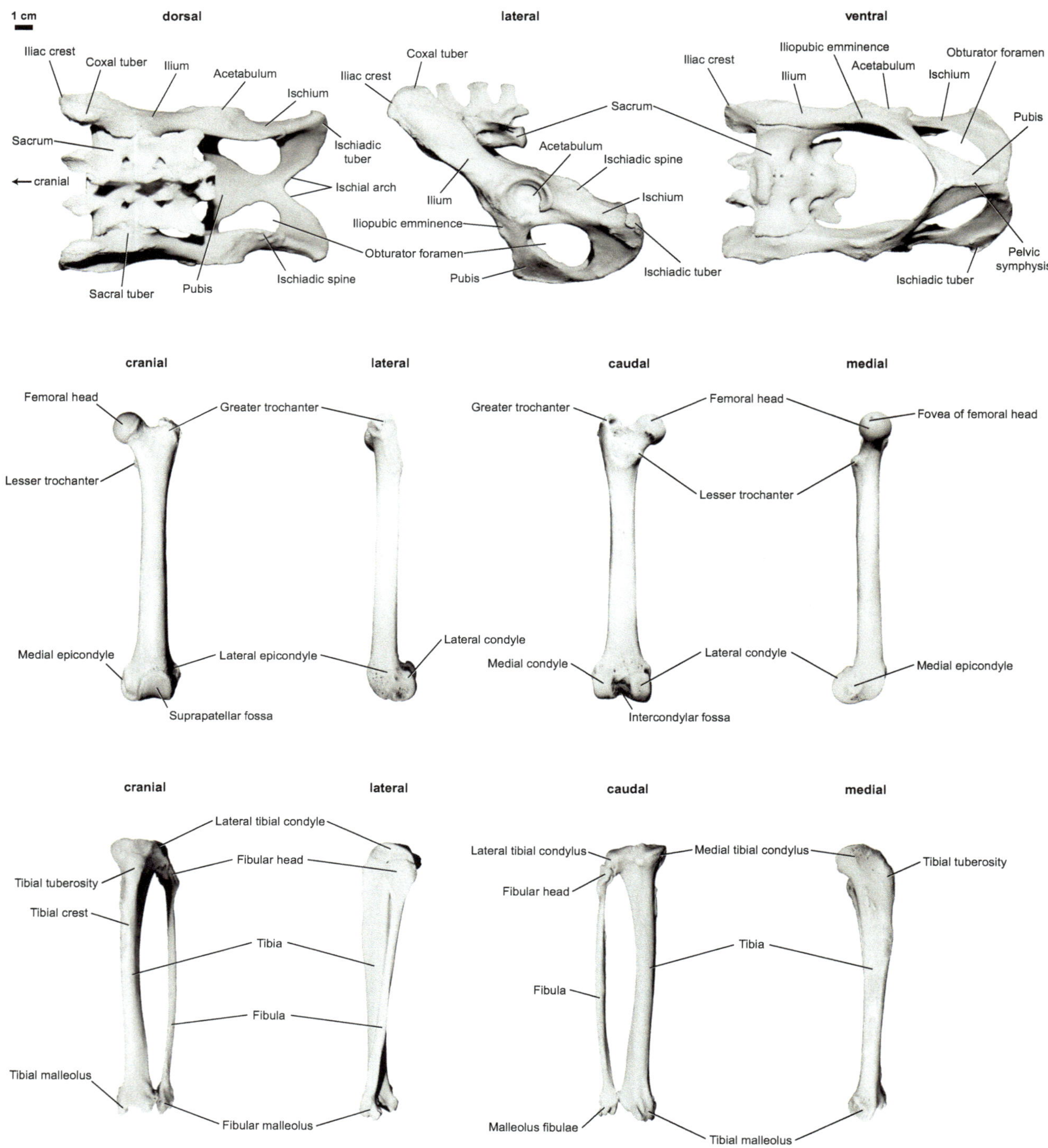

FIGURE I.2.4 **Hip and hind limb bones of the binturong (*Arctictis binturong*).** From top to bottom: pelvis (with sacrum) in dorsal, lateral and ventral view; femur, tibia and fibula in cranial, lateral, caudal and medial view.

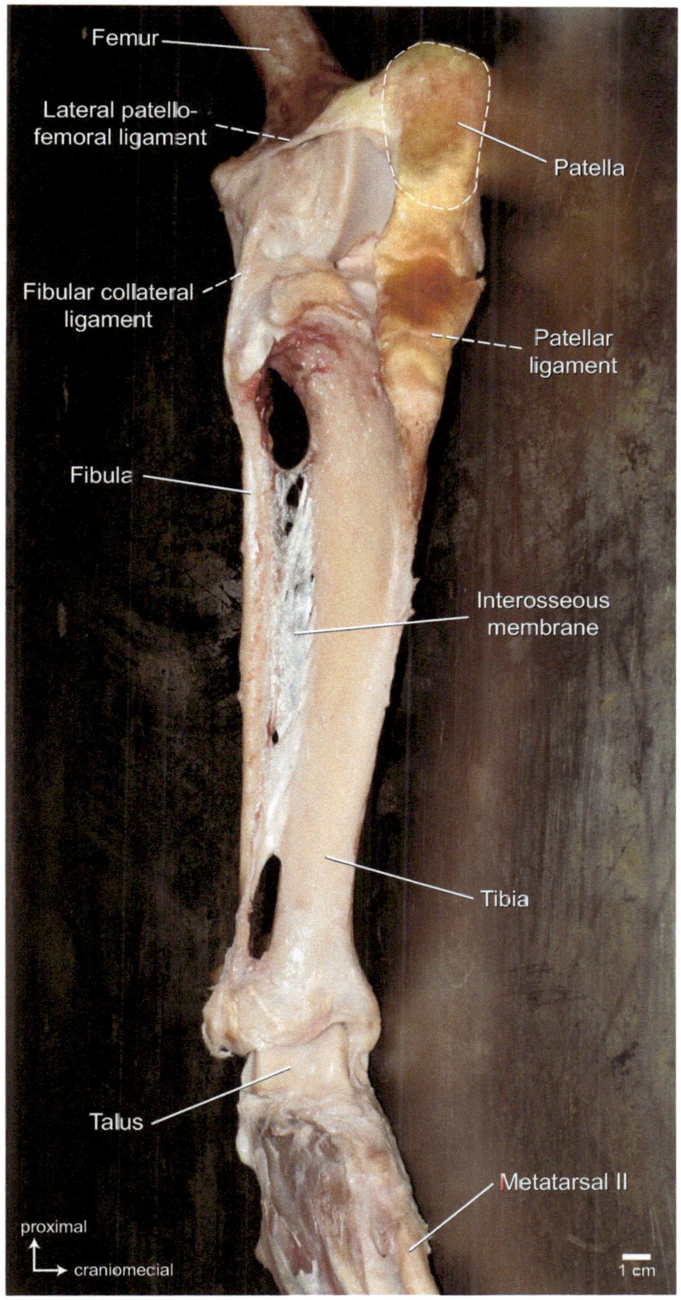

FIGURE I.2.5 Connection of the hind limb bones in the lion (*Panthera leo*). Knee joint, interosseous membrane between tibia and fibula and ankle joint in craniolateral view.

LIMB MUSCULATURE

Skeletal muscles are responsible for generating the forces that move the appendicular bones. They attach to bone at two or more places. The origin of a muscle is the attachment on a bone that remains immobile for an action. The insertion of a muscle is the attachment on a bone that moves during the action. The action of a muscle on a bone or joint depends on the distance and position of its attachment points on the bones. Several different movements can take place at the joints along three different planes. Movements that occur in sagittal plane include **extension** (increasing the joint angle) and **flexion** (decreasing the joint angle). Distal to the carpus and tarsus, these movements at the joints in the foot and hand are termed

dorsiflexion and **plantarflexion**. **Protraction** is moving a bone forward without changing the angle, and **retraction** is moving a bone backward without changing the angle. Movements that occur in horizontal plane are **adduction** (motion toward the body's midline), **abduction** (motion away from the body's midline), **elevation** (moving to a dorsal position), **depression** (moving to a ventral position), **inversion** (lifting the medial border of the foot) and **eversion** (lifting the lateral border of the foot). Movements that occur in transversal plane include **rotation** (inward or outward turning about the vertical axis), **pronation** (rotating the wrist joint medially) and **supination** (rotating the wrist joint laterally). According to their involvement in the abovementioned movements, muscles can be assigned to main functional groups. For example, the biceps muscle (BB, *M. biceps brachii*) acts as a shoulder joint extensor (increasing the angle between scapula and humerus) and elbow joint flexor (decreasing the angle between humerus and radius/ulna) (Figure I.2.6).

The following two tables provide an overview of the muscles in the limbs of mammals. The myological nomenclature essentially follows that of Nickel et al. (2003) and the *Nomina Anatomica Veterinaria* (NAV, 2017).

MUSCLE MAPS

In addition to the absence of certain muscles, the exact position and extent of a muscle's attachment area on the bone varies across taxa. The following illustrations (muscle maps) provide an overview of the musculature attaching to the shoulder and forelimb bones (Figures I.2.7–I.2.9) as well as the hip and hind limb bones (Figure I.2.10).

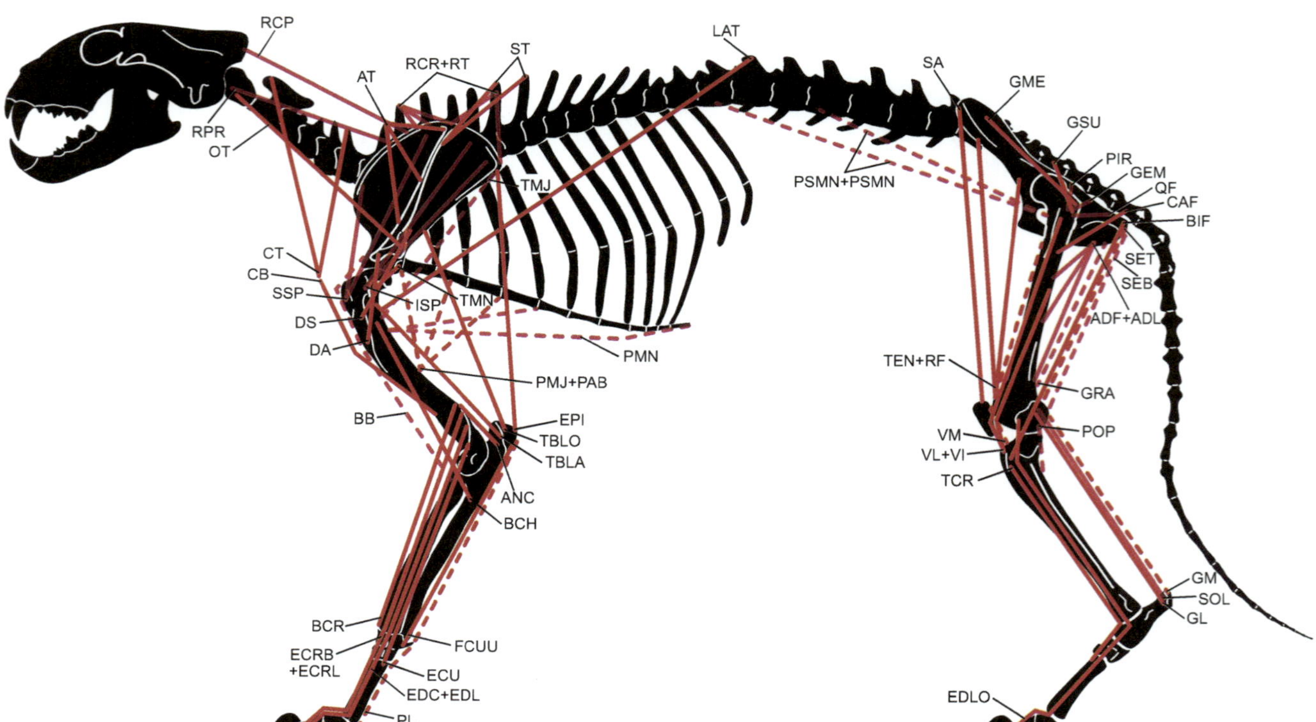

FIGURE I.2.6 Musculoskeletal system based on the lion (*Panthera leo*). The schematic illustration shows the muscles controlling the forelimb and hind limb (lateral view). The lines in red indicate the direction of pull of the superficial muscles (solid lines) and deeper muscles (dashed lines).

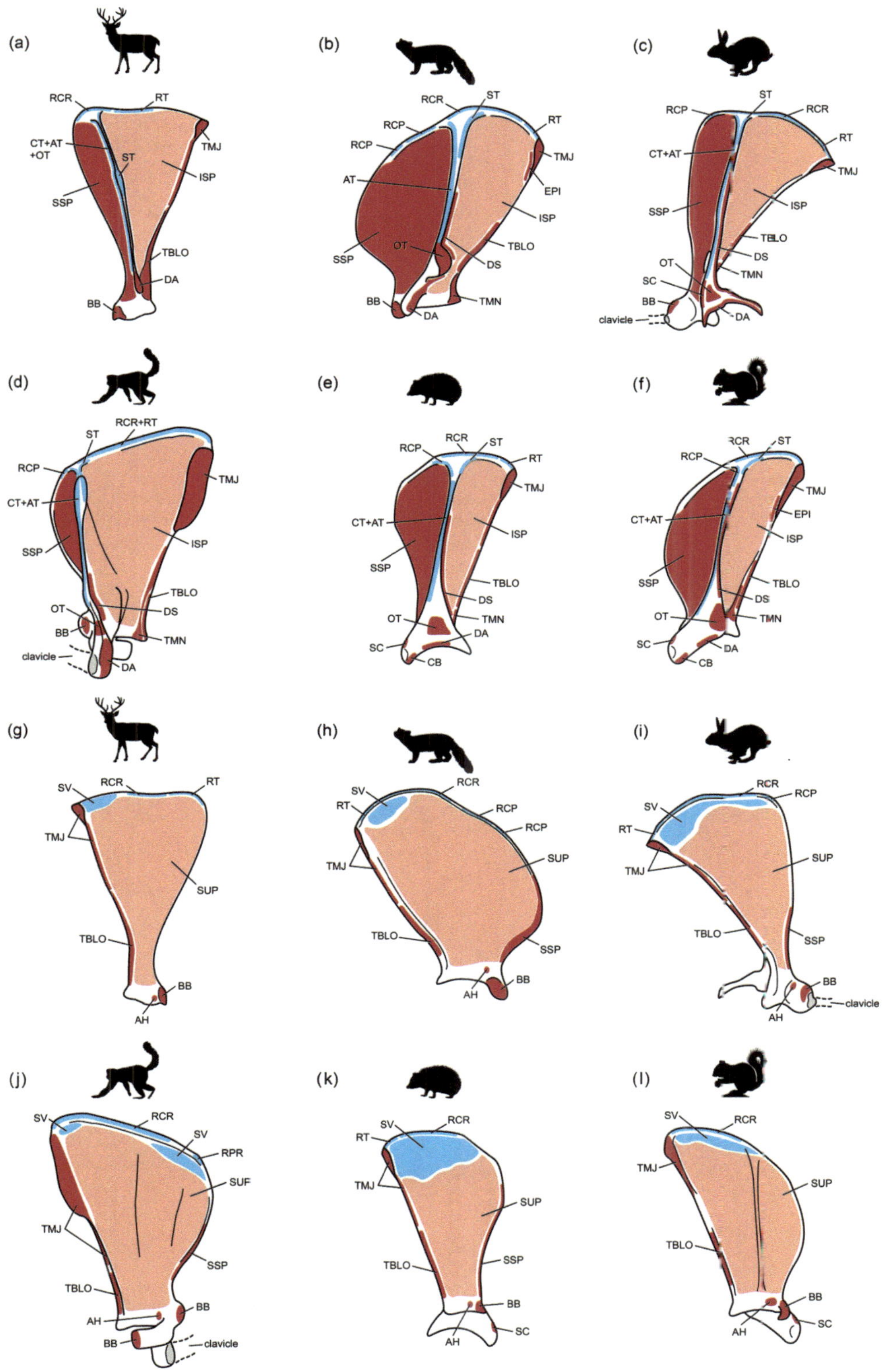

FIGURE I.2.7 **Comparative muscle maps of the musculature attaching to the scapula. (a–f)** Lateral view and **(g–l)** medial view. Red colors: muscle origins; blue colors: muscle insertions. Refer to Table I.2.1 for muscle abbreviations. Six phylogenetically different mammalian taxa representing different locomotor ecologies: **(a)** European roe deer (*Capreolus capreolus*, cursorial); **(b)** European stone marten (*Martes foina*, ambulatorial-scansorial); **(c)** European rabbit (*Oryctolagus cuniculus*, saltatorial); **(d)** Brown-bearded saki (*Chiropotes satanas*, scansorial); **(e)** European hedgehog (*Erinaceus europaeus*, ambulatorial); **(f)** Eurasian red squirrel (*Sciurus vulgaris*, scansorial-saltatorial). Not to scale.

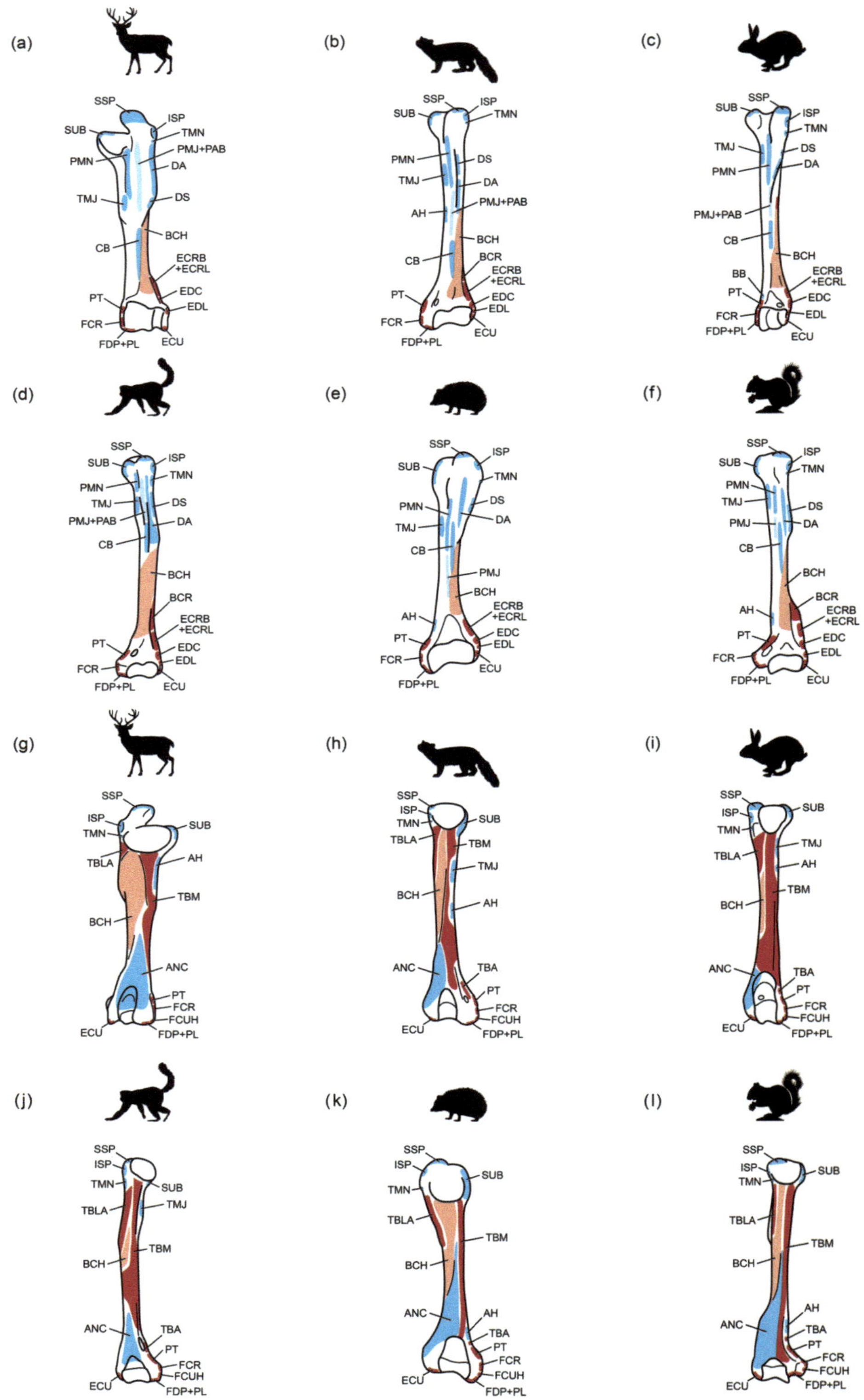

FIGURE I.2.8 **Comparative muscle maps of the musculature attaching to the humerus. (a–f)** Lateral view and **(g–l)** medial view. Red colors: muscle origins; blue colors: muscle insertions. Refer to Table I.2.1 for muscle abbreviations. Six phylogenetically different mammalian taxa representing different locomotor ecologies: **(a)** European roe deer (*Capreolus capreolus*, cursorial); **(b)** European stone marten (*Martes foina*, ambulatorial-scansorial); **(c)** European rabbit (*Oryctolagus cuniculus*, saltatorial); **(d)** Brown-bearded saki (*Chiropotes satanas*, scansorial); **(e)** European hedgehog (*Erinaceus europaeus*, ambulatorial); **(f)** Eurasian red squirrel (*Sciurus vulgaris*, scansorial-saltatorial). Not to scale.

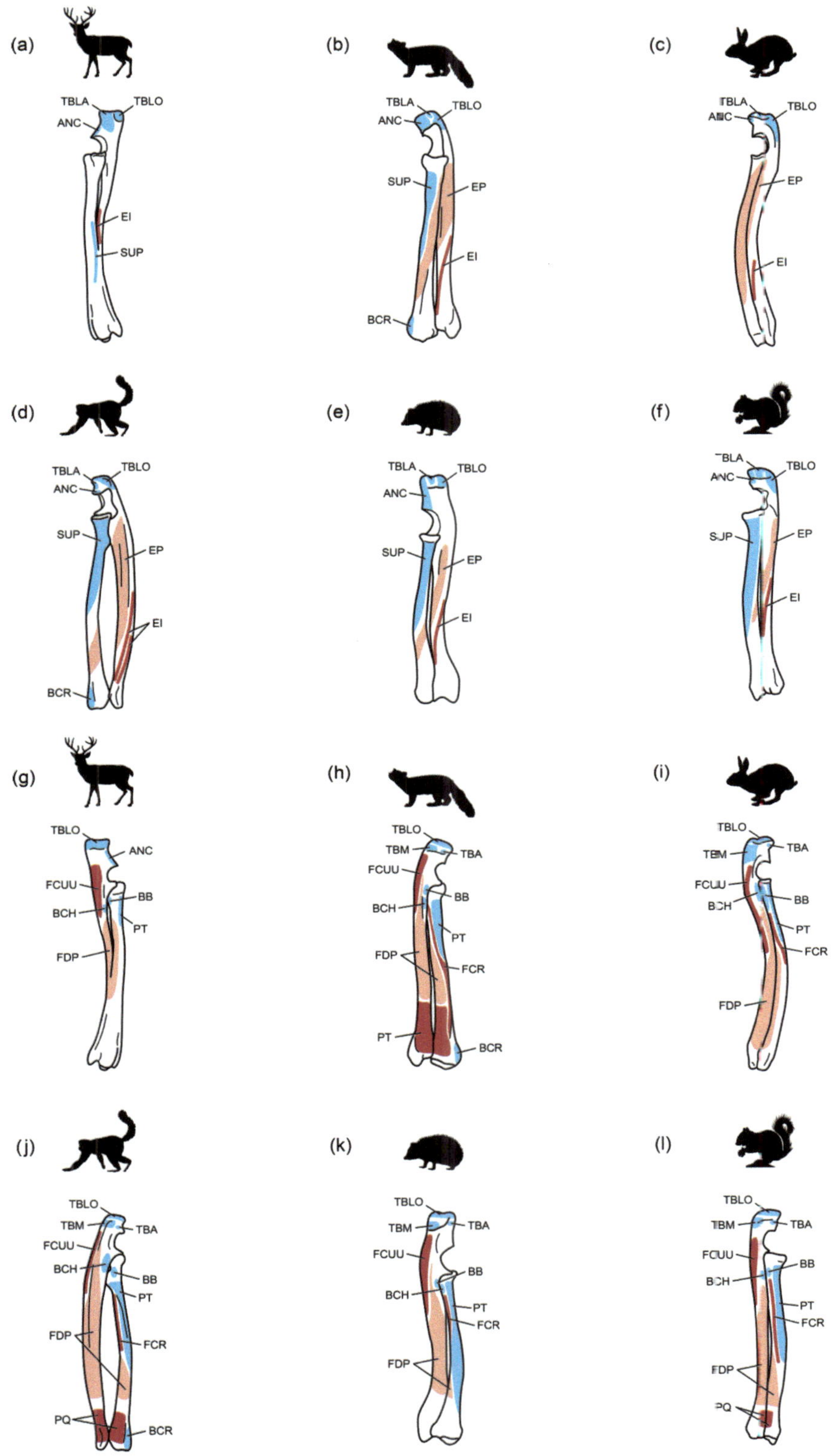

FIGURE I.2.9 Comparative muscle maps of the musculature attaching to the radius and ulna. (a–f) Lateral view and **(g–l)** medial view. Red colors: muscle origins; blue colors: muscle insertions. Refer to Table I.2.1 for muscle abbreviations. Six phylogenetically different mammalian taxa representing different locomotor ecologies: **(a)** European roe deer (*Capreolus capreolus*, cursorial); **(b)** European stone marten (*Martes foina*, ambulatorial-scansorial); **(c)** European rabbit (*Oryctolagus cuniculus*, saltatorial); **(d)** Brown-bearded saki (*Chiropotes satanas*, scansorial); **(e)** European hedgehog (*Erinaceus europaeus*, ambulatorial); **(f)** Eurasian red squirrel (*Sciurus vulgaris*, scansorial-saltatorial). Not to scale.

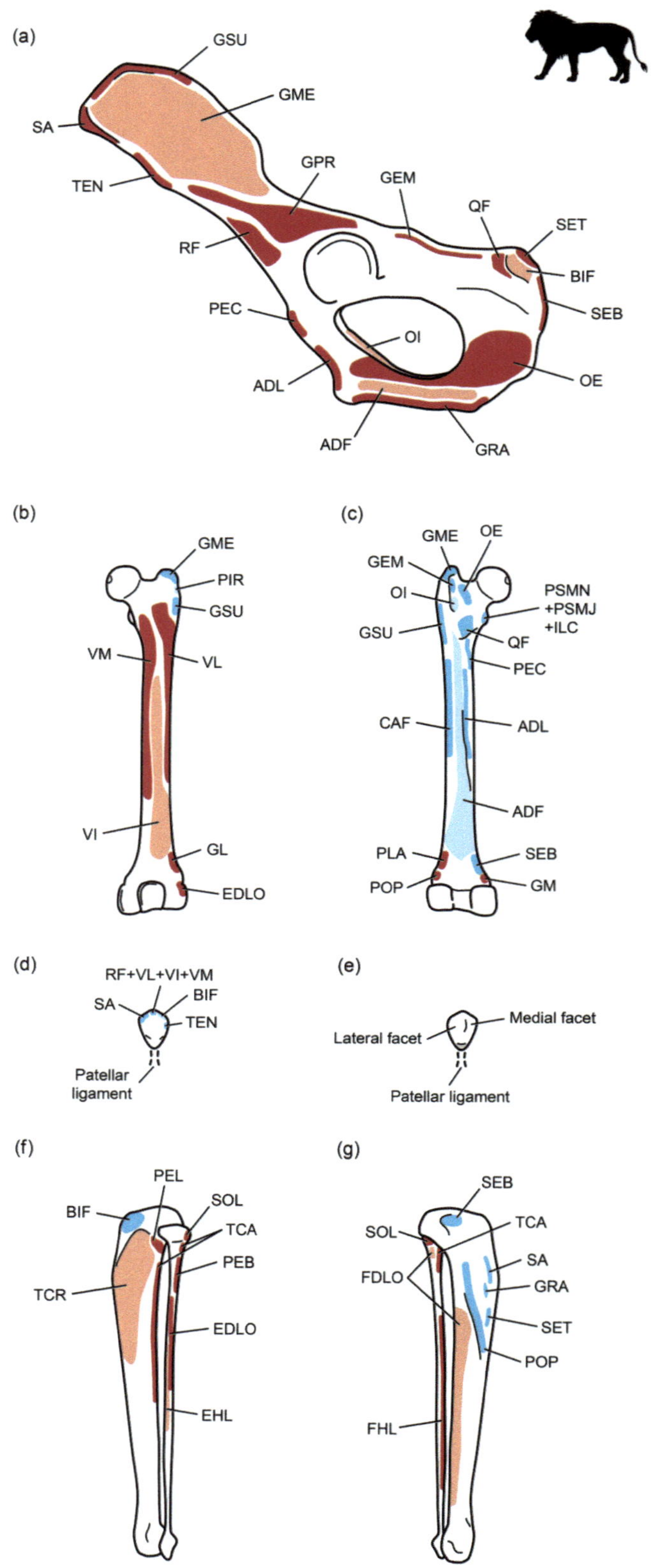

FIGURE I.2.10 Muscle maps of the musculature attaching to the pelvic limb bones. Lion (*Panthera leo*, ambulatorial-scansorial). **(a)** Pelvis in lateral view, **(b, c)** femur in cranial and caudal view, **(d, e)** patella in cranial and caudal view, **(f, g)** tibia and fibula in lateral and medial view. Red colors: muscle origins; blue colors: muscle insertions. Refer to Table I.2.2 for muscle abbreviations. Not to scale.

TABLE I.2.1
Muscles of the Shoulder and Forelimb with General Origin and Insertion

Muscle	Acronym	Origin	Insertion	Main Function
		Superficial dorsal neck and shoulder		
Mm. trapezii				
M. clavotrapezius	CT	Cervical vertebrae (nuchal and supraspinous ligament)	Clavicle (if present) and raphe with *M. clavobrachialis* and *M. cleidomastoideus*	Scapular abductor, humeral protractor
M. acromiotrapezius	AT	Cervical vertebrae (caudal nuchal ligament and supraspinous ligament)	Scapular spine and acromion	Scapular abductor and stabilizer, humeral protractor
M. spinotrapezius	ST	Thoracic vertebrae (thoracolumbar fascia)	Scapular spine	Scapular retractor and stabilizer
M. latissimus dorsi	LAT	Thoracic and lumbar vertebrae (thoracolumbar fascia and supraspinous ligament)	Medial proximal aspect of humeral diaphysis	Humeral retractor
		Ventral neck and shoulder		
M. sternomastoideus	SM	Mastoid process of temporal bone (skull)	Cranial portion of sternum (manubrium)	Skull depressor and rotator
M. cleidomastoideus	CM	Mastoid process of temporal bone (skull)	Clavicle (if present) and raphe with *M. clavobrachialis* and *M. clavotrapezius*	Skull depressor and rotator, humeral protractor
M. clavobrachialis	CB	Clavicle (if present) and raphe with *M. clavotrapezius* and *M. cleidomastoideus*	Cranial distal aspect of humeral diaphysis	Humeral protractor
Mm. pectorales				
M. pectoantebrachialis	PAB	Ventrolateral surface of cranial portion of sternum (manubrium)	Cranial proximal aspect of humeral diaphysis (superficial to insertion of *M. pectoralis minor*)	Humeral adductor
M. pectoralis major	PMJ	Ventral surface of cranial portion of sternum (manubrium) and body of sternum	Craniomedial middle of humeral diaphysis (superficial to insertion of *M. pectoralis minor*)	Humeral adductor and protractor
M. pectoralis minor	PMN	Ventral surface of body of sternum	Cranial proximal aspect of humeral diaphysis (deep to insertion of *M. pectoantebrachialis* and *M. pectoralis major*)	Humeral adductor and retractor
M. xiphihumeralis	XH	Ventrolateral surface of caudal portion of sternum (xiphoid process)	Craniomedial middle of humeral diaphysis	Humeral adductor and retractor
M. serratus ventralis	SV	Transverse processes of (all) cervical vertebrae and cranial thoracic ribs	Medial dorsal aspect of scapula (serratus muscle facet)	Scapular adductor and neck elevator
		Deep dorsal neck and shoulder		
Mm. rhomboidei		(Deep to *Mm. trapezii*)		
M. rhomboideus cervicis	RCR	Spinous processes of cervical vertebrae and occiput (if it reaches it)	Cranial dorsomedial border of scapula	Neck elevator and scapular adductor

(*Continued*)

TABLE I.2.1 ***(Continued)***

Muscles of the Shoulder and Forelimb with General Origin and Insertion

Muscle	Acronym	Origin	Insertion	Main Function
M. rhomboideus thoracis	RT	Spinous processes of thoracic vertebrae	Caudal dorsomedial border of scapula	Scapular adductor
M. rhomboideus capitis	RCP	Occiput	Cranial dorsal border of scapula (lateral to *M. rhomboideus cervicis*)	Neck elevator
M. rhomboideus profundus	RPR	Dorsolateral on atlas (cervical vertebra 1)	Cranial dorsolateral border of scapula (lateral to *M. rhomboideus capitis*)	Neck elevator
M. omotransversarius	OT	Atlantal transverse process (cervical vertebra 1)	Metacromion	Lateral neck rotator and scapular protractor
M. subclavius	SC	Ventral on first thoracic rib	Ventral surface of clavicle	Shoulder joint stabilizer
		Lateral shoulder and arm		
M. supraspinatus	SSP	Supraspinous fossa and scapular spine	Greater tubercle of humerus	Shoulder joint extensor and humeral protractor
M. infraspinatus	ISP	Infraspinous fossa and scapular spine	Lateral on greater tubercle of humerus (infraspinatus muscle facet)	Shoulder joint flexor and lateral humeral rotator
Mm. deltoidei				
M. spinodeltoideus	DS	Scapular spine (superficial to *M. infraspinatus*)	Deltoid crest of humerus	Shoulder joint flexor and humeral abductor
M. acromiodeltoideus	DA	Acromion	Deltoid crest of humerus (superficial to *M. spinodeltoideus*)	Shoulder joint flexor and humeral abductor
M. teres minor	TMN	Caudal border of scapula (near glenoid fossa)	Lateral on greater tubercle of humerus (distal to *M. infraspinatus*)	Shoulder joint flexor and lateral humeral rotator
M. epitrochlearis	EPI	Lateroventral on *M. teres major* and *M. latissimus dorsi* (caudal border of scapula)	Caudal aspect of olecranon tuber	Humeral retractor and elbow joint extensor
Mm. triceps brachii				
M. triceps brachii caput longum	TBLO	Caudal border of scapula (near glenoid fossa; medial to *M. teres minor*)	Caudal aspect of olecranon tuber (deep to *M. epitrochlearis*)	Shoulder joint flexor and elbow joint extensor
M. triceps brachii caput laterale	TBLA	Proximal aspect of deltoid crest of humerus	Caudolateral aspect of olecranon (lateral to *M. triceps brachii caput longum*)	Elbow joint extensor
M. triceps brachii caput mediale, short portion	TBMS	Medial on *M. triceps brachii caput longum* and *M. triceps brachii caput mediale, intermediate and long portion*	Medial aspect of olecranon	Elbow joint extensor
M. triceps brachii caput mediale, intermediate and long portion	TBM	Mediocaudal humeral diaphysis	Medial aspect of olecranon	Elbow joint extensor

(Continued)

TABLE I.2.1 (*Continued*)
Muscles of the Shoulder and Forelimb with General Origin and Insertion

Muscle	Acronym	Origin	Insertion	Main Function
M. brachialis	BCH	Proximal caudolateral humeral diaphysis	Tendon into bicipital tuberosity of radius or coronoid process of ulna (adjacent to insertion of *M. biceps brachii*)	Elbow joint flexor and forearm supinator
M. anconeus	ANC	Distal caudal aspect of humeral diaphysis (along lateral epicondylar crest)	Lateral aspect of olecranon (deep to *M. triceps brachii caput laterale*)	Elbow joint extensor and forearm pronator
		Medial shoulder and arm		
M. subscapularis	SUB	Subscapular fossa	Lesser tubercle of humerus	Scapular adductor
M. teres major	TMJ	Caudal border of scapula (teres major muscle facet)	Craniomedial on humeral diaphysis (teres major tuberosity; near pectoral crest)	Shoulder joint flexor and humeral retractor
M. articularis humeri	AH	Coracoid process of scapula	Medial proximal aspect of humeral diaphysis	Shoulder joint stabilizer and humeral adductor
M. triceps brachii caput accessorium	TBA	Distal caudomedial aspect of humeral diaphysis (along medial epicondylar crest)	Medial aspect of olecranon	Elbow joint extensor
M. biceps brachii	BB	Tendon from the supraglenoid tubercle of scapula (and from coracoid process of scapula if second head present)	Tendon into bicipital tuberosity of radius or coronoid process of ulna	Shoulder joint extensor and elbow joint flexor
		Laterodorsal distal arm		
M. brachioradialis	BCR	Proximal lateral epicondylar crest of humerus (proximal to origin of *M. extensor carpi radialis*)	Distal medial aspect of radius	Elbow joint flexor and forearm supinator
M. extensor carpi radialis longus	ECRL	Proximal lateral epicondylar crest of humerus (distal to origin of *M. brachioradialis* – if present)	Base of metacarpal II	Elbow joint flexor and wrist joint extensor
M. extensor carpi radialis brevis	ECRB	Proximal lateral epicondylar crest of humerus (together with or distal to origin of *M. extensor carpi radialis longus*)	Base of metacarpal III	Elbow joint flexor and wrist joint extensor
M. extensor pollicis	EP	Lateral ulnar and radial diaphysis (and interosseus membrane between radius and ulna)	Base of metacarpal I (and sesamoid proximal to metacarpal I)	Wrist joint extensor and extensor of digit I
M. extensor digitorum communis	EDC	Lateral epicondylar crest of humerus (distal to origin of *M. extensor carpi radialis*)	Tendons into distal phalanges of digits II–V	Elbow joint flexor, wrist joint extensor and digital extensor
M. extensor digitorum lateralis	EDL	Lateral epicondylar crest of humerus (distal to origin of *M. extensor digitorum communis*	Tendons into distal phalanges of digits IV–V	Elbow joint flexor, wrist joint extensor and digital extensor
M. extensor digiti I and II	EI	Lateral middle of ulnar diaphysis	Tendons into distal phalanges of digits I and II	Extensor of digits I and II

(*Continued*)

TABLE I.2.1 (*Continued*)

Muscles of the Shoulder and Forelimb with General Origin and Insertion

Muscle	Acronym	Origin	Insertion	Main Function
M. extensor carpi ulnaris	ECU	Lateral epicondylar crest of humerus (distal to origin of *M. extensor digitorum lateralis*)	Base of metacarpal V	Elbow joint flexor and wrist joint extensor
		Medioventral distal arm		
M. flexor carpi radialis	FCR	Medial epicondyle of humerus	Base of metacarpal II and III	Wrist joint flexor
M. flexor carpi ulnaris, caput ulnare	FCUU	Medial aspect of olecranon	Sesamoid proximal to metacarpal V (pisiform)	Wrist joint flexor
M. flexor carpi ulnaris, caput humerale	FCUH	Medial epicondyle of humerus	Sesamoid proximal to metacarpal V (pisiform)	Wrist joint flexor
M. palmaris longus	PL	Medial epicondyle of humerus	Tendons into distal phalanges of digits II–V or palmar aponeurosis	Wrist joint flexor and digital flexor
M. flexor digitorum profundus (4 heads)	FDP	Medial epicondyle of humerus and medial aspect of olecranon	Tendons into distal phalanges of digits II–V	Wrist joint flexor and digital flexor
		Deep distal arm		
M. supinator	SUP	Lateral epicondyle of humerus	Dorsomedial aspect of radial diaphysis	Forearm supinator
M. pronator teres	PT	Medial epicondyle of humerus (dorsal to origin of *M. flexor carpi radialis*)	Medial aspect of radial diaphysis	Forearm pronator
M. pronator quadratus	PQ	Distal medioventral surface of ulna	Distal medioventral surface of radius	Forearm pronator

Synonyms are listed in Table III.1.1. Diaphysis refers to the shaft (main section) of a long bone

TABLE I.2.2

Muscles of the Hip and Hind Limb with General Origin and Insertion

Muscle	Acronym	Origin	Insertion	Main Function
		Superficial hip		
M. sartorius	SA	Cranioventral iliac crest and medioventral border of ilium	Fascia of knee and medial aspect of tibial crest	Hip joint flexor and femoral adductor
M. tensor fasciae latae	TEN	Ventral border of ilium and tendinous raphe with *M. gluteus superficialis*	End of fascia lata on craniolateral aspect of thigh	Hip joint flexor and femoral abductor
M. caudofemoralis	CAF	Transverse process of first caudal vertebrae	Caudolateral femoral diaphysis	Hip joint extensor and femoral abductor
M. biceps femoris	BIF	Ischial tuberosity	Lateral fascia of knee and cranial border of tibia	Hip joint extensor and knee joint flexor
M. semitendinosus	SET	Dorsal aspect of ischial tuberosity	Medial fascia of knee and proximal craniomedial aspect of tibia	Hip joint extensor and knee joint flexor
M. semimembranosus	SEB	Caudal border of tuber ischiadicum (deep to *M. semitendinosus*)	Caudomedial distal femur and proximomedial aspect of tibia	Femoral adductor and knee joint flexor

(*Continued*)

TABLE I.2.2 (*Continued*)
Muscles of the Hip and Hind Limb with General Origin and Insertion

Muscle	Acronym	Origin	Insertion	Main Function
Mm. glutei				
M. gluteus superficialis	GSU	Iliac crest and tips of transverse processes of last sacral vertebrae and first caudal vertebrae	Caudolateral aspect of greater trochanter of femur	Hip joint flexor and femoral abductor
M. gluteus medius	GME	Crest and lateral surface of ilium, transverse processes of last sacral vertebrae and first caudal vertebrae (deep to *M. gluteus superficialis*)	Proximal aspect of greater trochanter of femur	Hip joint extensor and femoral abductor
M. gluteus profundus	GPR	Lateral surface of ilium (caudal to *M. gluteus medius*)	Craniolateral aspect of greater trochanter of femur	Hip joint extensor and femoral abductor
		Deep ventral hip		
M. psoas minor	PSMN	Ventral bodies and transverse processes of lumbar vertebrae	Lesser trochanter of femur and medial aspect of neck of femur	Hip joint flexor
		M. iliopsoas		
M. psoas major	PSMJ	Ventral bodies and transverse processes of (last thoracic and) lumbar vertebrae	Lesser trochanter of femur and medial aspect of neck of femur	Hip joint flexor
M. iliacus	ILC	Ventral surface of ilium (more or less fused with *M. psoas major*)	Lesser trochanter of femur and medial aspect of neck of femur	Hip joint flexor
		Deep medial hip		
M. gracilis	GRA	Ventral pubis near symphysis of ischium and pubis	Medial fascia of knee and proximal tibia (often continuous with *M. sartorius*)	Femoral adductor and knee joint flexor
M. adductor femoris (magnus et brevis)	ADF	Ventral pubis near symphysis of ischium and pubis (deep to *M. gracilis*)	Proximal aspect of ventral femoral diaphysis	Hip joint extensor and femoral adductor
M. obturator externus	OE	Caudoventral margins of obturator foramen and dorsoventral surfaces of pubis and ischium	Proximal aspect of trochanteric fossa	Femoral rotator
M. obturator internus	OI	Medial surface of obturator foramen and dorsal surface of ischium along symphysis	Proximal aspect of trochanteric fossa	Femoral rotator and abductor
		Deep lateral hip		
M. piriformis	PIR	Transverse processes of last two sacral vertebrae and first caudal vertebrae	Lateral aspect of greater trochanter of femur	Hip joint extensor and femoral abductor
Mm. gemelli (M. gemellus superioris, M. gemellus inferioris)	GEM	Dorsal aspect of acetabulum and dorsocaudal surface of ischium	Proximal aspect of trochanteric fossa	Femoral rotator and abductor
M. quadratus femoris	QF	Ventral ischial tuberosity	Caudal aspect of lesser trochanter for femur	Hip joint extensor and lateral hip joint rotator

(*Continued*)

TABLE I.2.2 (*Continued*)
Muscles of the Hip and Hind Limb with General Origin and Insertion

Muscle	Acronym	Origin	Insertion	Main Function
M. tenuissimus	TE	Transverse process of first caudal vertebra	Lateral fascia of knee	Hip joint extensor and knee joint flexor
M. adductor longus	ADL	Cranioventral surface of pubis	Caudal femoral diaphysis	Hip joint extensor and femoral adductor
M. pectineus	PEC	Cranial surface of pubis near pelvic symphysis (iliopubic eminence)	Caudomedial aspect of proximal femoral diaphysis (distal to lesser trochanter)	Hip joint extensor and femoral adductor
Mm. quadriceps femoris				
M. rectus femoris	RF	Distal ilium near acetabulum	Tendon onto patella (common with *Mm. vasti*)	Hip joint flexor and knee joint extensor
M. vastus lateralis	VL	Cranial aspect of great trochanter of femur and dorsolateral femoral diaphysis	Tendon onto patella (common with *M. rectus femoris*)	Knee joint extensor
M. vastus intermedius	VI	Cranial aspect of femoral diaphysis	Tendon onto patella (common with *M. rectus femoris*)	Knee joint extensor
M. vastus medialis	VM	Craniomedial femoral diaphysis	Tendon onto patella (common with *M. rectus femoris*)	Knee joint extensor
		Craniolateral distal leg		
M. tibialis cranialis	TCR	Lateral condyle of tibia and proximal craniolateral surface of tibia	Tendon into medial base of metatarsal I	Ankle joint dorsiflexor and foot inverter
M. extensor digitorum longus	EDLO	Lateral surface of lateral epicondyle of femur	Tendons into distal phalanges of digits II-V	Ankle joint dorsiflexor and digital extensor
M. extensor digitorum lateralis	EDLA	Proximal lateral on fibula	Tendon together with *M. extensor digitorum longus* into distal phalanx of digit V	Ankle joint dorsiflexor and digital extensor
M. extensor hallucis longus	EHL	Craniolateral surface of fibula	Tendon into distal phalanx of digit I	Ankle joint dorsiflexor and digital extensor
Mm. peronei				
M. peroneus longus	PEL	Caudolateral condyle of tibia (and lateral fibular head)	Tendon into lateroplantar base of metatarsal I	Ankle joint flexor and foot everter
M. peroneus tertius	PET	Craniolateral diaphysis of fibula	Tendon into distal tarsus	Ankle joint flexor and foot everter
M. peroneus brevis	PEB	Proximal caudolateral aspect of fibula	Tendon into lateral base of metatarsal V	Ankle joint flexor and foot everter
		Caudal distal leg		
Mm. gastrocnemii				
M. gastrocnemius caput laterale	GL	Lateral epicondylar ridge of femur (distal to *M. plantaris*)	Tendon into calcaneus (together with *M. gastrocnemius caput mediale*)	Knee joint flexor and ankle joint flexor
M. gastrocnemius caput mediale	GM	Medial epicondylar ridge of femur	Tendon into calcaneus (together with *M. gastrocnemius caput laterale*)	Knee joint flexor and ankle joint flexor

(*Continued*)

TABLE I.2.2 (*Continued*)

Muscles of the Hip and Hind Limb with General Origin and Insertion

Muscle	Acronym	Origin	Insertion	Main Function
M. plantaris	PLA	Lateral epicondylar ridge of femur (proximal to *M. gastrocnemius caput laterale*)	Tendon into calcaneus (medial and deep to tendon of *Mm. gastrocnemii*)	Ankle joint flexor and digital flexor
M. soleus	SOL	Caudal head of fibula	Tendon into calcaneus (lateral and deep to tendon of *Mm. gastrocnemii*)	Ankle joint flexor
M. popliteus	POP	Lateral epicondyle of femur	Proximal caudomedial aspect of tibia	Knee joint flexor and tibial rotator
M. flexor digitorum longus	FDLO	Caudomedial aspect of fibular head and caudomedial diaphysis of tibia (interosseous membrane)	Tendons into plantar aspect of distal phalanges of digits I–V	Ankle joint extensor and digital flexor
M. flexor hallucis longus	FHL	Caudal aspect of fibula (interosseous membrane)	Tendons into plantar aspect of distal phalanges of digits I–V (joining that of *M. flexor digitorum longus*)	Ankle joint plantaflexor and digital flexor
M. tibialis caudalis	TCA	Caudal aspect of the fibular head and proximal diaphysis of tibia	Tendon into medioplantar aspect of navicular bone and medial cuneiforme bone (tarsal bones)	Ankle joint plantaflexor and foot inverter

Synonyms are listed in Table III.1.2. Diaphysis refers to the shaft (main section) of a long bone.

REFERENCES

Nickel R, Schummer A, Seiferle E. 2003. *Lehrbuch der Anatomie der Haustiere I: Bewegungsapparat.* Parey: Stuttgart.

World Association of Veterinary Anatomists. 2017. *Nomina anatomica veterinaria.* International Committee on Veterinary Anatomical Nomenclature: Vienna.

Part II

Martes foina (European stone marten)

Classification:	Carnivora, Canoidea, Mustelidae
Mean body mass:	1.1–2.0 kg
Habitat:	Open forest, forest edge, (sub) urban
Locomotor ecology:	Ambulatorial – scansorial
Peculiarities:	Digitigrade

FIGURE II.1.1 **Superficial muscles of forelimb and hind limb in lateral view. (a) Without labels. (b) With labels and muscle outlines.** In the neck and shoulder region, the *Mm. trapezii* are clearly divided into three parts (CT, *M. clavotrapezius*; AT, *M. arcomiotrapezius*; and ST, *M. spinotrapezius*). The EPI, *M. epitrochlearis* is well developed, and the BCR, *M. brachioradialis* is present. The OEA, *M. obliquus externus abdominis* originates from the thoracic ribs and inserts into the linea alba at the midline and on the pubis near the symphysis. The RA, *M. rectus abdominis* originates from the costal cartilages and inserts into the prebubic tendon. In the distal hind limb, the four tendons of the EDLO, *M. extensor digitorum longus* insert on the distal phalanges of digits II through V. The EHL, *M. hallucis longus* lies deep to the TCR, *M. tibialis cranialis* and the EDLO, *M. extensor digitorum longus*. The PET, *M. peroneus tertius* is absent. Note that the retinacula (ligaments) spanning the wrist and ankle joints have been largely removed.

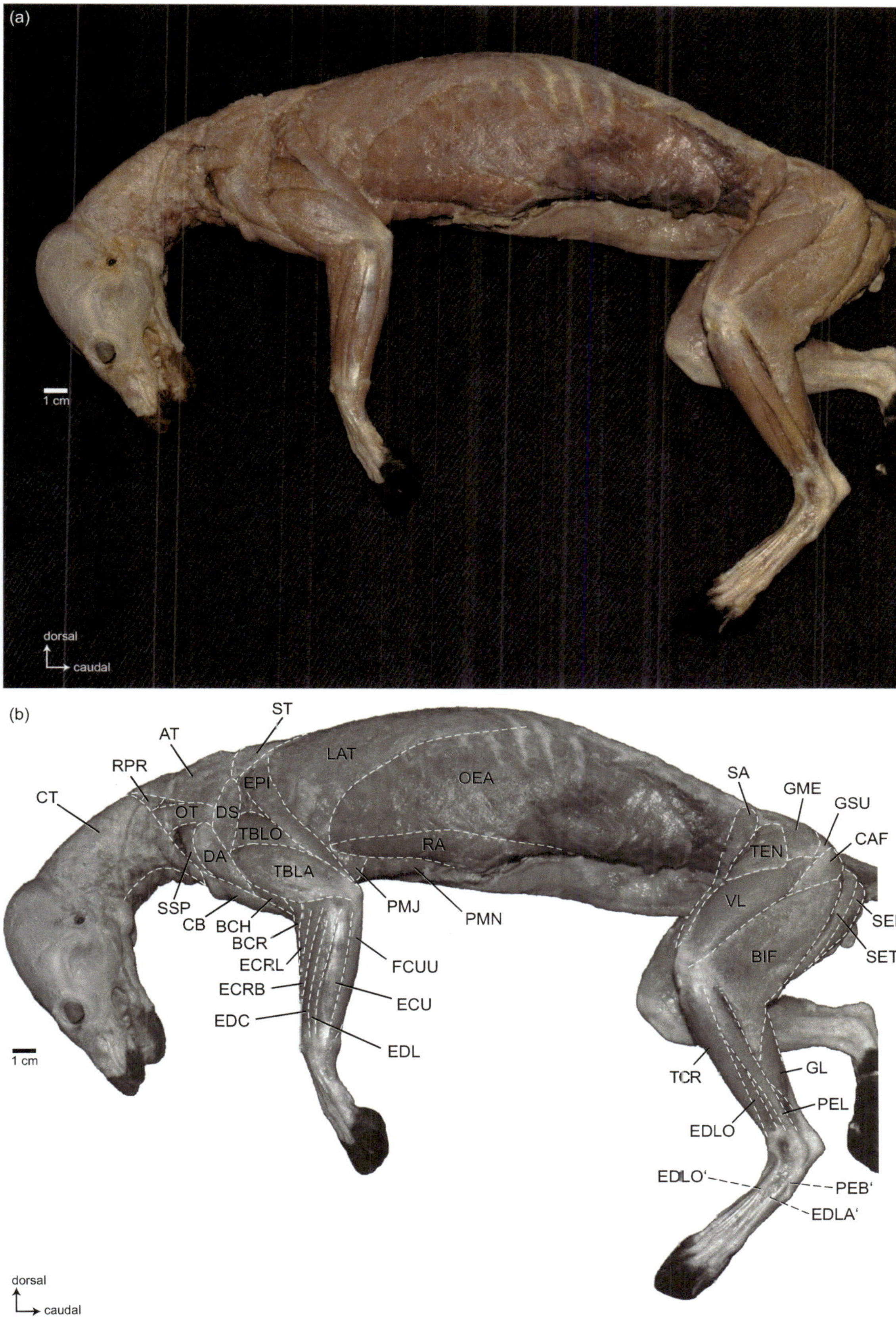
(a)
1 cm
dorsal
caudal
(b)
ST
AT
LAT
RPR
EPI
OEA
CT
OT
DS
TBLO
DA
RA
TBLA
SSP
CB
BCH
BCR
ECRL
ECRB
EDC
PMJ
PMN
FCUU
ECU
EDL
SA
GME
GSU
CAF
TEN
VL
SEB
SET
BIF
1 cm
TCR
GL
PEL
EDLO
EDLO‘
PEB‘
EDLA‘
dorsal
caudal

FIGURE II.1.2 **Pectoral muscles in ventral view. (a) Superficial view.** The clavicle is absent in this specimen. **(b) Deep view.** The SM, *M. sternomastoideus*, the PMJ, *M. pectoralis major* and the PAB, *M. pectoantebrachialis* have been removed. The XH, *M. xiphihumeralis* is clearly separated from the PMN, *M. pectoralis minor.* A rudimentary clavicle is sometimes present in martens (embedded into the raphe between the CM, *M. cleidomastoideus* and the CB, *M. clavobrachialis*). The SC, *M. subclavius* is absent.

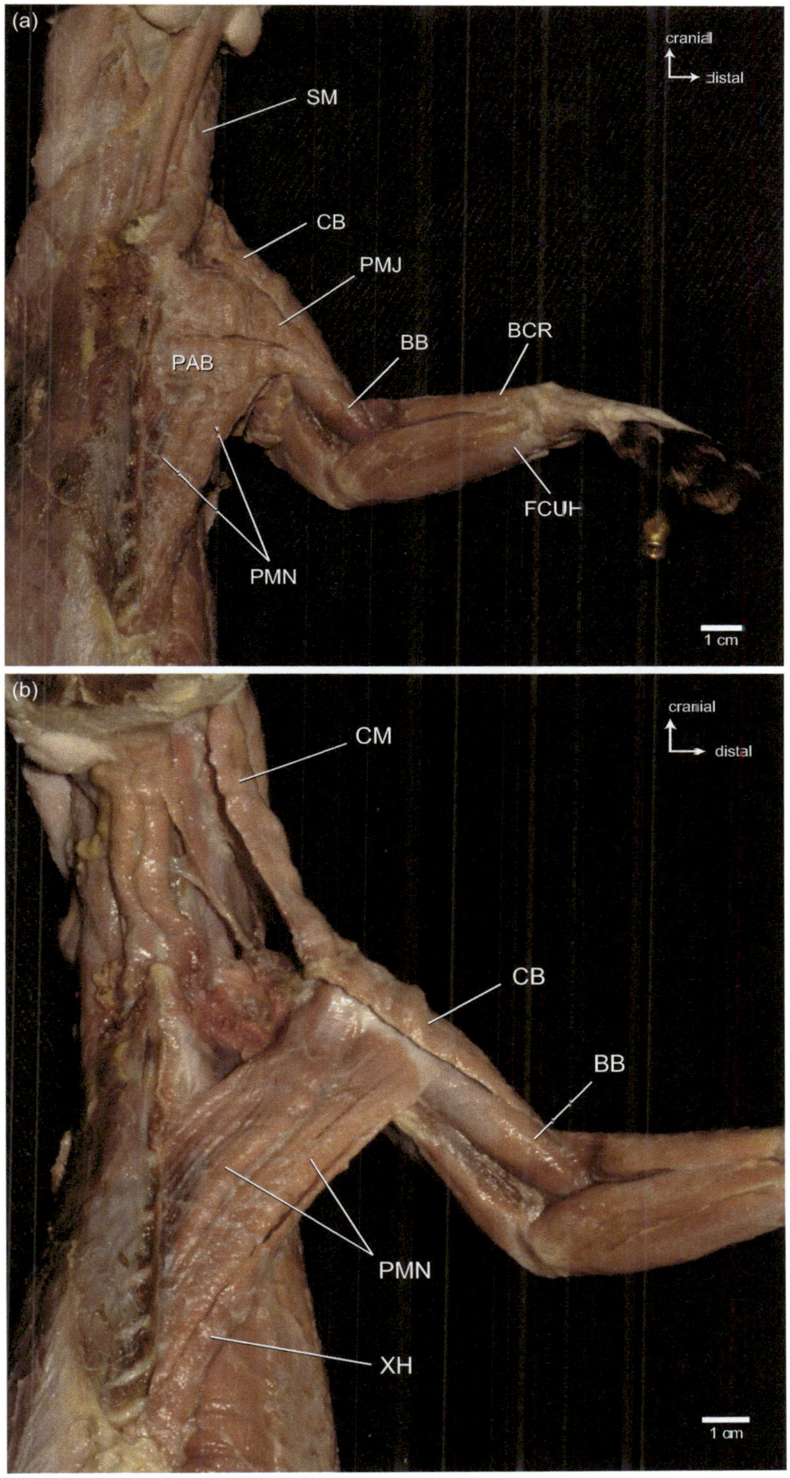
(a)
cranial
distal
SM
CB
PMJ
BB
BCR
PAB
FCUH
PMN
1 cm
(b)
cranial
distal
CM
CB
BB
PMN
XH
1 cm

FIGURE II.1.3 **Deep neck and shoulder muscles. (a) In lateral view.** In the neck region, all four *Mm. rhomboidei* are present (RCR, *M. rhomboideus cervicis*; RT, *M. rhomboideus thoracis*; RCP, *M. rhomboideus capitis*; and RPR, *M. rhomboideus profundus*). The *Mm. deltoidei* and the EPI, *M. epitrochlearis* have been removed. The clavicle is absent in this specimen. **(b) In dorsal view.** The RCR, *M. rhomboideus cervicis* does not reach the occiput.

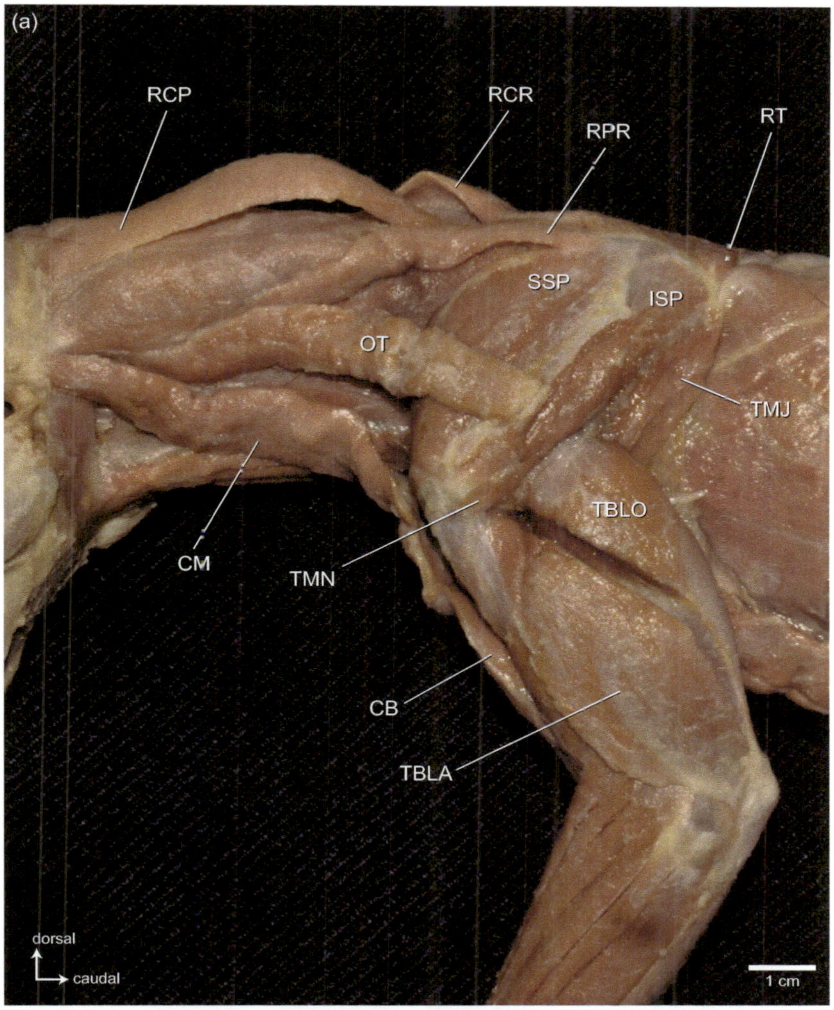
(a)
RCP
RCR
RPR
RT
SSP
ISP
OT
TMJ
TBLO
CM
TMN
CB
TBLA
dorsal
caudal
1 cm

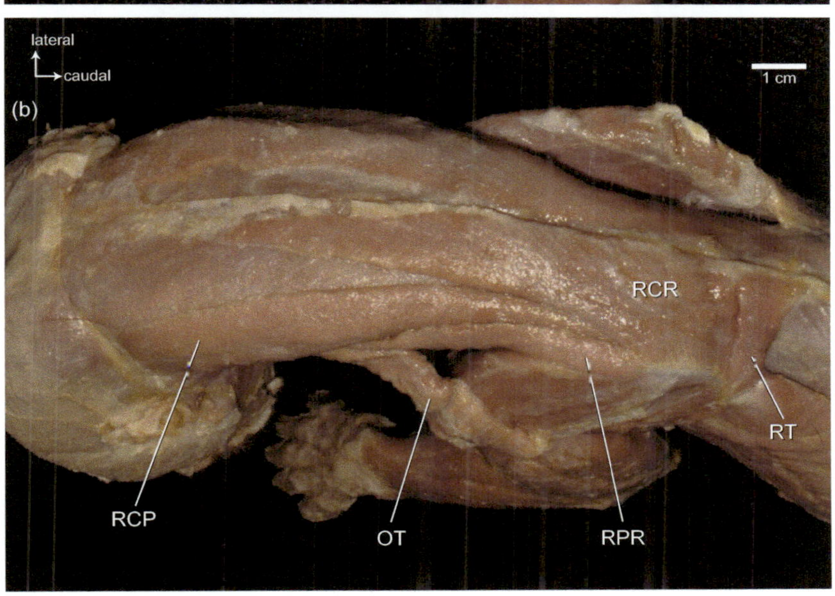
lateral
caudal
1 cm
(b)
RCR
RCP
OT
RPR
RT

FIGURE II.1.4 **Intrinsic forelimb muscles in (a) lateral view and (b) medial view. Distal intrinsic forelimb muscles in (c) lateral view and (d) medial view.**

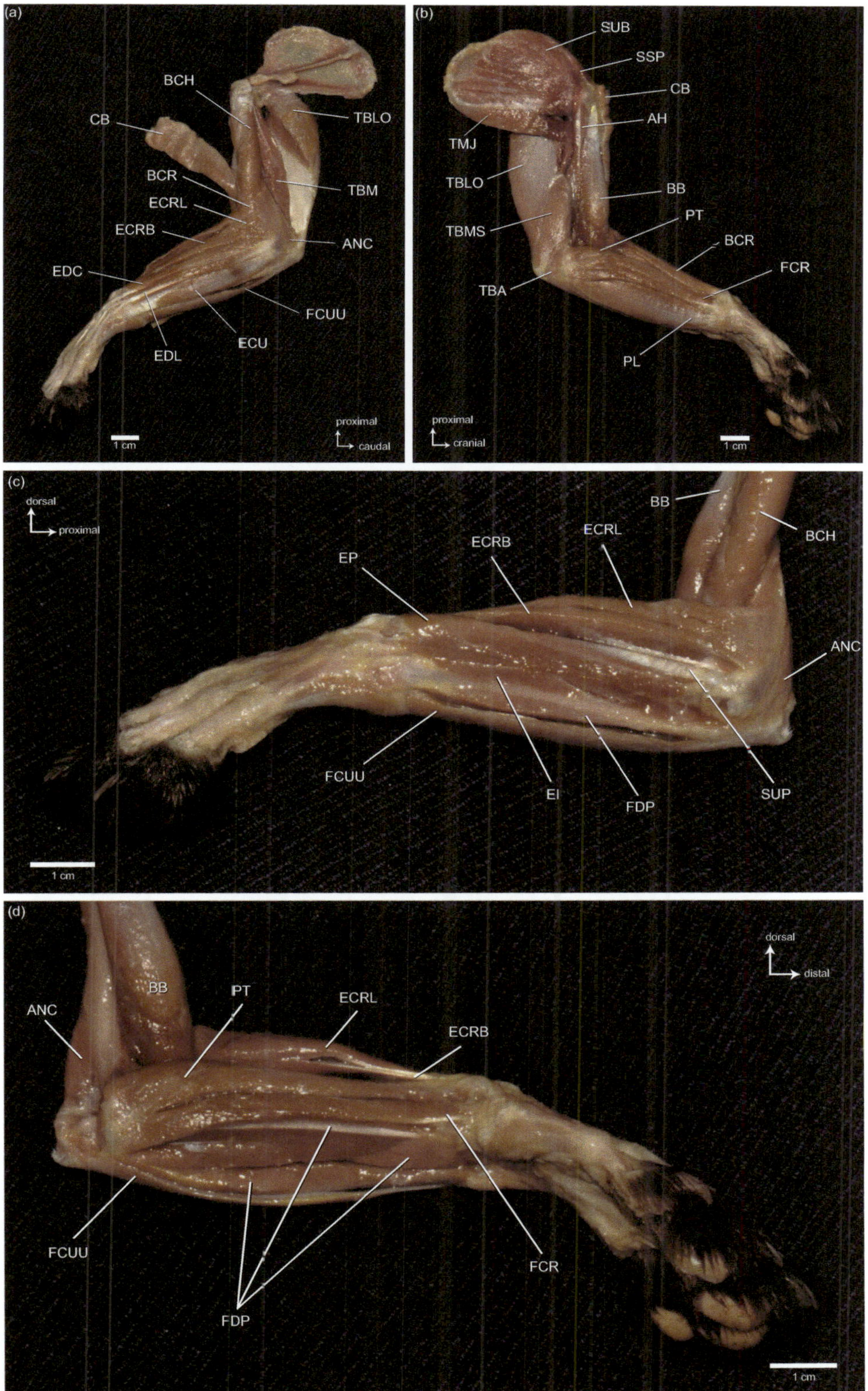

(a)
BCH
CB
TBLO
BCR
ECRL
TBM
ECRB
ANC
EDC
FCUU
EDL
ECU
1 cm
proximal
caudal
(b)
SUB
SSP
CB
AH
TMJ
TBLO
BB
TBMS
PT
BCR
FCR
TBA
PL
proximal
cranial
1 cm
(c)
dorsal
proximal
BB
BCH
ECRB
ECRL
EP
ANC
FCUU
EI
FDP
SUP
1 cm
(d)
dorsal
distal
BB
PT
ECRL
ANC
ECRB
FCUU
FCR
FDP
1 cm

FIGURE II.1.5 **Distal forelimb muscles. (a) In dorsal view.** The EDC, *M. extensor digitorum communis* inserts with five tendons on the distal phalanges of digits I through V. The EDL, *M. extensor digitorum lateralis* inserts with three tendons on the distal phalanges of digits III through V. **(b) In ventral view.** Not visible in the picture, but the PL, *M. palmaris longus* inserts with five tendons on the distal phalanges of digits I through V.

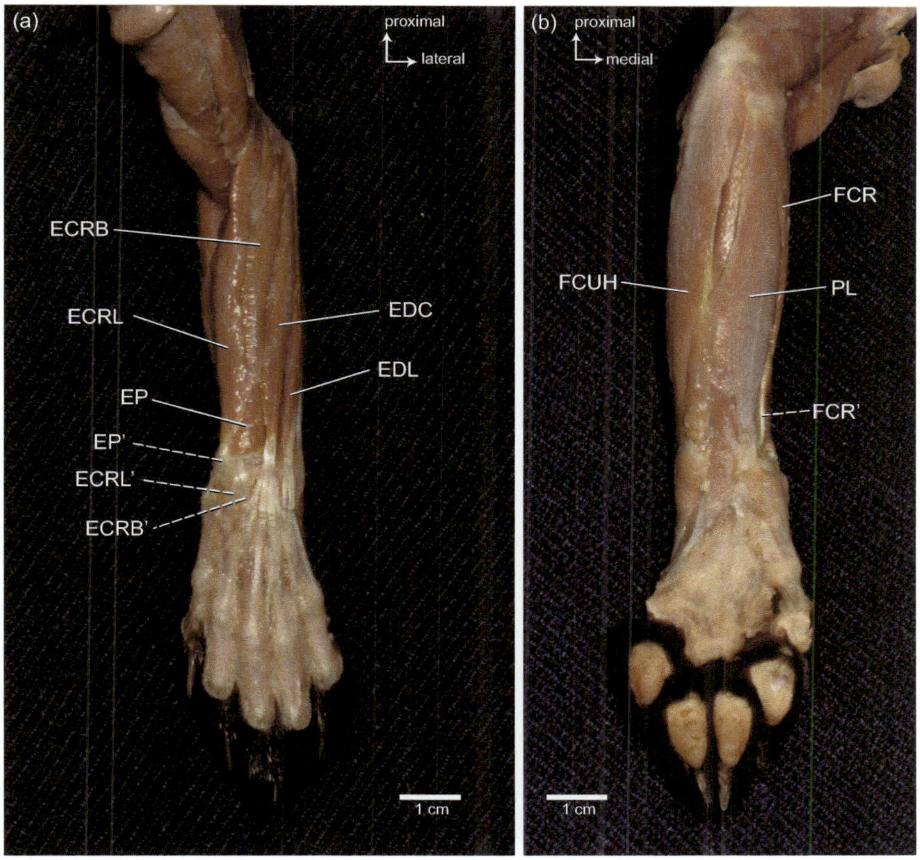
(a)
proximal
lateral
ECRB
ECRL
EDC
EDL
EP
EP'
ECRL'
ECRB'
1 cm
(b)
proximal
medial
FCR
FCUH
PL
FCR'
1 cm

FIGURE II.1.6 **Deep rotator muscles. (a) Distal rotator muscles in dorsal view.** The EI, *M. extensor digiti I and II* inserts with two tendons on the distal phalanges of digits I and II. **(b) Distal rotator muscles in ventral view. (c) Proximal rotator muscles in ventral view.**

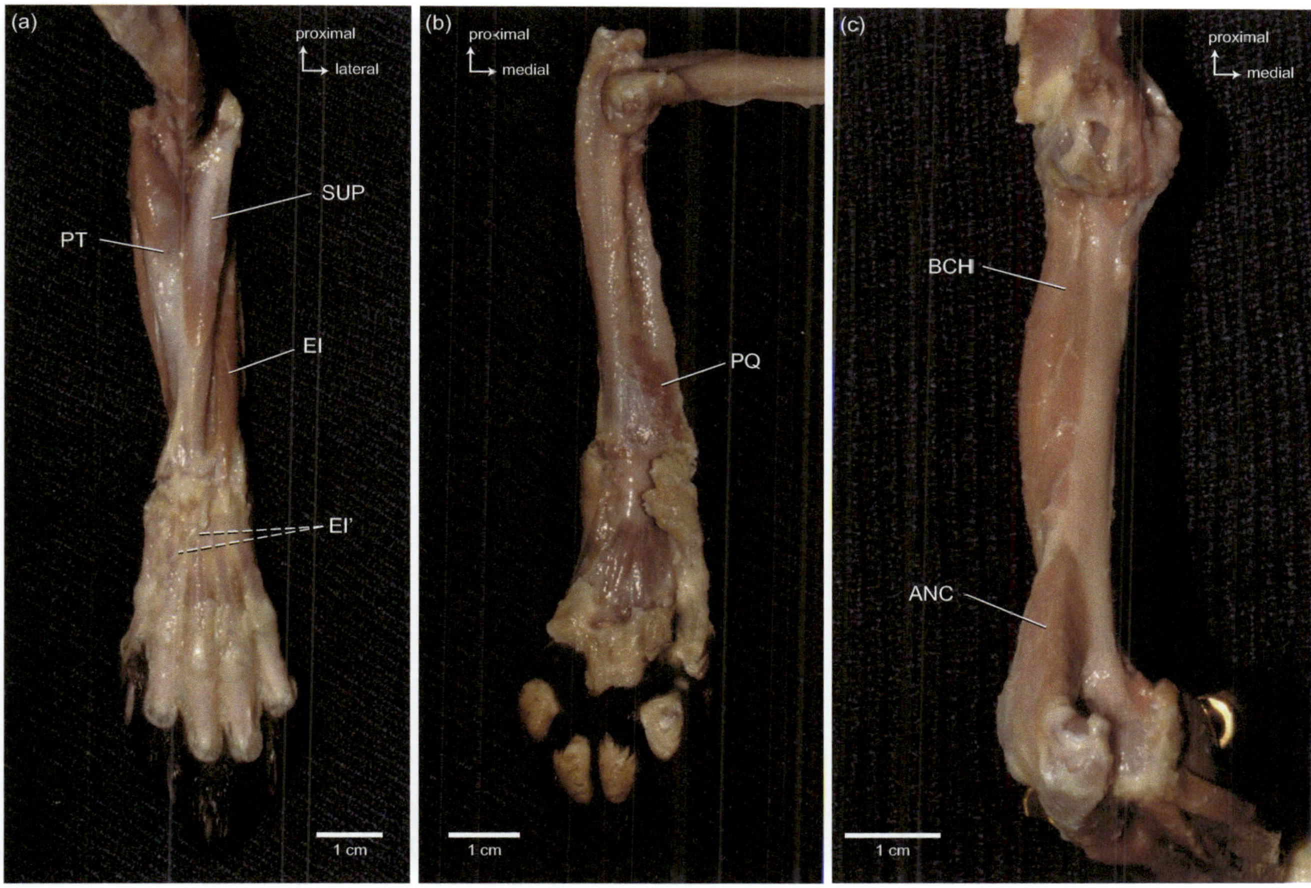
(a)
proximal
lateral
SUP
PT
EI
EI'
1 cm
(b)
proximal
medial
PQ
1 cm
(c)
proximal
medial
BCH
ANC
1 cm

FIGURE II.1.7 **Hip muscles. (a) In ventral view.** The GRA, *M. gracilis* and the SET, *M. semitendinosus* have been removed. **(b) In lateral view.** The SA, *M. sartorius*, the BIF, *M. biceps femoris* and the SET, *M. semitendinosus* have been removed.

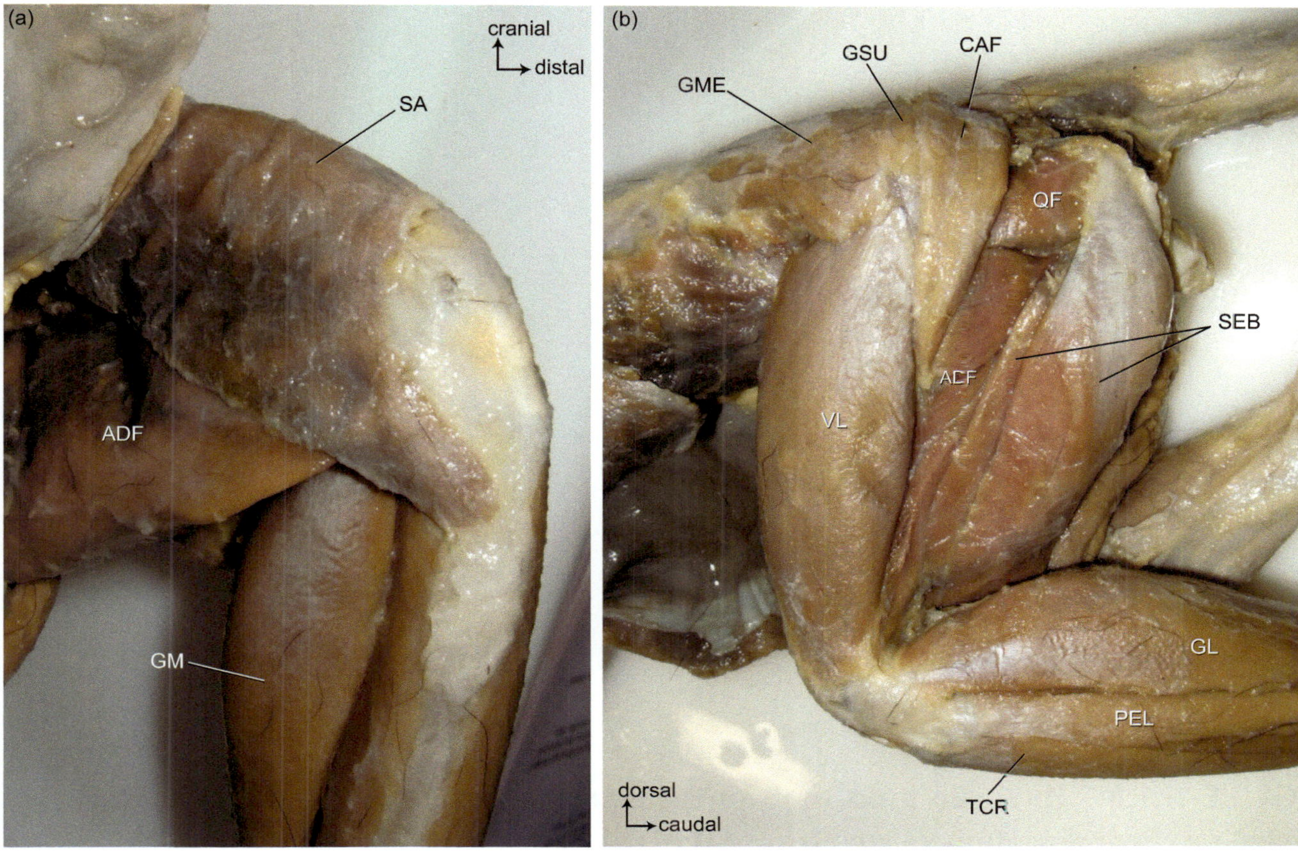
(a)
cranial
distal
SA
ADF
GM
(b)
GSU
CAF
GME
QF
SEB
ACF
VL
GL
PEL
TCR
dorsal
caudal

FIGURE II.1.8 Intrinsic hind limb muscles in (a) lateral view and (b) medial view. Distal intrinsic forelimb muscles in (c) lateral view and (d) medial view. The EDLA, *M. extensor digitorum lateralis* inserts on the distal phalanges of digits IV and V, joining the EDLO, *M. extensor digitorum longus* that inserts into the distal phalanges of digits II–V.

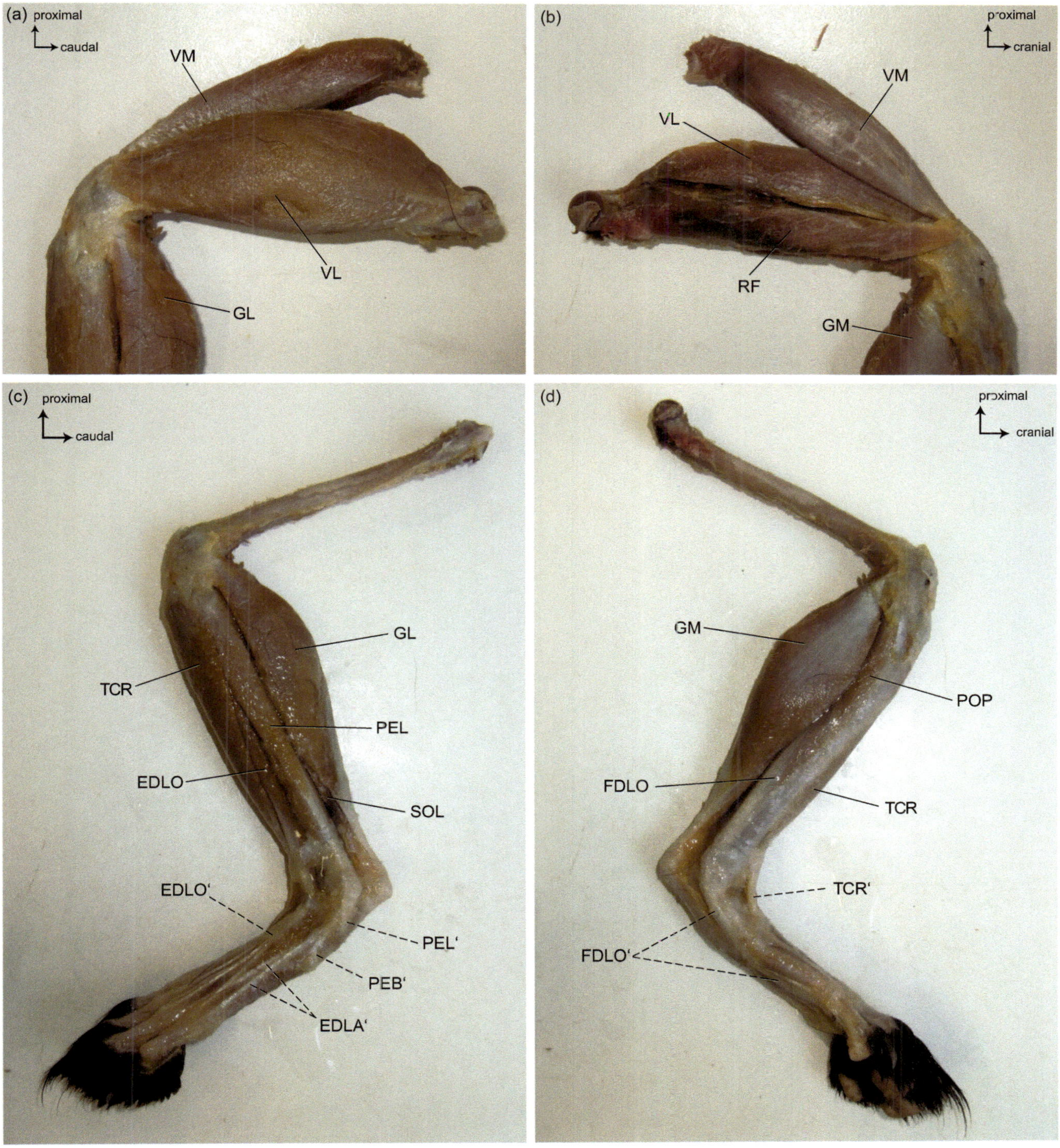
(a) proximal
caudal
VM
VL
GL
(b) proximal
cranial
VM
VL
RF
GM
(c) proximal
caudal
GL
TCR
PEL
EDLO
SOL
EDLO'
PEL'
PEB'
EDLA'
(d) proximal
cranial
GM
POP
FDLO
TCR
TCR'
FDLO'

Meles meles (Eurasian badger)

Classification:	Carnivora, Canoidea, Mustelidae
Mean body mass:	6.6–16.7 kg
Habitat:	Open forest, forest edge, (sub) urban
Locomotor ecology:	Ambulatorial, semifossorial
Peculiarities:	Plantigrade; strong, non-retractable claws

FIGURE II.2.1 Superficial muscles of forelimb and hind limb in lateral view. (a) Without labels. (b) With labels and muscle outlines. In the neck and shoulder region, the *Mm. trapezii* are clearly divided into three parts (CT, *M. clavotrapezius*; AT, *M. arcomiotrapezius*; and ST, *M. spinotrapezius*). The EPI, *M. epitrochlearis* is well developed, and the BCR, *M. brachioradialis* is present. In the hip region, the GME, *M. gluteus medius*, the GSU, *M. gluteus superficialis* and the CAF, *M. caudofemoralis* lie deep to the TEN, *M. tensor fasciae latae*. The SEB, *M. semimembranosus* lies medially to the SET, *M. semitendinosus*. In the distal hind limb, the four tendons of the EDLO, *M. extensor digitorum longus* insert on the distal phalanges of digits II through V. The EHL, *M. hallucis longus* lies deep to the TCR, *M. tibialis cranialis* and the EDLO, *M. extensor digitorum longus*. The PET, *M. peroneus tertius* is absent. Note that the retinacula (ligaments) spanning the wrist and ankle joints have been partially removed.

(a)

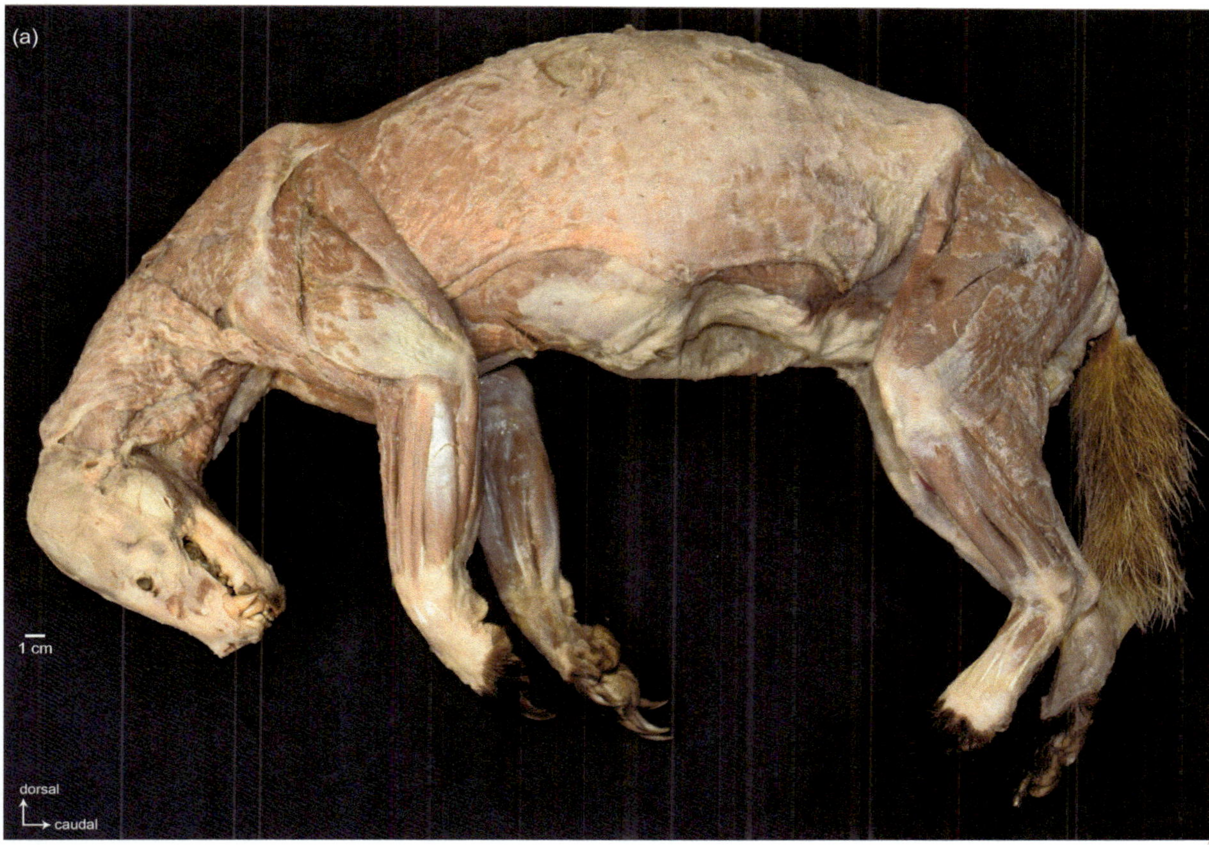
1 cm
dorsal
caudal

(b)

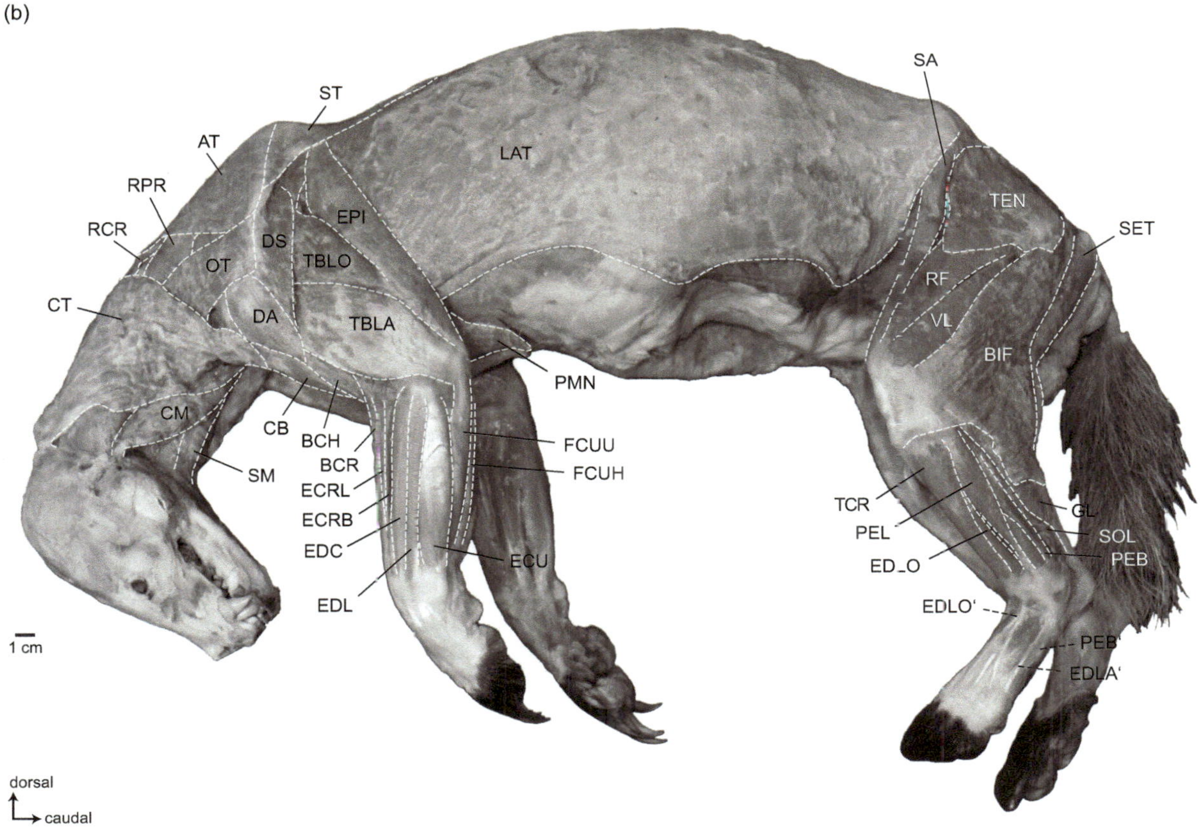
SA
ST
AT
LAT
RPR
EPI
TEN
SET
RCR
DS
TBLO
OT
RF
CT
DA
TBLA
VL
BIF
PMN
CM
CB
BCH
BCR
FCUU
SM
ECRL
FCUH
ECRB
TCR
GL
PEL
SOL
EDC
ECU
ED_O
PEB
EDL
EDLO'
PEB
EDLA'
1 cm
dorsal
caudal

FIGURE II.2.2 **Pectoral muscles in ventral view. (a) Superficial view.** A rudimentary clavicle is present (embedded into the raphe between the CM, *M. cleidomastoideus* and the CB, *M. clavobrachialis*). The SC, *M. subclavius* is absent. The XH, *M. xiphihumeralis* is not clearly developed as a separate muscle, but part of the PMN, *M. pectoralis minor.* **(b) Deep view.** The PAB, *M. pectoantebrachialis* has been removed. The PMJ, *M. pectoralis major* has been detached from its origin to display the PMN, *M. pectoralis minor.*

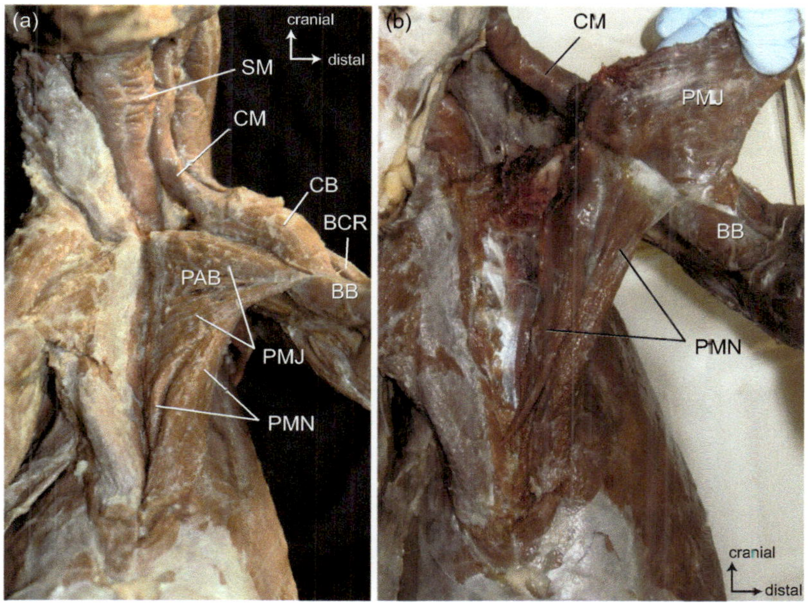
(a)
cranial
distal
SM
CM
CB
BCR
PAB
BB
PMJ
PMN
(b)
CM
PMJ
BB
PMN
cranial
distal

FIGURE II.2.3 **Deep neck and shoulder muscles. (a) In lateral view.** In the neck region, all four *Mm. rhomboidei* are present (RCR, *M. rhomboideus cervicis*; RT, *M. rhomboideus thoracis*; RCP, *M. rhomboideus capitis*; and RPR, *M. rhomboideus profundus*). The EPI, *M. epitrochlearis* and the LAT, *M. latissimus dorsi* have been removed. The clavicle is embedded into the raphe between the CM, *M. cleidomastoideus* and the CB, *M. clavobrachialis*. **(b) In dorsal view.** The RCR, *M. rhomboideus cervicis* reaches the occiput.

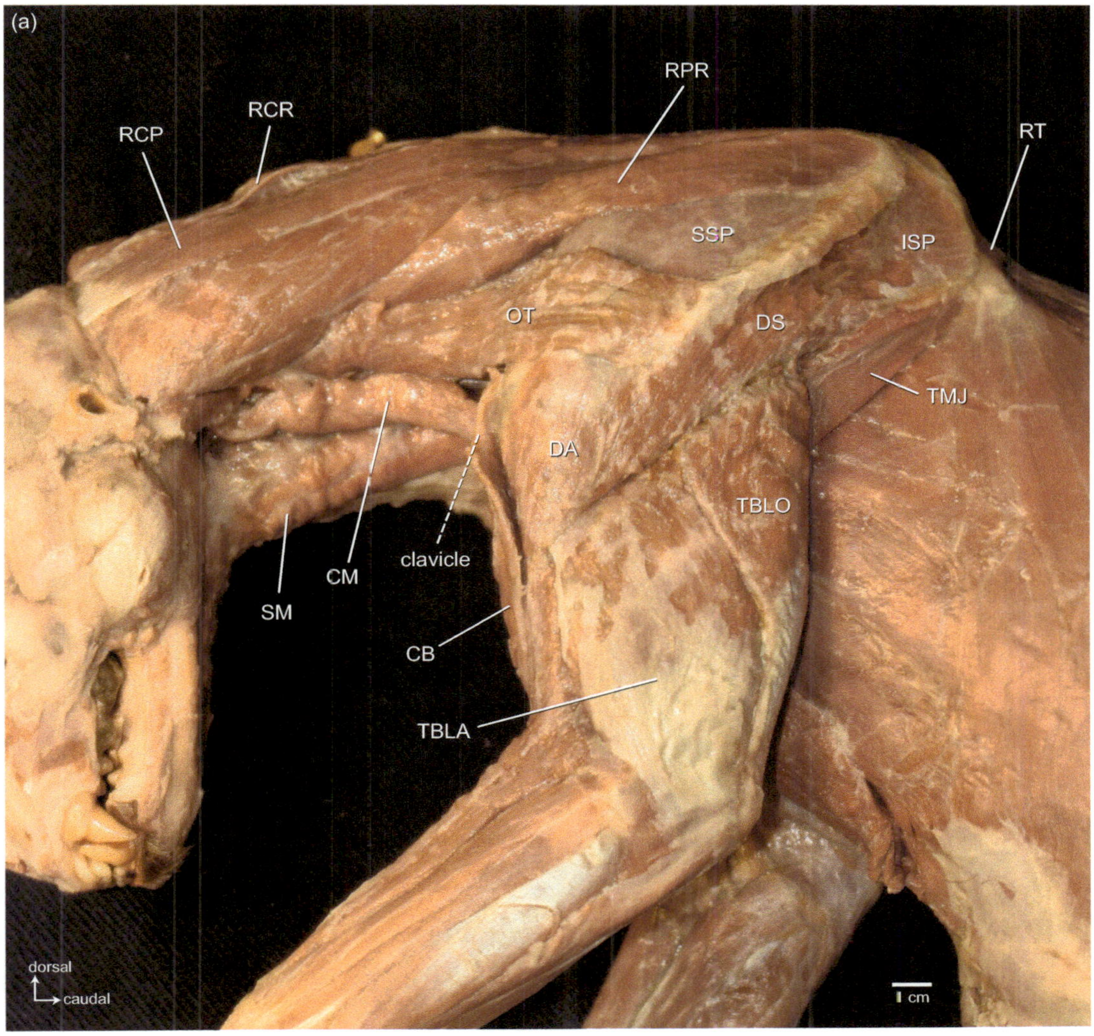
(a)
RPR
RCR
RCP
RT
SSP
ISP
OT
DS
TMJ
DA
TBLO
clavicle
CM
SM
CB
TBLA
dorsal
caudal
1 cm

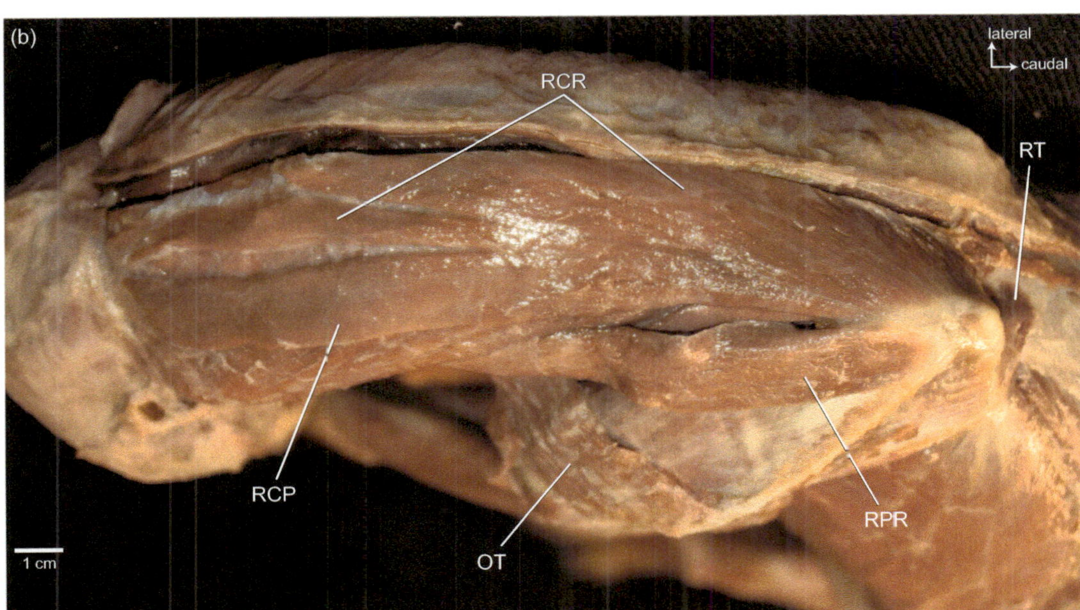
(b)
lateral
caudal
RCR
RT
RCP
RPR
OT
1 cm

FIGURE II.2.4 **Intrinsic forelimb muscles in (a) lateral view and (b) medial view.** The *Mm. deltoidei* (DS, *M. spinodeltoideus* and DA, *M. acromiodeltoideus)* and the TMJ, *M. teres major* have been removed. The BB, *M. biceps brachii* has one distinct origin. **Distal intrinsic forelimb muscles in (c) lateral view and (d) medial view.** The TMN, *M. teres minor* has been removed. **Distal intrinsic forelimb muscles in (e) lateral view and (f) medial view.** The EDL, *M. extensor digitorum lateralis* inserts with three tendons on the distal phalanges of digits III through V.

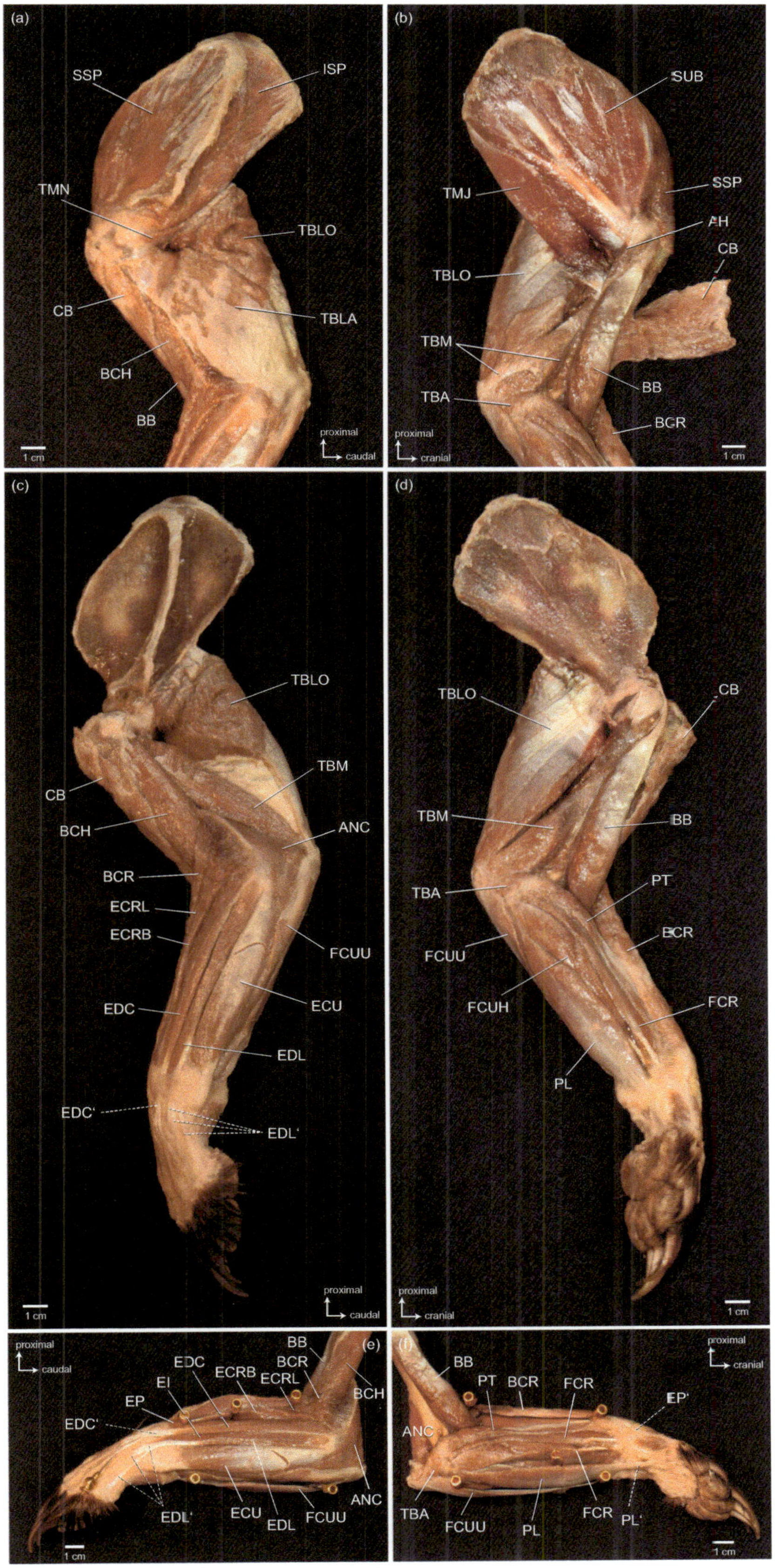
(a)
SSP
ISP
TMN
TBLO
CB
TBLA
BCH
BB
1 cm
proximal
caudal
(b)
SUB
SSP
TMJ
AH
CB
TBLO
TBM
BB
TBA
BCR
proximal
cranial
1 cm
(c)
TBLO
TBM
CB
BCH
ANC
BCR
ECRL
ECRB
FCUU
EDC
ECU
EDL
EDC'
EDL'
1 cm
proximal
caudal
(d)
TBLO
CB
TBM
BB
TBA
PT
BCR
FCUU
FCUH
FCR
PL
proximal
cranial
1 cm
(e)
proximal
caudal
BB
BCR
ECRB
ECRL
BCH
EDC
EI
EP
EDC'
ANC
EDL'
ECU
EDL
FCUU
1 cm
(f)
BB
PT
BCR
FCR
EP'
ANC
proximal
cranial
TBA
FCUU
PL
FCR
PL'
1 cm

FIGURE II.2.5 **Distal forelimb muscles. (a) In dorsal view.** The EDC, *M. extensor digitorum communis* inserts with four tendons on the distal phalanges of digits II through V. **(b, c) In ventral view.** The PL, *M. palmaris longus* and the FDP, *M. flexor digitorum profundus* insert with five tendons on the distal phalanges of digits I through V.

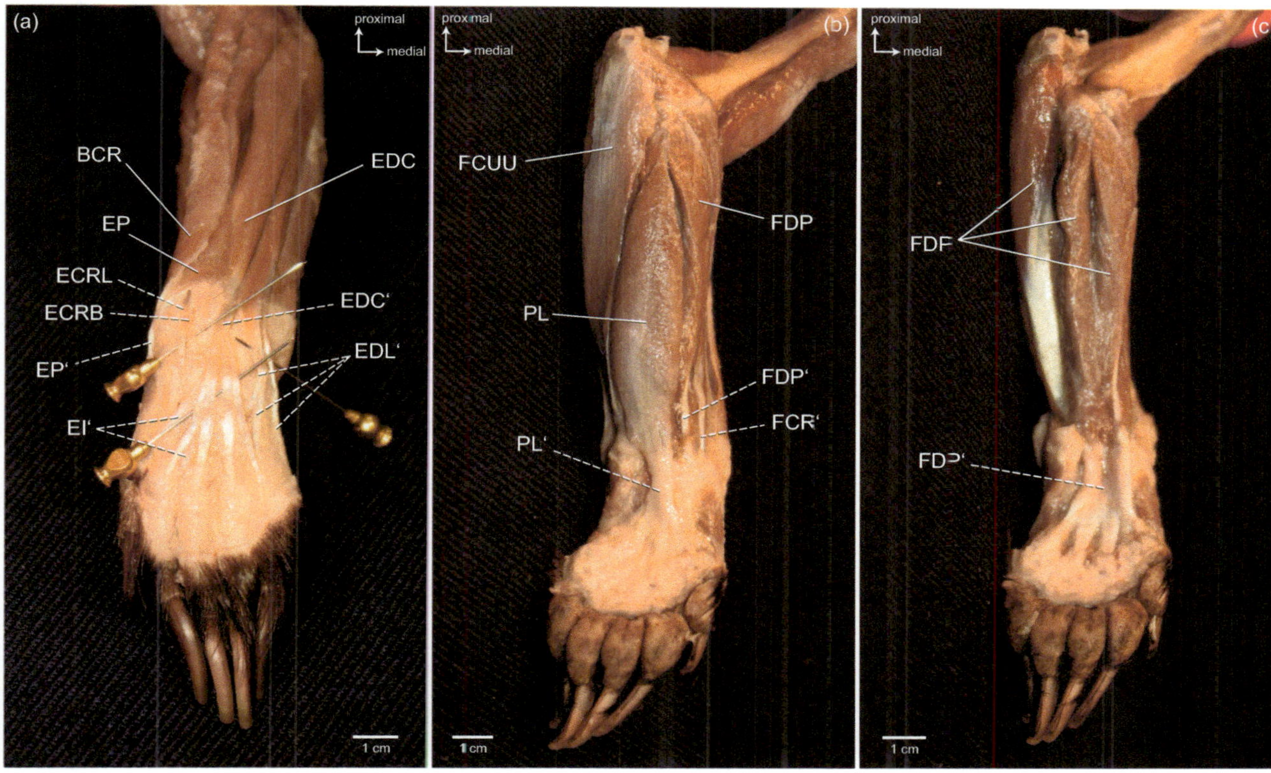
(a)
proximal
medial
BCR
EDC
EP
ECRL
ECRB
EDC'
EP'
EDL'
EI'
1 cm
proximal
medial
(b)
FCUU
FDP
PL
FDP'
FCR'
PL'
1 cm
proximal
medial
(c)
FDF
1 cm

FIGURE II.2.6 Deep rotator muscles. (a) Distal rotator muscles in dorsal view. (b) Distal rotator muscles in ventral view. Not visible in the picture, but the tendon of the BB, *M. biceps brachii* inserts into bicipital tuberosity of the radius. The tendon of the BCH, *M. brachioradialis* inserts into the coronoid process of the ulna. **(c) Proximal rotator muscles in ventral view.**

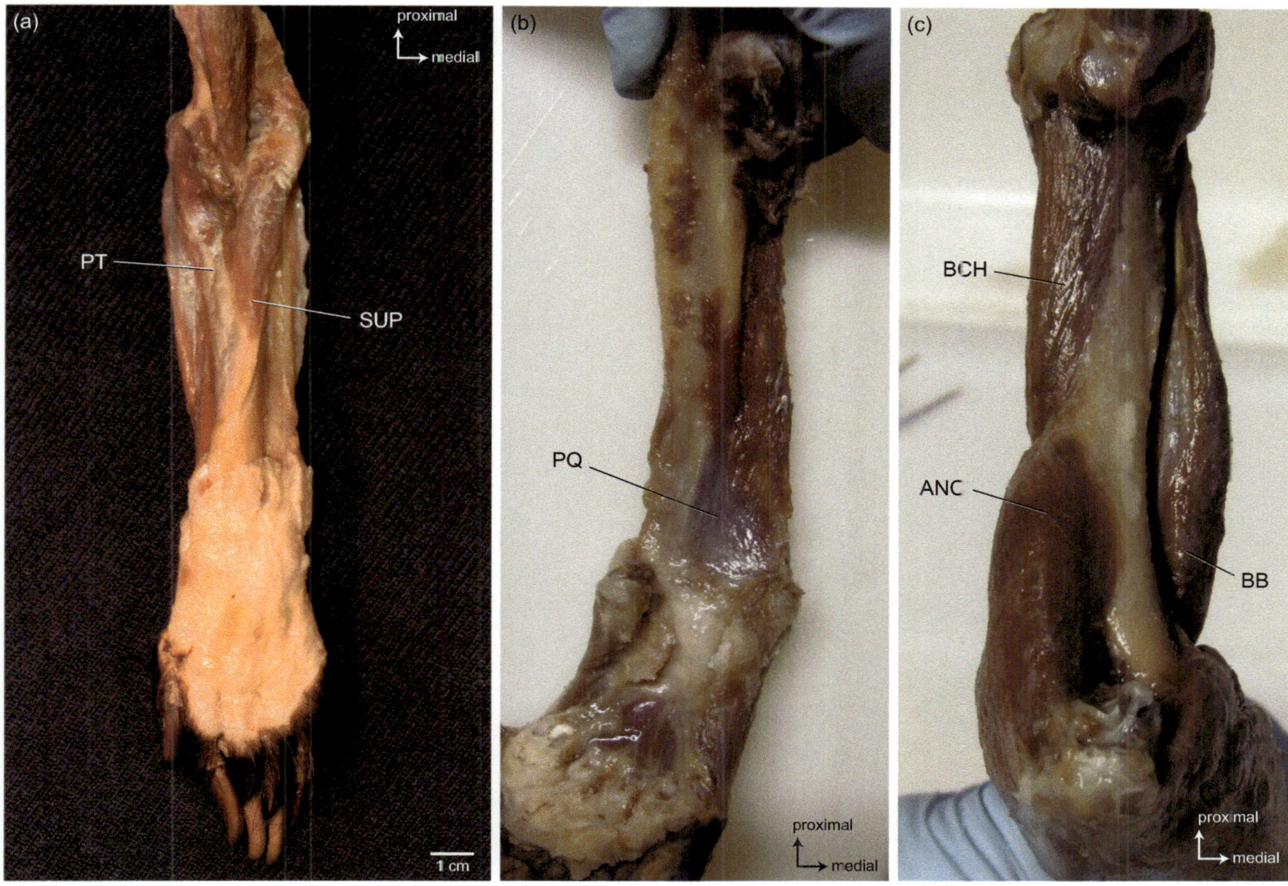
(a)
proximal
medial
PT
SUP
1 cm
(b)
PQ
proximal
medial
(c)
BCH
ANC
BB
proximal
medial

Nasua nasua (South American coati)

Classification:	Carnivora, Canoidea, Procyonidae
Mean body mass:	3.5–6.0 kg
Habitat:	Forest, mountains
Locomotor ecology:	Scansorial
Peculiarities:	Reversible hind feet

FIGURE II.3.1 Superficial muscles of forelimb and hind limb in lateral view. (a) Without labels. (b) With labels and muscle outlines. In the neck and shoulder region, the *Mm. trapezii* are clearly divided into three parts (CT, *M. clavotrapezius* – not visible in this specimen; AT, *M. arcomiotrapezius*; and ST, *M. spinotrapezius*). The EPI, *M. epitrochlearis* is well developed, and the BCR, *M. brachioradialis* is present. The OEA, *M. obliquus externus abdominis* originates from the thoracic ribs and inserts into the linea alba at the midline and on the pubis near the symphysis. In the distal hind limb, the four tendons of the EDLO, *M. extensor digitorum longus* insert on the distal phalanges of digits II through V. The EHL, *M. hallucis longus* lies deep to the TCR, *M. tibialis cranialis* and the EDLO, *M. extensor digitorum longus*. The PET, *M. peroneus tertius* is absent. Note that the retinacula (ligaments) spanning the wrist and ankle joints have been partially removed.

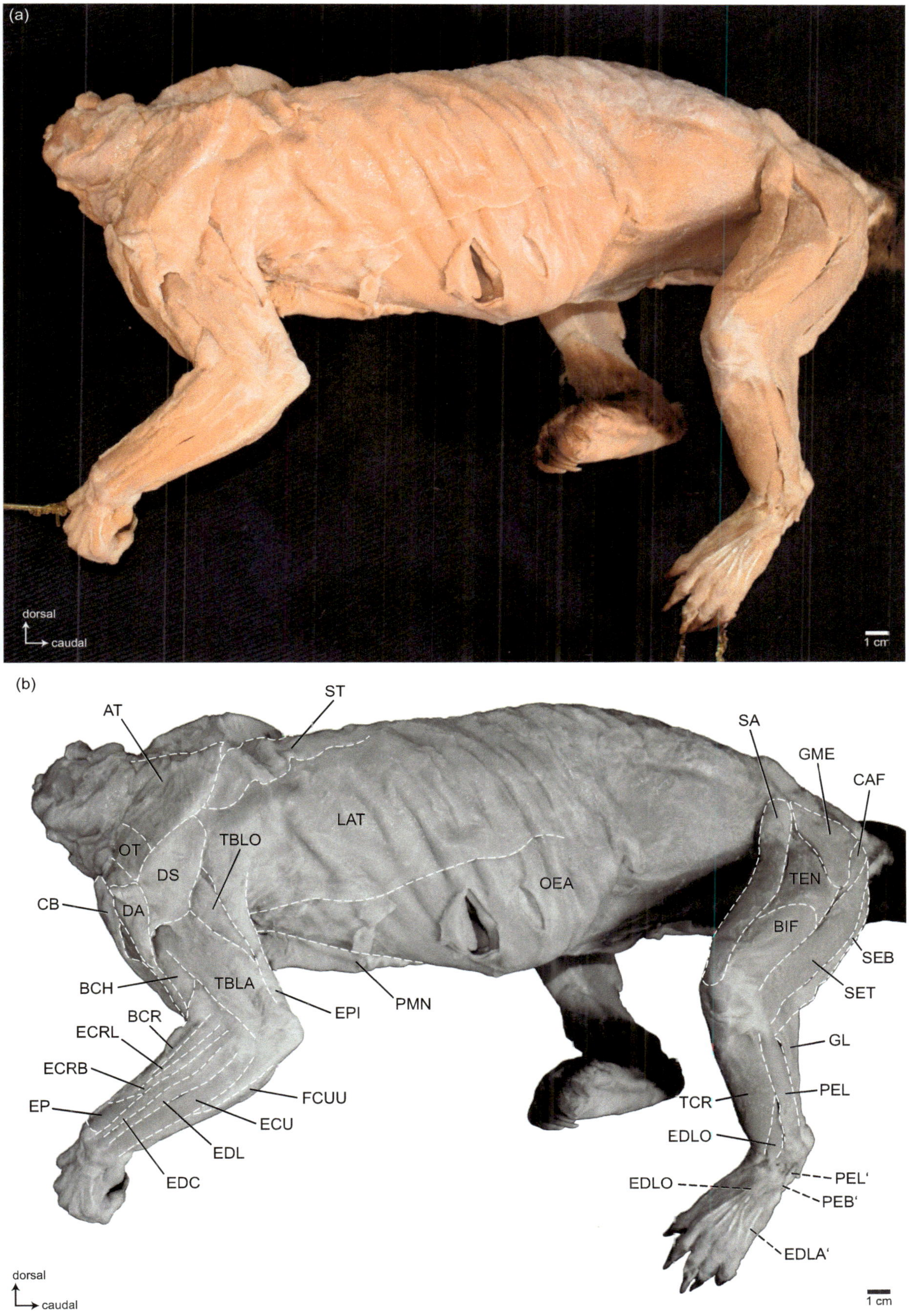
(a)
dorsal
caudal
1 cm
(b)
AT
ST
SA
GME
CAF
LAT
TBLO
OT
DS
OEA
TEN
CB
DA
BIF
SEB
BCH
TBLA
SET
EPI
PMN
BCR
ECRL
GL
ECRB
FCUU
PEL
EP
TCR
ECU
EDLO
EDL
EDC
EDLO
PEL'
PEB'
EDLA'
dorsal
caudal
1 cm

FIGURE II.3.2 **Pectoral muscles in ventral view. (a) Superficial view.** The clavicle is absent. The XH, *M. xiphihumeralis* is not clearly developed as a separate muscle, but part of the PMN, *M. pectoralis minor.* **(b) Deep view.** The PAB, *M. pectoantebrachialis* and the PMJ, *M. pectoralis major* have been removed.

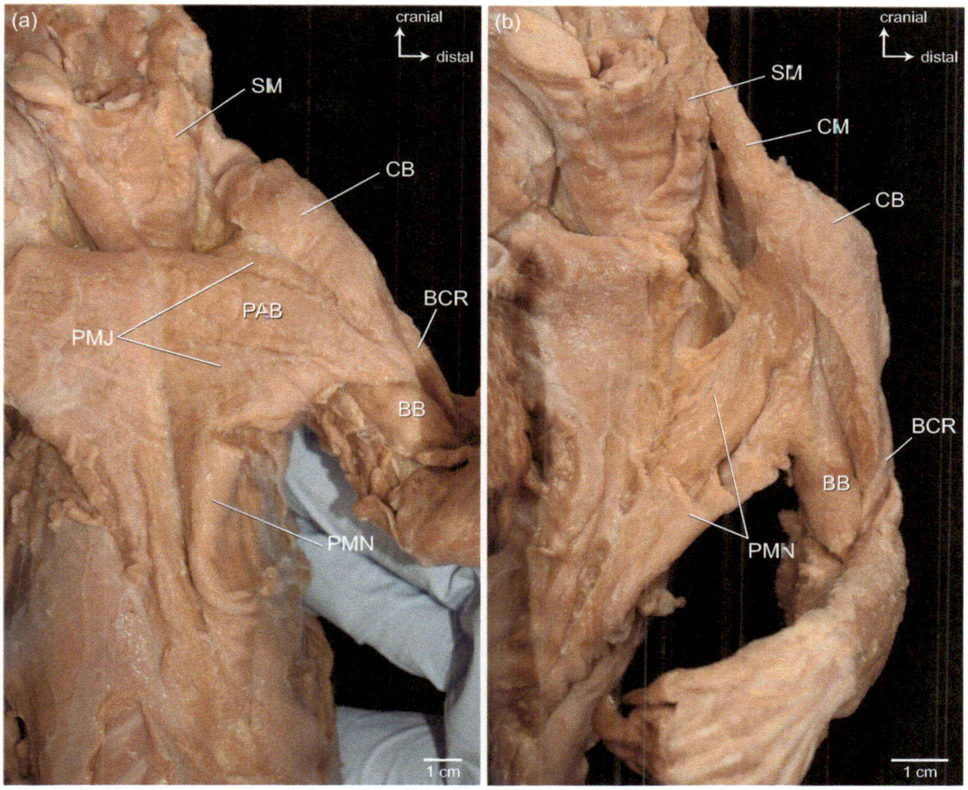
(a)
cranial
distal
SM
CB
PAB
PMJ
BCR
BB
PMN
1 cm
(b)
cranial
distal
SM
CM
CB
BCR
BB
PMN
1 cm

FIGURE II.3.3 Deep neck and shoulder muscles. (a) In lateral view. In the neck region, three *Mm. rhomboidei* are present (RCR, *M. rhomboideus cervicis*; RT, *M. rhomboideus thoracis*; and RCP, *M. rhomboideus capitis*). The RCR, *M. rhomboideus cervicis* does not reach the occiput. The RPR, *M. rhomboideus profundus* is absent. **(b) In dorsal view.**

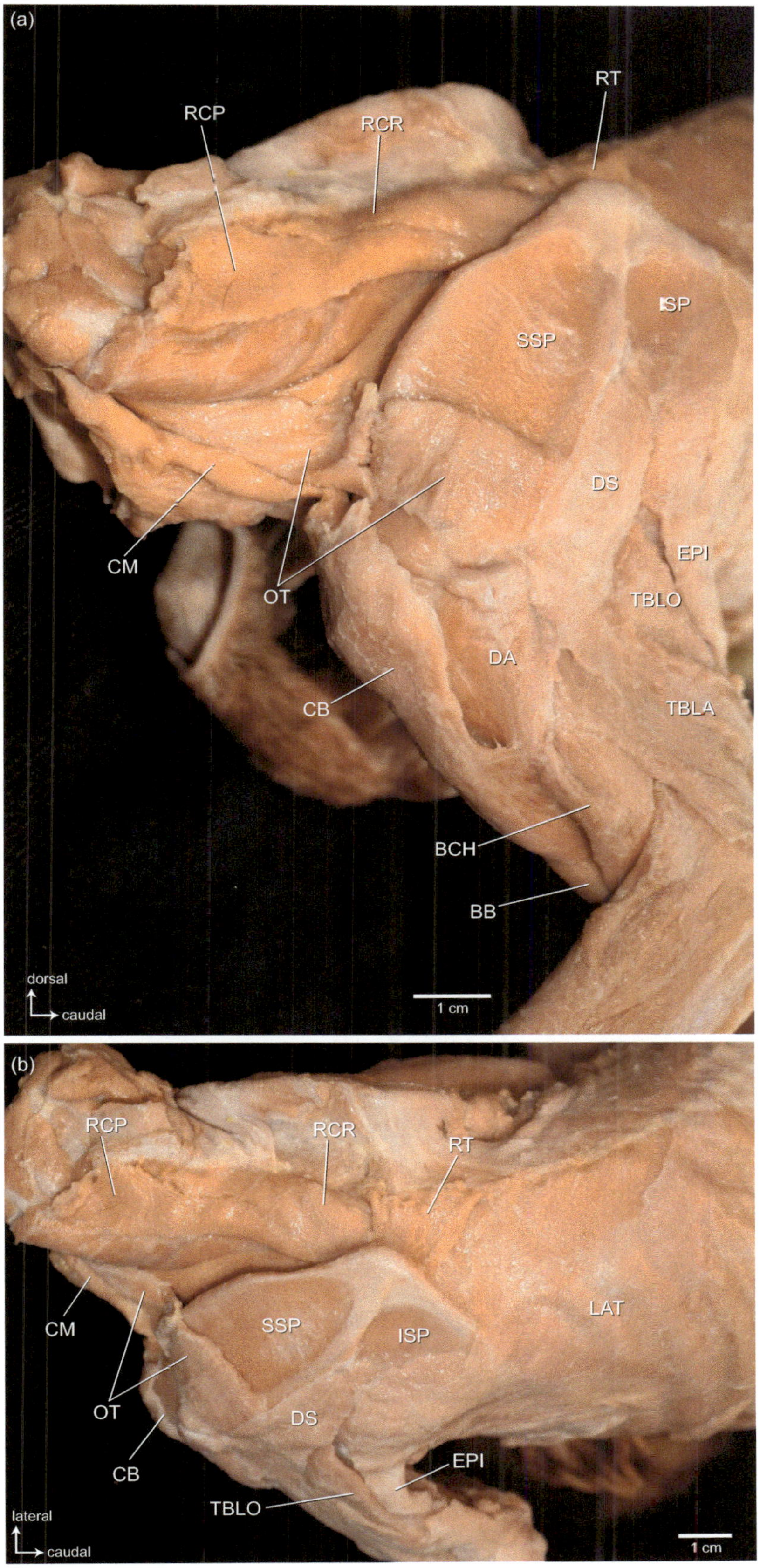
(a)
RT
RCP
RCR
ISP
SSP
DS
EPI
CM
OT
TBLO
DA
CB
TBLA
BCH
BB
dorsal
caudal
1 cm
(b)
RCP
RCR
RT
CM
SSP
ISP
LAT
OT
DS
CB
EPI
TBLO
lateral
caudal
1 cm

FIGURE II.3.4 **Intrinsic forelimb muscles in (a) proximolateral view and (b) proximomedial view.** The *Mm. deltoidei* (DS, *M. spinodeltoideus* and DA, *M. acromiodeltoideus)* have been removed. The BB, *M. biceps brachii* has one distinct origin. **Intrinsic forelimb muscles in (c) lateral view and (d) medial view.** The EDL, *M. extensor digitorum lateralis* inserts with three tendons on the distal phalanges of digits III through V. The PL, *M. palmaris longus* inserts into the palmar aponeurosis.

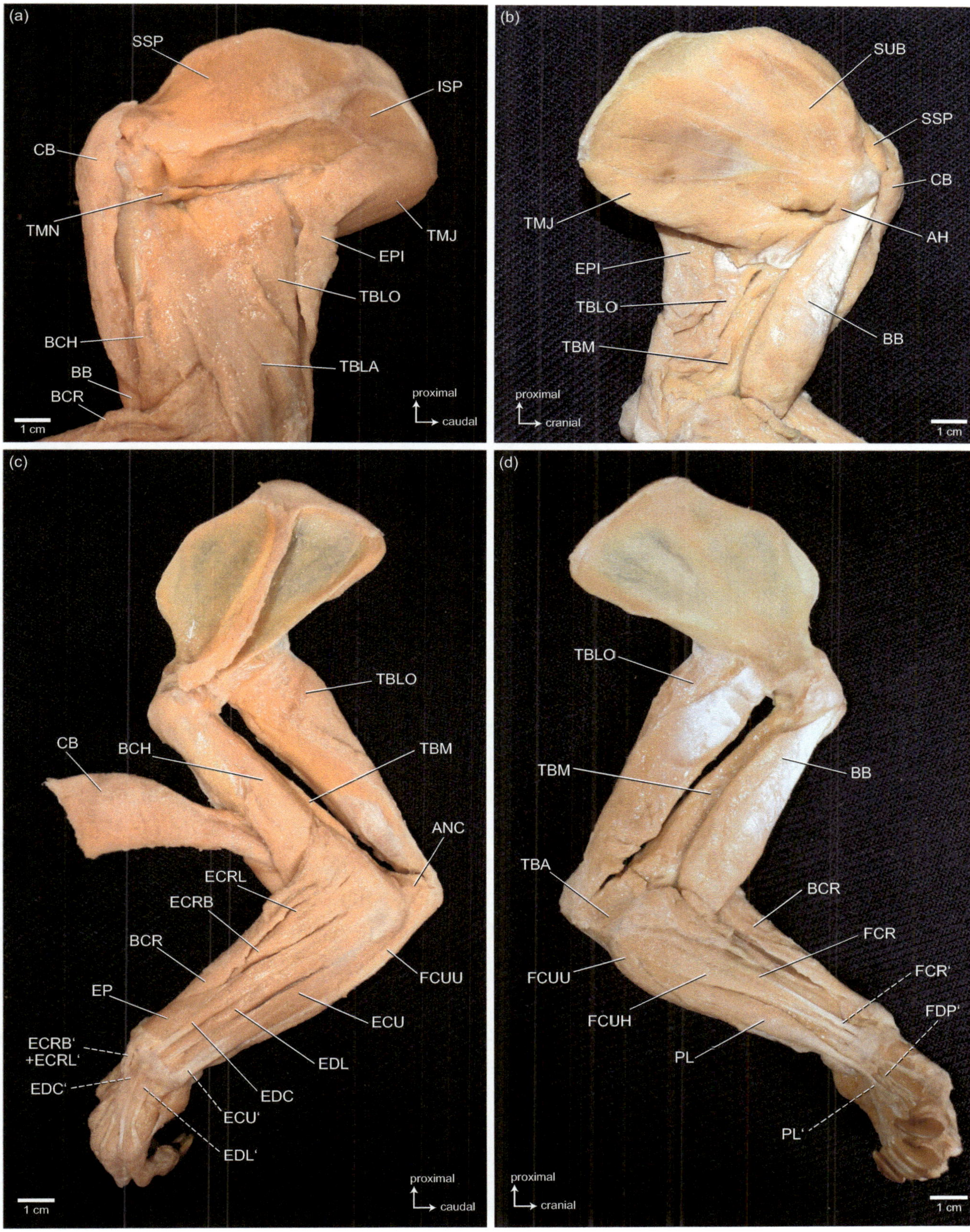
(a)
SSP
ISP
CB
TMN
TMJ
EPI
TBLO
BCH
TBLA
BB
BCR
1 cm
proximal
caudal
(b)
SUB
SSP
CB
TMJ
AH
EPI
TBLO
BB
TBM
proximal
cranial
1 cm
(c)
TBLO
CB
BCH
TBM
ANC
ECRL
ECRB
BCR
FCUU
EP
ECU
ECRB‘
+ECRL‘
EDL
EDC‘
EDC
ECU‘
EDL‘
1 cm
proximal
caudal
(d)
TBLO
TBM
BB
TBA
BCR
FCR
FCUU
FCR‘
FCUH
FDP‘
PL
PL‘
proximal
cranial
1 cm

FIGURE II.3.5 **Distal forelimb muscles. (a, b) In dorsal view.** The EDC, *M. extensor digitorum communis* inserts with four tendons on the distal phalanges of digits II through V. The ECRB, *M. extensor carpi radialis brevis* inserts with one, but very broad tendon into the base of metacarpal III. The EI, *M. extensor digiti I and II* inserts with two tendons on the distal phalanges of digits I and II. **(c) In ventral view.** The FDP, *M. flexor digitorum profundus* (four muscle bellies) inserts with five tendons on the distal phalanges of digits I through V. The LUM, *Mm. lumbricales* originate from the four tendons of the FDP, *M. flexor digitorum profundus* and insert into the proximal phalanges of the digits. **Hand muscles. (d, e) In dorsal and ventral views.** The ADQ, *M. abductor digiti V* originates from the Os pisiforme (carpal bone) and inserts into the proximal phalanx of digit V. The FDQB, *M. flexor digiti V brevis* originates from the Os hamatum (carpal bone, adjacent to the Os pisiforme) and inserts distal to the ADQ, *M. abductor digiti V* into the proximal phalanx of digit V. The APB, *M. abductor pollicis brevis* originates from the sesamoid and inserts into proximal phalanx of digit I. The FBP, *Mm. flexores breves profundi* serve every digit. They originate from the palmar aspect of the base of the metacarpals and insert into the proximal phalanges of the digits.

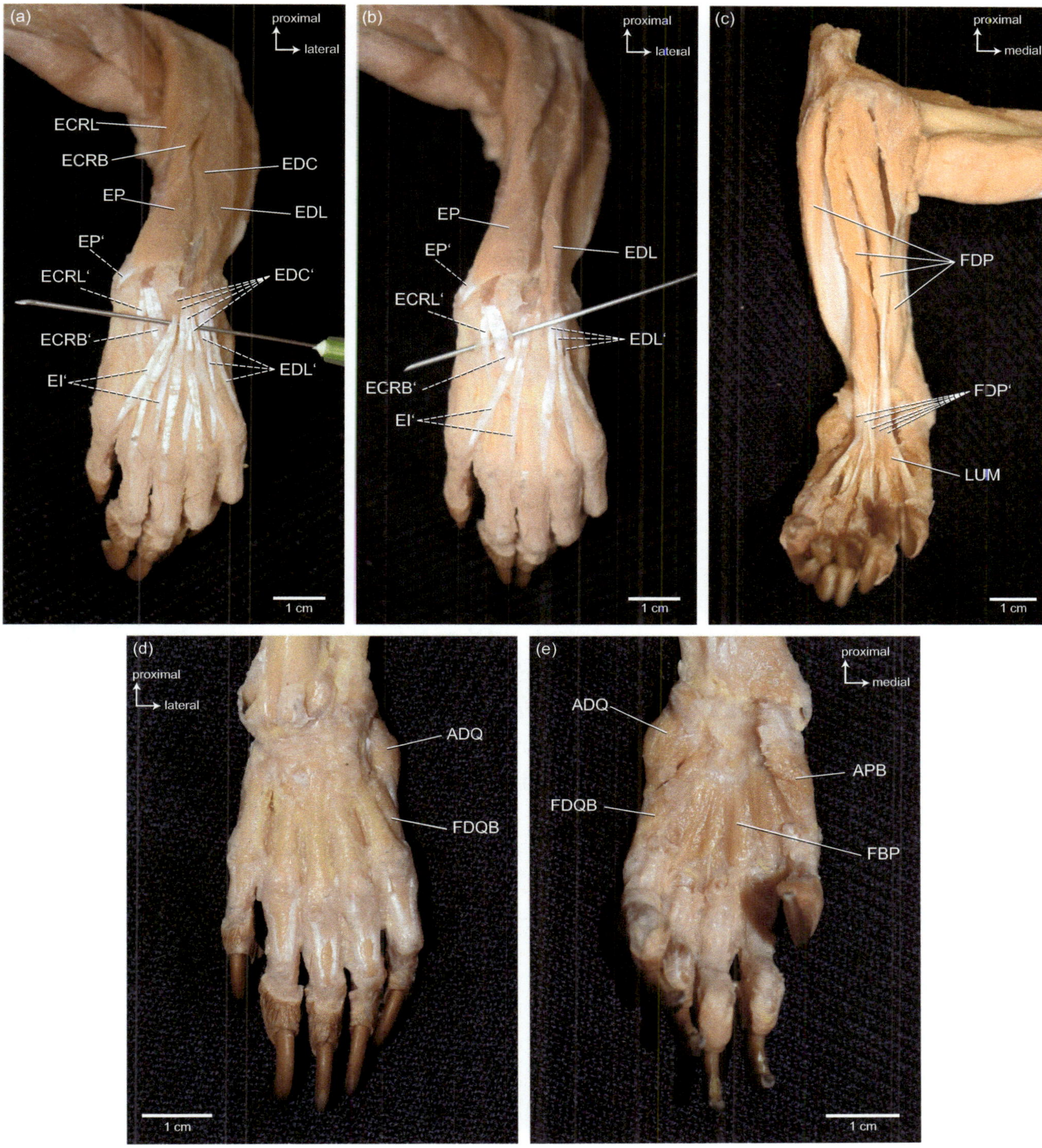
(a)
proximal
lateral
ECRL
ECRB
EDC
EP
EDL
EP'
ECRL'
EDC'
ECRB'
EI'
EDL'
1 cm
(b)
proximal
lateral
EP
EP'
EDL
ECRL'
EDL'
ECRB'
EI'
1 cm
(c)
proximal
medial
FDP
FDP'
LUM
1 cm
(d)
proximal
lateral
ADQ
FDQB
1 cm
(e)
proximal
medial
ADQ
APB
FDQB
FBP
1 cm

FIGURE II.3.6 Deep rotator muscles. (a) Distal rotator muscles in dorsal view. (b) Distal rotator muscles in ventral view. (c) Proximal rotator muscles in dorsomedial view. The tendon of the BB, *M. biceps brachii* inserts distally to the tendon of the BCH, *M. brachioradialis* into the radius. The tendon of the BCH, *M. brachioradialis* inserts distally into the coronoid process into the ulna. **(d) Proximal rotator muscles in ventral view.**

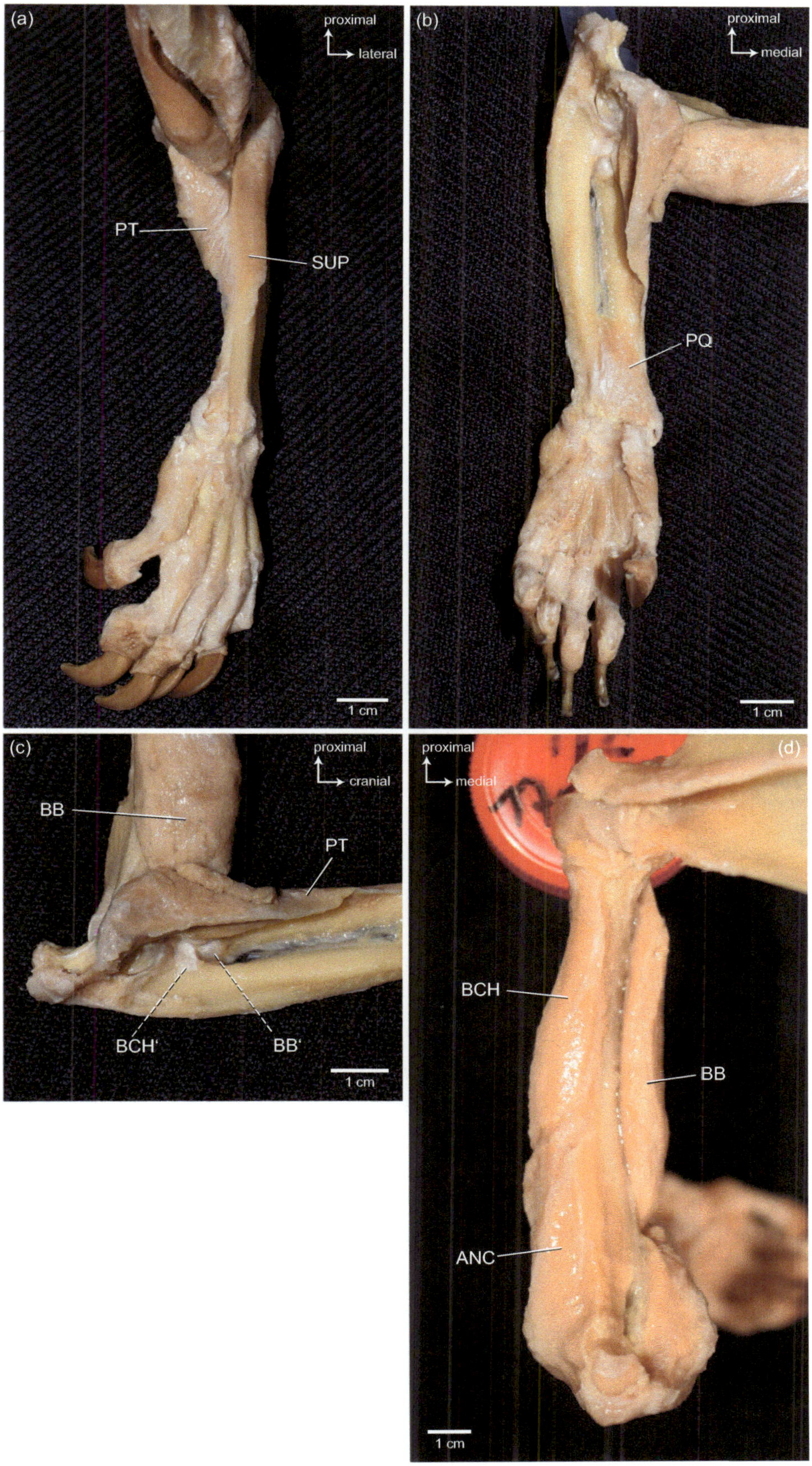
(a)
proximal
lateral
PT
SUP
1 cm
(b)
proximal
medial
PQ
1 cm
(c)
proximal
cranial
BB
PT
BCH'
BB'
1 cm
(d)
proximal
medial
BCH
BB
ANC
1 cm

Potos flavus (Kinkajou)

Classification:	Carnivora, Canoidea, Procyonidae
Mean body mass:	2.0–4.6 kg
Habitat:	Forest, rainforest
Locomotor ecology:	Scansorial
Peculiarities:	Reversible hind feet; prehensile tail; high spinal flexibility

FIGURE II.4.1 Superficial muscles of forelimb and hind limb in lateral view. (a) Without labels. (b) With labels and muscle outlines. In the neck and shoulder region, the *Mm. trapezii* are clearly divided into three parts (CT, *M. clavotrapezius*; AT, *M. arcomiotrapezius*; and ST, *M. spinotrapezius*). A separate muscle runs parallel to the SM, *M. sternomastoideus.* It originates from the occiput and inserts into the sternum: the SO, *M. sternooccipitalis.* The EPI, *M. epitrochlearis* is well developed, and the BCR, *M. brachioradialis* is present. The OEA, *M. obliquus externus abdominis* originates from the thoracic ribs and inserts into the linea alba at the midline and on the pubis near the symphysis. The RA, *M. rectus abdominis* originates from the costal cartilages and inserts into the prebubic tendon. The PIR, *M. piriformis* lies deep to the GME, *M. gluteus medius* and GSU, *M. gluteus superficialis.* In the distal hind limb, the four tendons of the EDLO, *M. extensor digitorum longus* insert on the distal phalanges of digits II through V. The PET, *M. peroneus tertius* is absent. Note that the retinacula (ligaments) spanning the wrist and ankle joints have been largely removed.

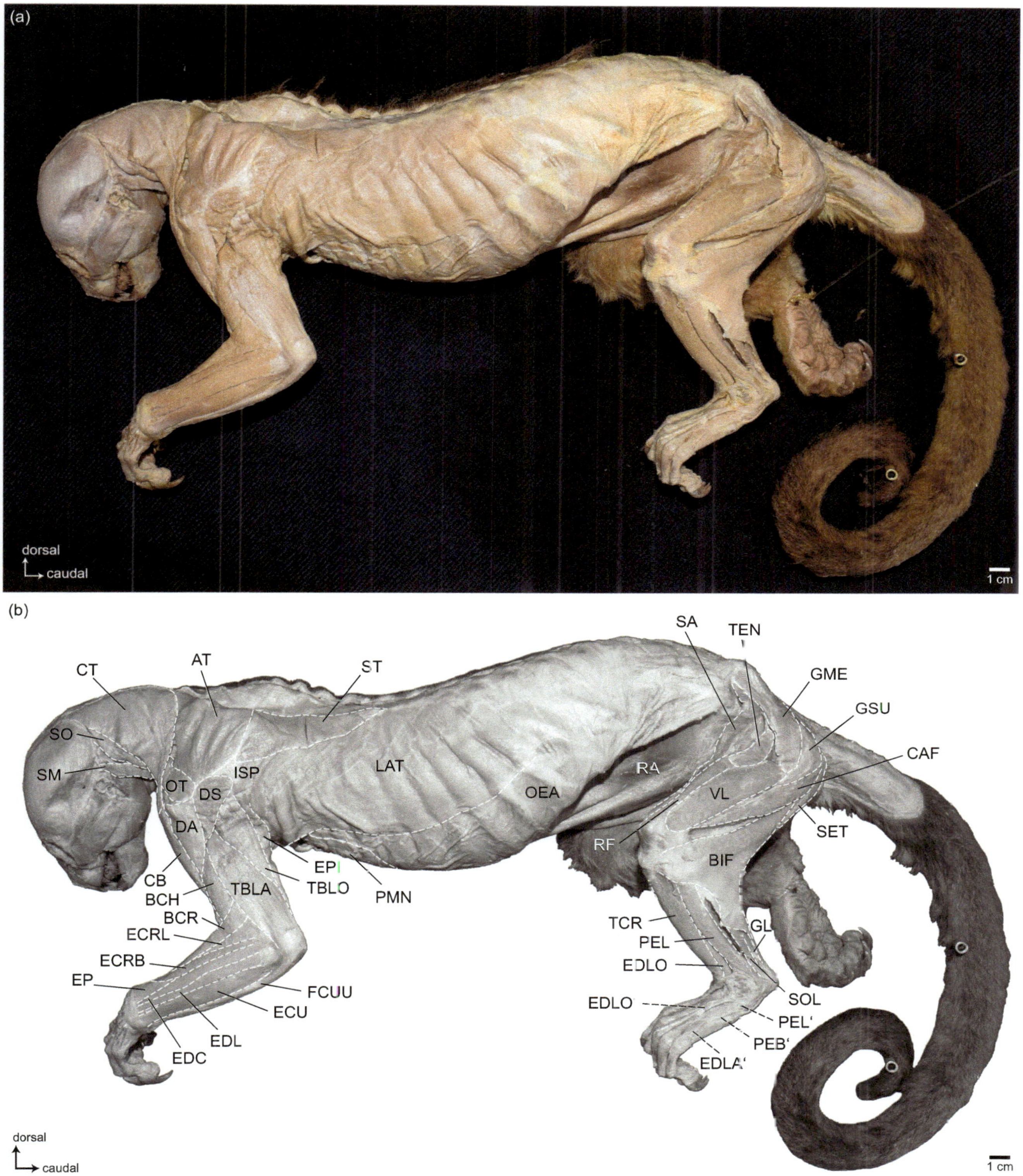
(a)
dorsal
caudal
1 cm
(b)
CT
AT
ST
SO
SM
OT
ISP
DS
DA
LAT
OEA
RA
SA
TEN
GME
GSU
CAF
VL
SET
RF
BIF
CB
BCH
TBLA
EPI
TBLO
PMN
BCR
ECRL
ECRB
EP
FCUU
ECU
EDL
EDC
TCR
PEL
GL
EDLO
SOL
EDLO'
PEL'
PEB'
EDLA'
dorsal
caudal
1 cm

FIGURE II.4.2 **Pectoral muscles in ventral view. (a) Superficial view.** A separate muscle runs parallel to the SM, *M. sternomastoideus*. It originates from the occiput and inserts into the sternum: the SO, *M. sternooccipitalis*. The clavicle is absent. The XH, *M. xiphihumeralis* is separated from the PMN, *M. pectoralis minor.* **(b) Deep view.** The PMJ, *M. pectoralis major*; the PAB, *M. pectoantebrachialis*; the XH, *M. xiphihumeralis*; and the LAT, M. *latissimus dorsi* have been removed.

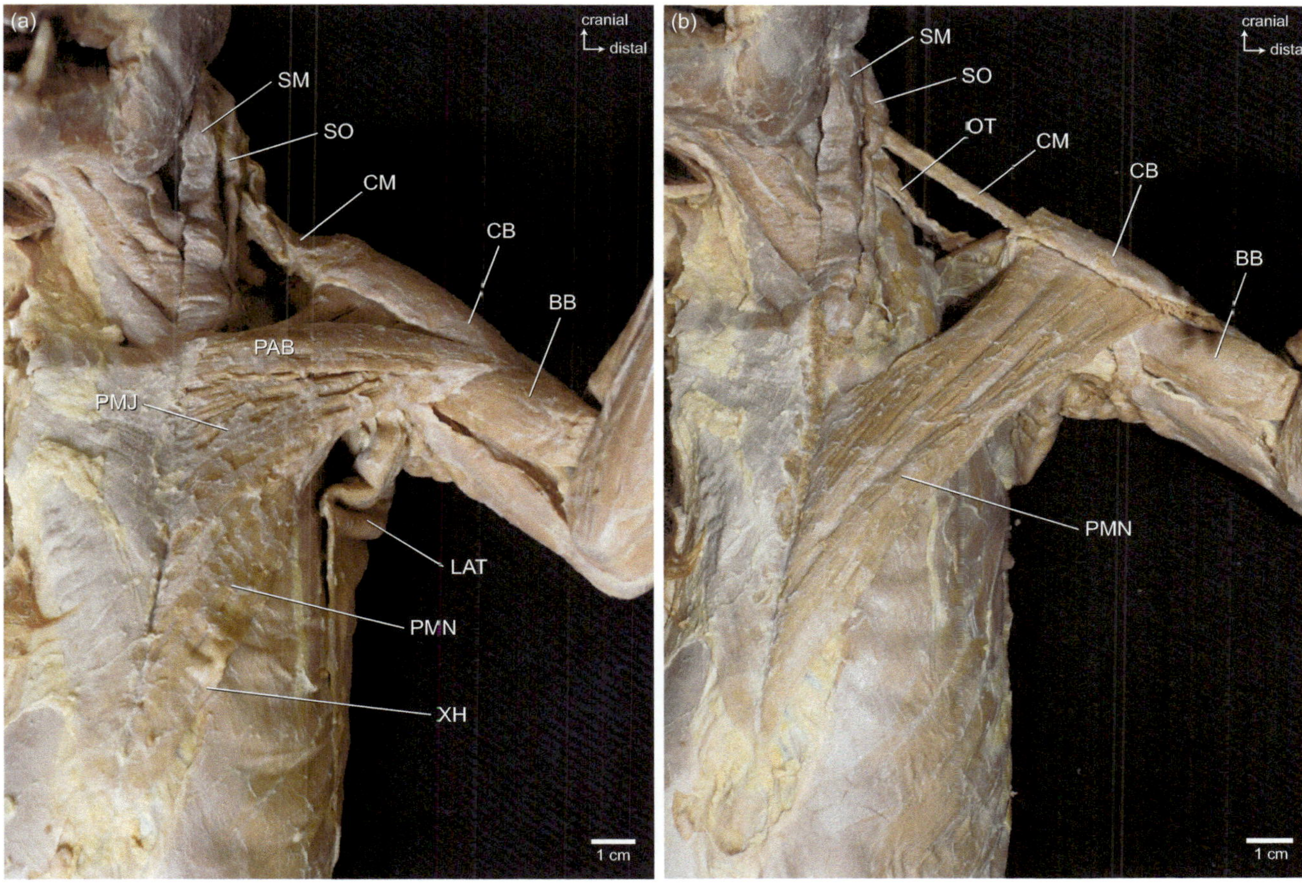
(a)
cranial
distal
SM
SO
CM
CB
BB
PAB
PMJ
LAT
PMN
XH
1 cm
(b)
cranial
distal
SM
SO
OT
CM
CB
BB
PMN
1 cm

FIGURE II.4.3 **Deep neck and shoulder muscles. (a) In lateral view.** In the neck region, all four *Mm. rhomboidei* are present (RCR, *M. rhomboideus cervicis*; RT, *M. rhomboideus thoracis*; RCP, *M. rhomboideus capitis*; and RPR, *M. rhomboideus profundus*). The SO, *M. sternooccipitalis* originates from the occiput and inserts into the sternum. **(b) In dorsal view.** The RCR, *M. rhomboideus cervicis* does not reach the occiput.

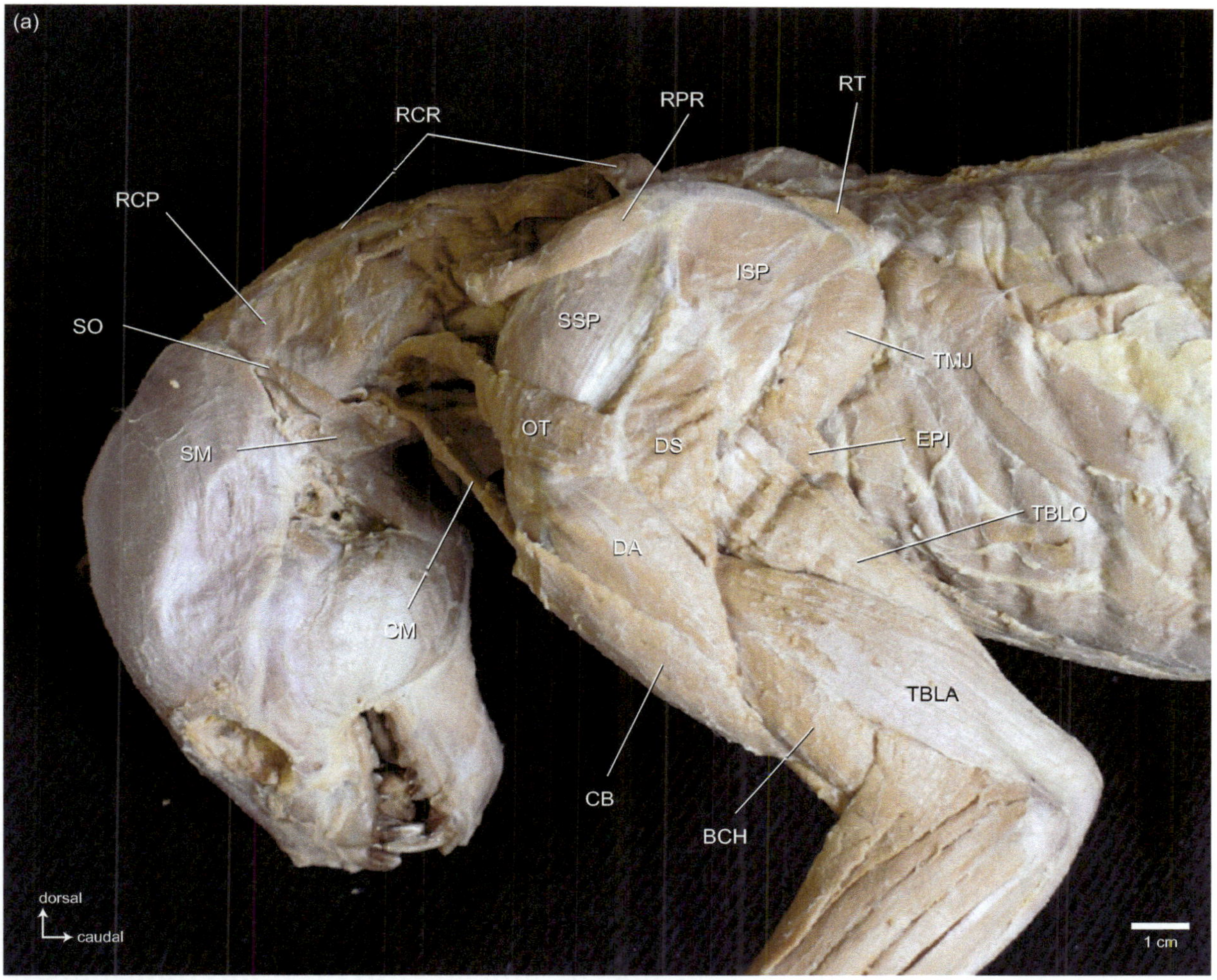
(a)
RCR
RPR
RT
RCP
ISP
SSP
SO
TMJ
OT
SM
DS
EPI
TBLO
DA
CM
TBLA
CB
BCH
dorsal
caudal
1 cm

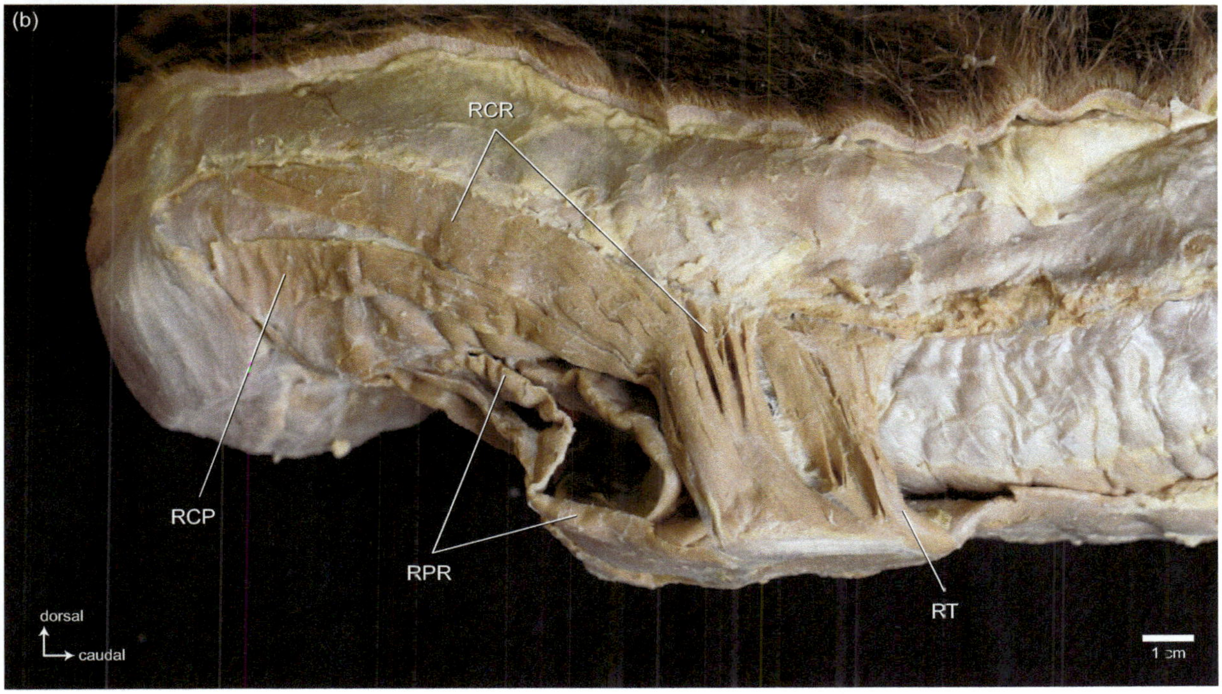
(b)
RCR
RCP
RPR
RT
dorsal
caudal
1 cm

FIGURE II.4.4 **Intrinsic muscles in (a) proximolateral view and (b) proximomedial view.** The *Mm. deltoidei* (DS, *M. spinodeltoideus* and DA, *M. acromiodeltoideus*) have been removed. Note the very small attachment area for the *M. serratus ventralis* medial on the scapula. The distal tendon of the AH, *M. articularis humeri* gives origin to a muscle belly that distally fuse with the BB, *M. biceps brachii*. **Intrinsic forelimb muscles in (c) lateral view and (d) medial view.** The AH, *M. articularis humeri* and the TBMS, *M. triceps brachii caput mediale, short portion* have been removed. The PL, *M. palmaris longus* inserts into the palmar aponeurosis.

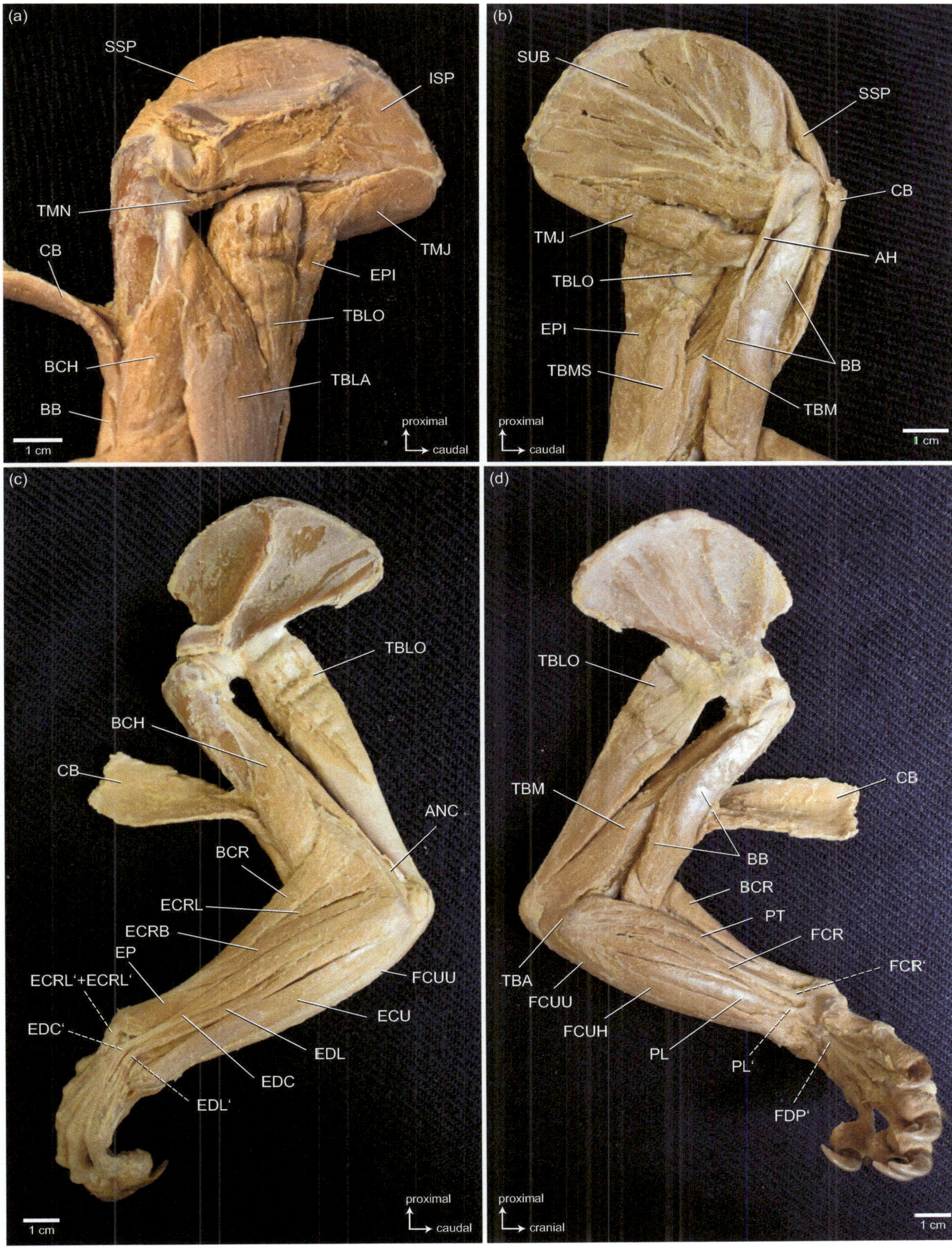
(a)
SSP
ISP
TMN
CB
TMJ
EPI
TBLO
BCH
TBLA
BB
1 cm
proximal
caudal
(b)
SUB
SSP
CB
TMJ
AH
TBLO
EPI
BB
TBMS
TBM
proximal
caudal
1 cm
(c)
TBLO
BCH
CB
ANC
BCR
ECRL
ECRB
EP
FCUU
ECRL'+ECRL'
ECU
EDC'
EDL
EDC
EDL'
1 cm
proximal
caudal
(d)
TBLO
TBM
CB
BB
BCR
PT
FCR
FCR'
TBA
FCUU
FCUH
PL
PL'
FDP'
proximal
cranial
1 cm

FIGURE II.4.5 **Distal forelimb muscles. (a, b) In dorsal view.** Note the strong tendon of ECRB, *M. extensor carpi radialis brevis* that inserts into metacarpal III. The EI, *M. extensor digiti I and II* inserts with one tendon on the distal phalanges of digits II. The EDC, *M. extensor digitorum communis* inserts with four tendons on the distal phalanges of digits II through V. The EDL, *M. extensor digitorum lateralis* inserts with three tendons on the distal phalanges of digits III through V. Note the proximal origin of the EI, *M. extensor digiti I and II.* **(c) In ventral view.** The FDP, *M. flexor digitorum profundus* (four muscle bellies) inserts with five tendons on the distal phalanges of digits I through V.

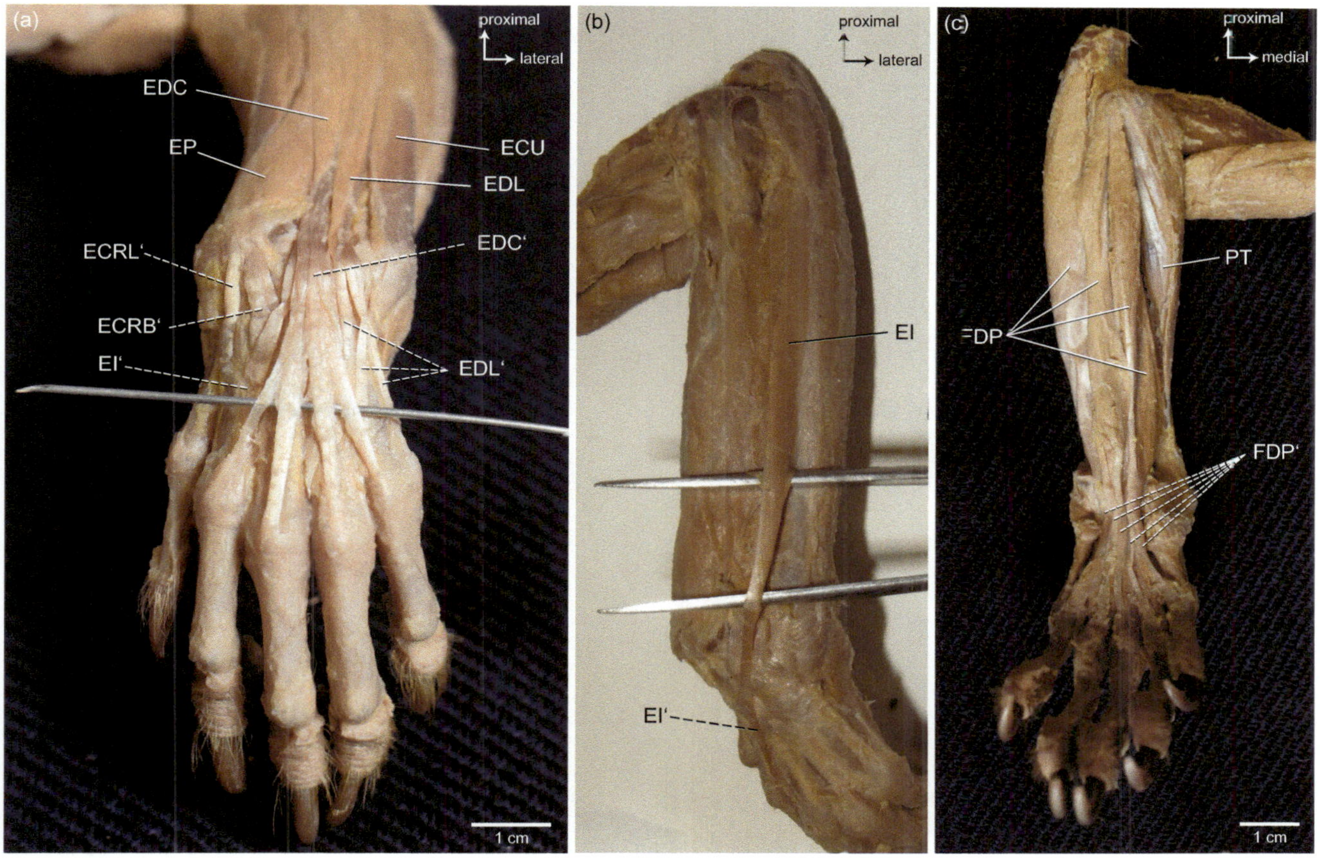
(a)
proximal
lateral
EDC
EP
ECU
EDL
ECRL'
EDC'
ECRB'
EI'
EDL'
1 cm
(b)
proximal
lateral
EI
EI'
(c)
proximal
medial
PT
FDP
FDP'
1 cm

FIGURE II.4.6 Deep rotator muscles. (a) Distal rotator muscles in dorsal view. (b) Distal rotator muscles in ventral view. The tendon of the BB, *M. biceps brachii* inserts distally to the tendon of the BCH, *M. brachioradialis* into the radius. The tendon of the BCH, *M. brachioradialis* inserts distally into the coronoid process into the ulna. Note the well-developed interosseus membrane between radius and ulna. **(c) Proximal rotator muscles in ventromedial view.**

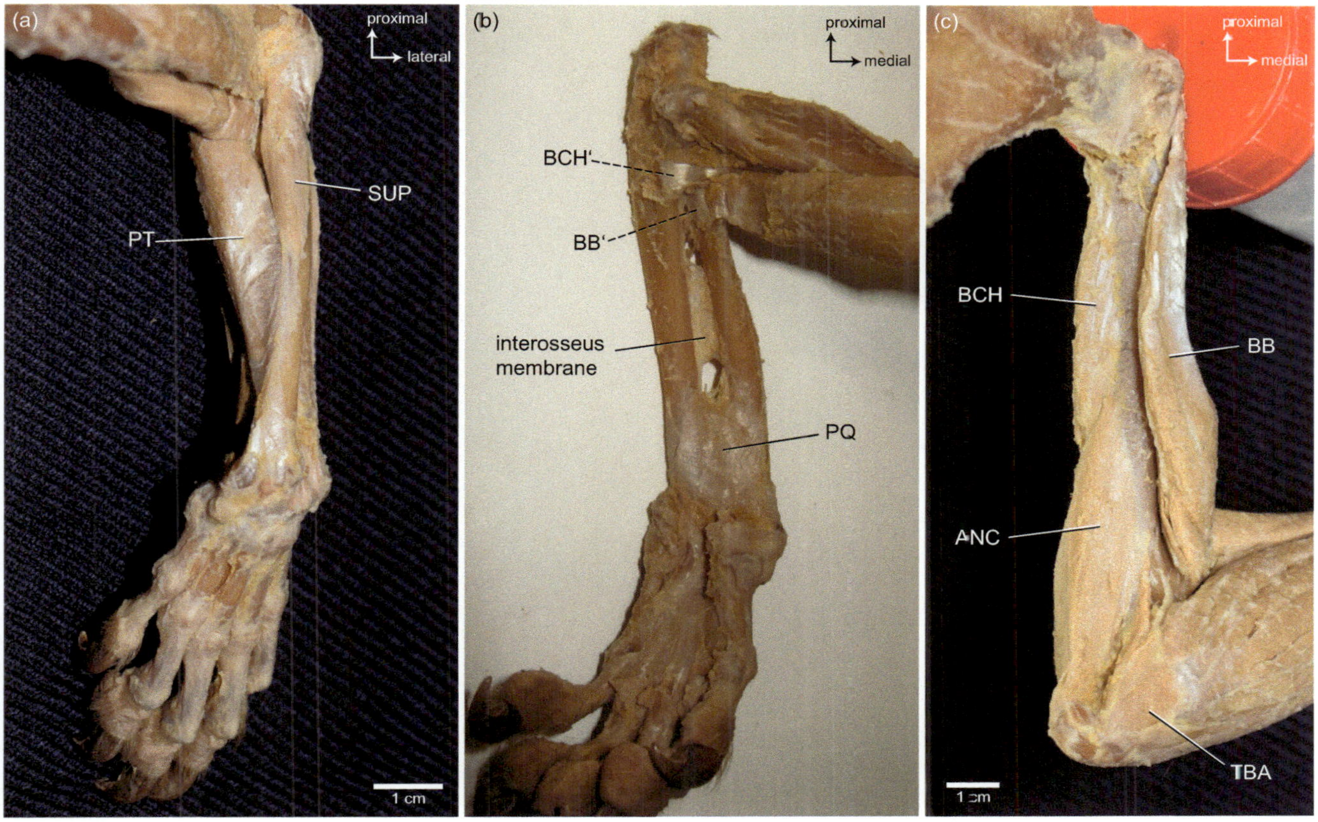
(a)
proximal
lateral
SUP
PT
1 cm
(b)
proximal
medial
BCH'
BB'
interosseus
membrane
PQ
(c)
proximal
medial
BCH
BB
ANC
TBA
1 cm

Cuon alpinus (Dhole)

Classification:	Carnivora, Canoidea, Canidae
Mean body mass:	17.0–21.0 kg
Habitat:	Forest, forest steppes
Locomotor ecology:	Cursorial
Peculiarities:	Digitigrade

FIGURE II.5.1 **Superficial muscles of forelimb and hind limb in lateral view. (a) Without labels. (b) With labels and muscle outlines.** In the neck and shoulder region, the *Mm. trapezii* are clearly divided into three parts (CT, *M. clavotrapezius* – not visible in this specimen; AT, *M. arcomiotrapezius*; and ST, *M. spinotrapezius*). The EPI, *M. epitrochlearis* and the BCR, *M. brachioradialis* are absent. The OEA, *M. obliquus externus abdominis* originates from the thoracic ribs and inserts into the linea alba at the midline and on the pubis near the symphysis. The RA, *M. rectus abdominis* originates from the costal cartilages and inserts into the prebubic tendon. In the hip region, the CAF, *M. caudofemoralis* lies deep to the BIF, *M. biceps femoris*. In the distal hind limb, the four tendons of the EDLO, *M. extensor digitorum longus* insert on the distal phalanges of digits II through V. The PET, *M. peroneus tertius* is absent. Note that the retinacula (ligaments) spanning the wrist and ankle joints have been partially removed.

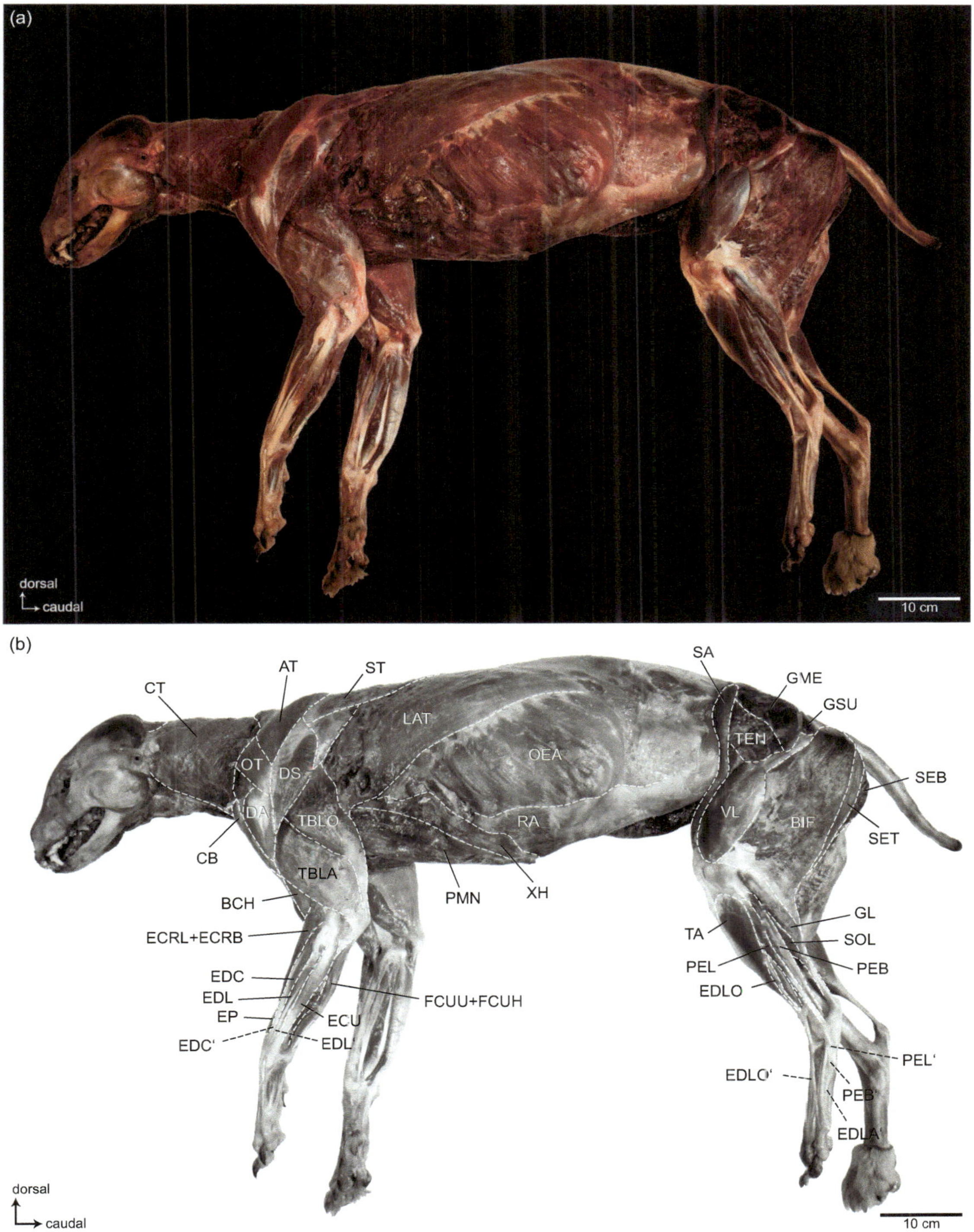
(a)
dorsal
caudal
10 cm
(b)
CT
AT
ST
LAT
OT
DS
DA
TBLO
TBLA
OEA
RA
CB
BCH
PMN
XH
ECRL+ECRB
EDC
EDL
EP
EDC'
EDL'
ECU
FCUU+FCUH
SA
GME
GSU
TEN
SEB
VL
BIF
SET
GL
TA
SOL
PEL
PEB
EDLO
PEL'
EDLO'
PEB'
EDLA'
dorsal
caudal
10 cm

FIGURE II.5.2 **Pectoral muscles in ventral view. Superficial view.** The clavicle is absent. The four pectoral muscles, PMJ, *M. pectoralis major*, PAB, *M. pectoantebrachialis*, PMN, *M. pectoralis minor* and XH, *M. xiphihumeralis* are clearly separated each other.

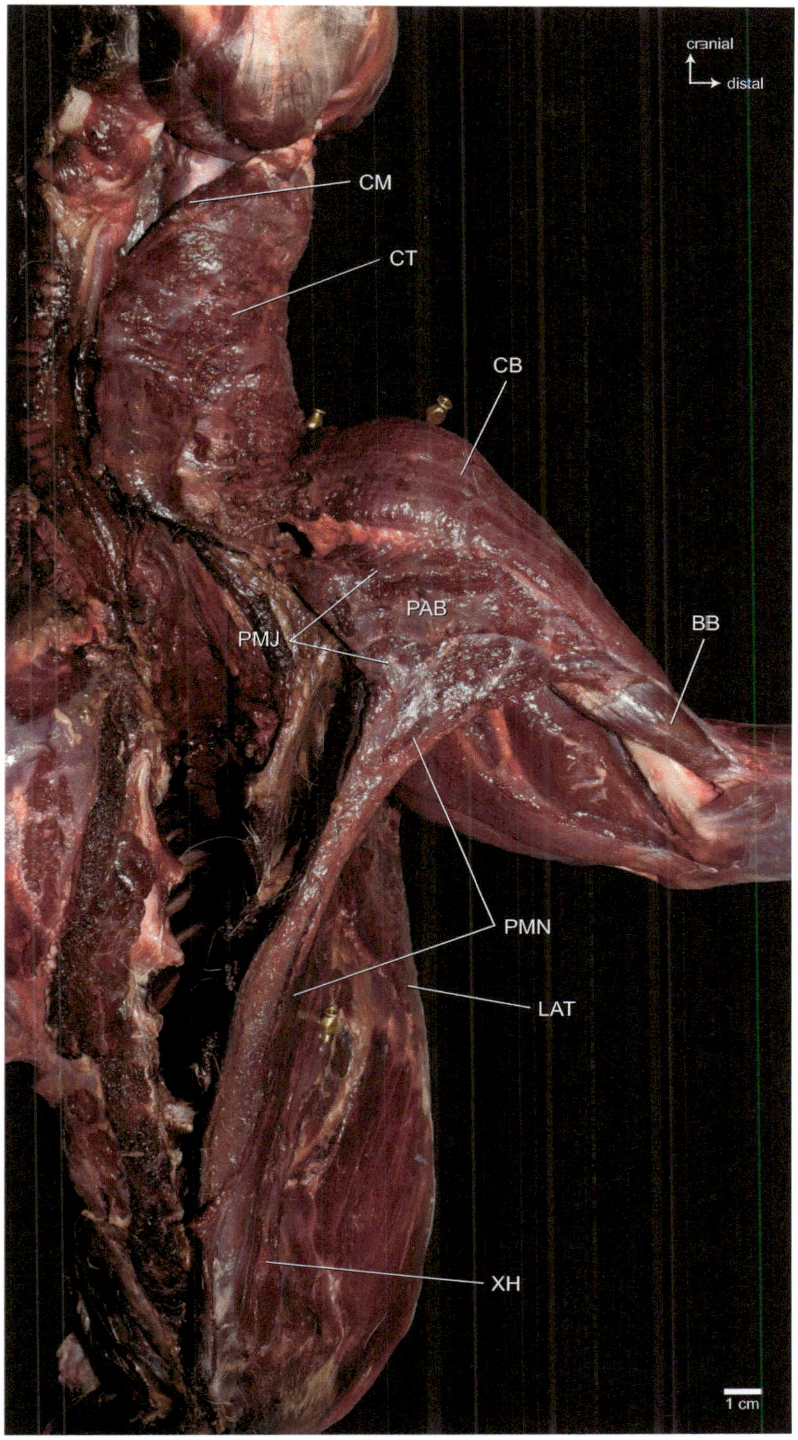
cranial
distal
CM
CT
CB
PAB
BB
PMJ
PMN
LAT
XH
1 cm

FIGURE II.5.3 Deep neck and shoulder muscles. (a) In lateral view. In the neck region, three *Mm. rhomboidei* are present (RCR, *M. rhomboideus cervicis*; RT, *M. rhomboideus thoracis*; and RCP, *M. rhomboideus capitis*). The RPR, *M. rhomboideus profundus* is absent. **(b) In dorsal view.** The RCR, *M. rhomboideus cervicis* does not reach the occiput.

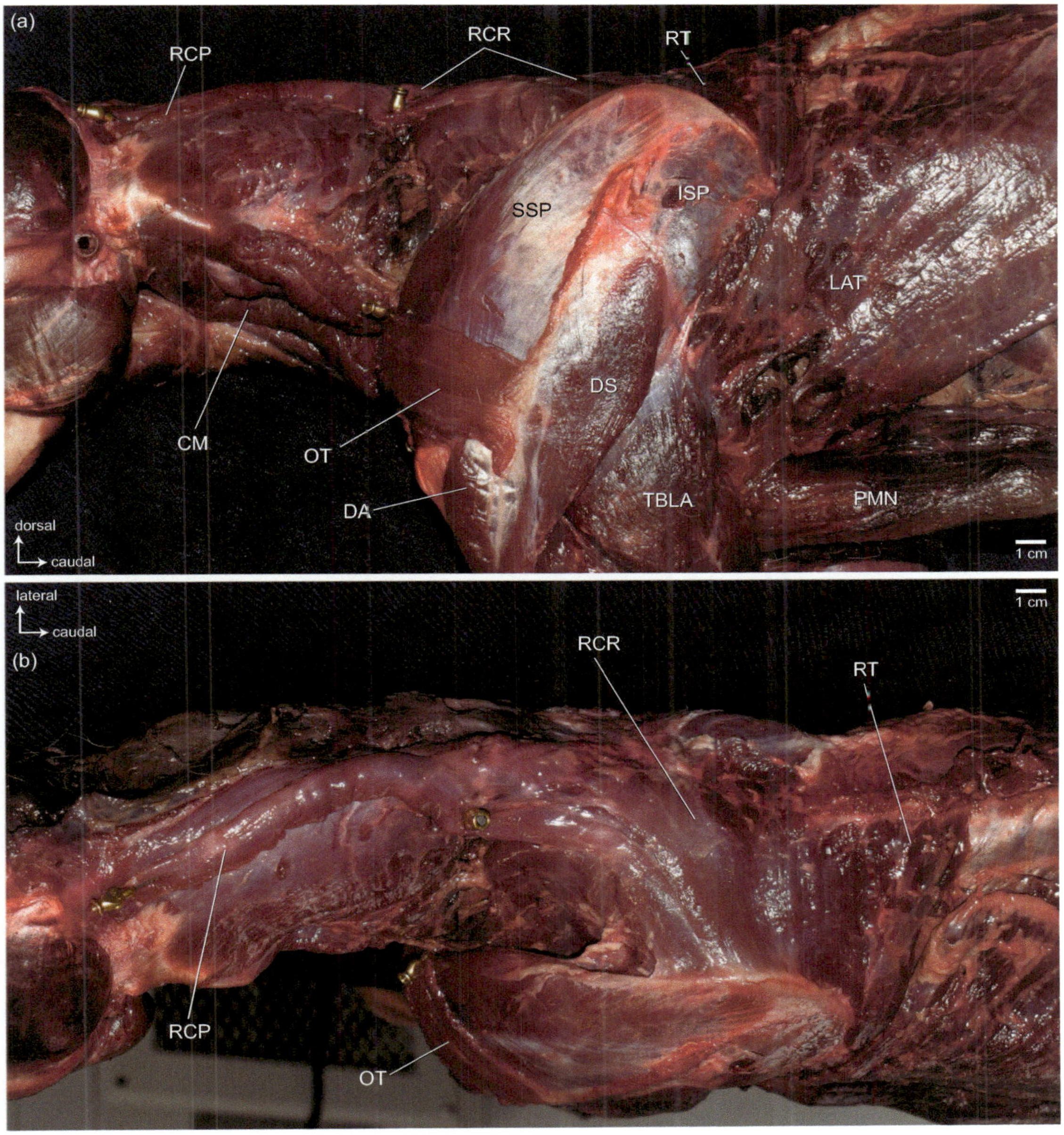
(a)
RCP
RCR
RT
SSP
ISP
LAT
CM
OT
DS
DA
TBLA
PMN
dorsal
caudal
1 cm
lateral
caudal
1 cm
(b)
RCR
RT
RCP
OT

FIGURE II.5.4 **Intrinsic forelimb muscles in (a) proximolateral view and (b) proximomedial view.** The *Mm. deltoidei* (DS, *M. spinodeltoideus* and DA, *M. acromiodeltoideus*) have been removed. The BB, *M. biceps brachii* has one distinct origin. **Intrinsic forelimb muscles in (c) lateral view and (d) medial view.** The PL, *M. palmaris longus* inserts into the palmar aponeurosis. **Distal intrinsic forelimb muscles in (e) lateral view.** The EDC, *M. extensor digitorum communis* and the ECU, *M. extensor carpi ulnaris* have been removed. The EDL, *M. extensor digitorum lateralis* inserts with two tendons on the distal phalanges of digits IV through V. The EI, *M. extensor digiti I and II* inserts with two tendons on the distal phalanges of digits I and II. **(f) In medial view.** The ulnar and humeral head of the *M. flexor carpi ulnaris* (FCUU, FCUH) have been removed.

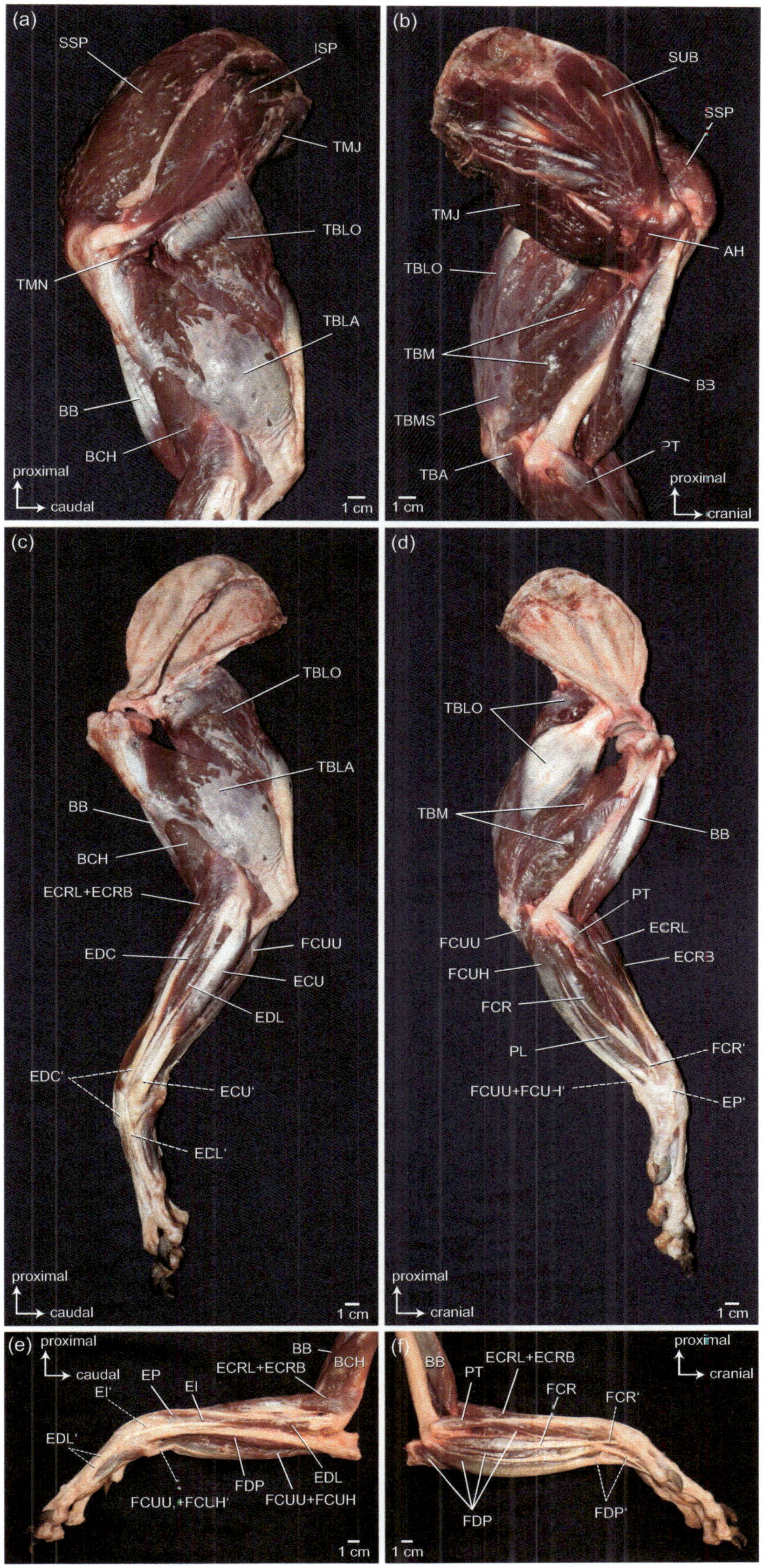
(a)
SSP
ISP
TMJ
TBLO
TMN
TBLA
BB
BCH
proximal
caudal
1 cm
(b)
SUB
SSP
TMJ
AH
TBLO
TBM
BB
TBMS
PT
TBA
proximal
cranial
1 cm
(c)
TBLO
TBLA
BB
BCH
ECRL+ECRB
FCUU
EDC
ECU
EDL
EDC'
ECU'
EDL'
proximal
caudal
1 cm
(d)
TBLO
TBM
BB
PT
FCUU
ECRL
ECRB
FCUH
FCR
PL
FCR'
FCUU+FCUH'
EP'
proximal
cranial
1 cm
(e)
proximal
caudal
BB
BCH
EP
EI
ECRL+ECRB
EI'
EDL'
FDP
EDL
FCUU+FCUH'
FCUU+FCUH
1 cm
(f)
BB
PT
ECRL+ECRB
proximal
cranial
FCR
FCR'
FDP
FDP'
1 cm

FIGURE II.5.5 **Distal forelimb muscles. (a) In dorsal view.** The EDC, *M. extensor digitorum communis* inserts with four tendons on the distal phalanges of digits II through V. The EDL, *M. extensor digitorum lateralis* inserts with two tendons on the distal phalanges of digits IV through V. **(b) In ventral view.** The FDP, *M. flexor digitorum profundus* (four muscle bellies) inserts with five tendons on the distal phalanges of digits I through V.

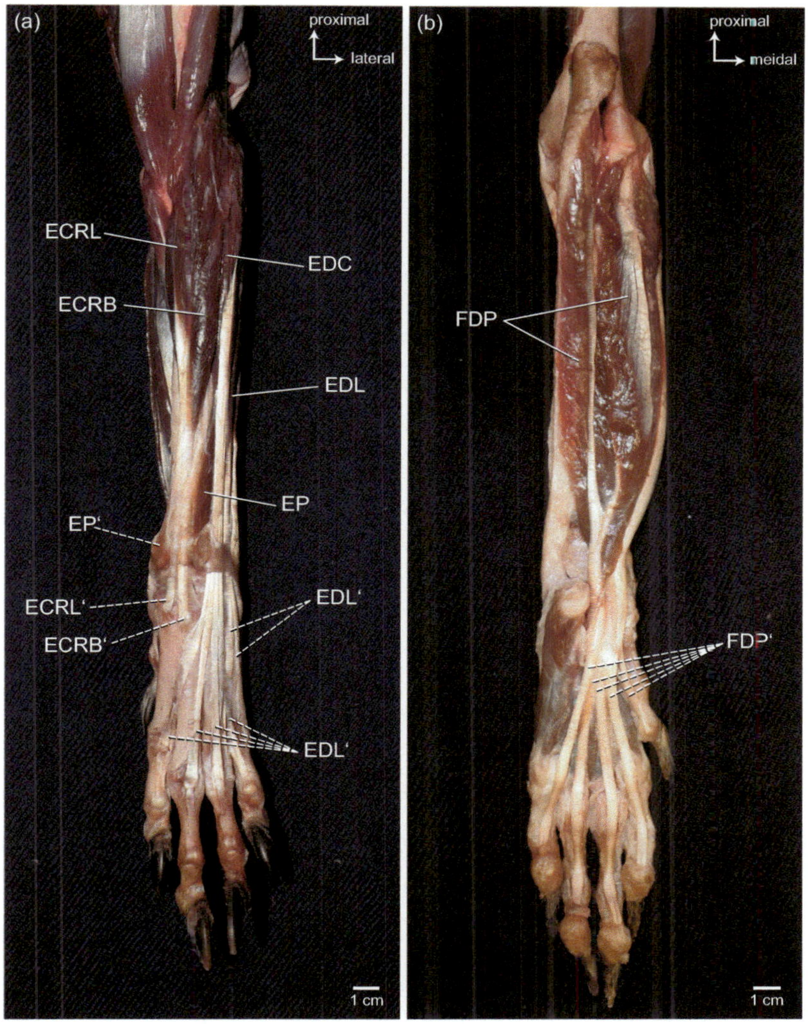
(a)
proximal
lateral
ECRL
EDC
ECRB
EDL
EP
EP'
ECRL'
ECRB'
EDL'
EDL'
1 cm
(b)
proximal
meidal
FDP
FDP'
1 cm

FIGURE II.5.6 **Deep rotator muscles. (a) Distal rotator muscles in dorsal view.** The BB, *M. biceps brachii* inserts with two tendons into the coronoid process of the ulna. The tendon of the BCH, *M. brachioradialis* inserts proximally into the bicipital tuberosity of the radius. **(b) Distal rotator muscles in ventral view.** The ADQ, *M. abductor digiti V* originates from the Os pisiforme (carpal bone) and inserts into the proximal phalanx of digit V. The FDQB, *M. flexor digiti V brevis* originates from the Os hamatum (carpal bone, adjacent to the Os pisiforme) and inserts distal to the ADQ, *M. abductor digiti V* into the proximal phalanx of digit V. The APB, *M. abductor pollicis brevis* originates from the sesamoid and inserts into proximal phalanx of digit I. The FBP, *Mm. flexores breves profundi* serve every digit. They originate from the palmar aspect of the base of the metacarpals and insert into the proximal phalanges of the digits. **(c) Proximal rotator muscles in ventral view.**

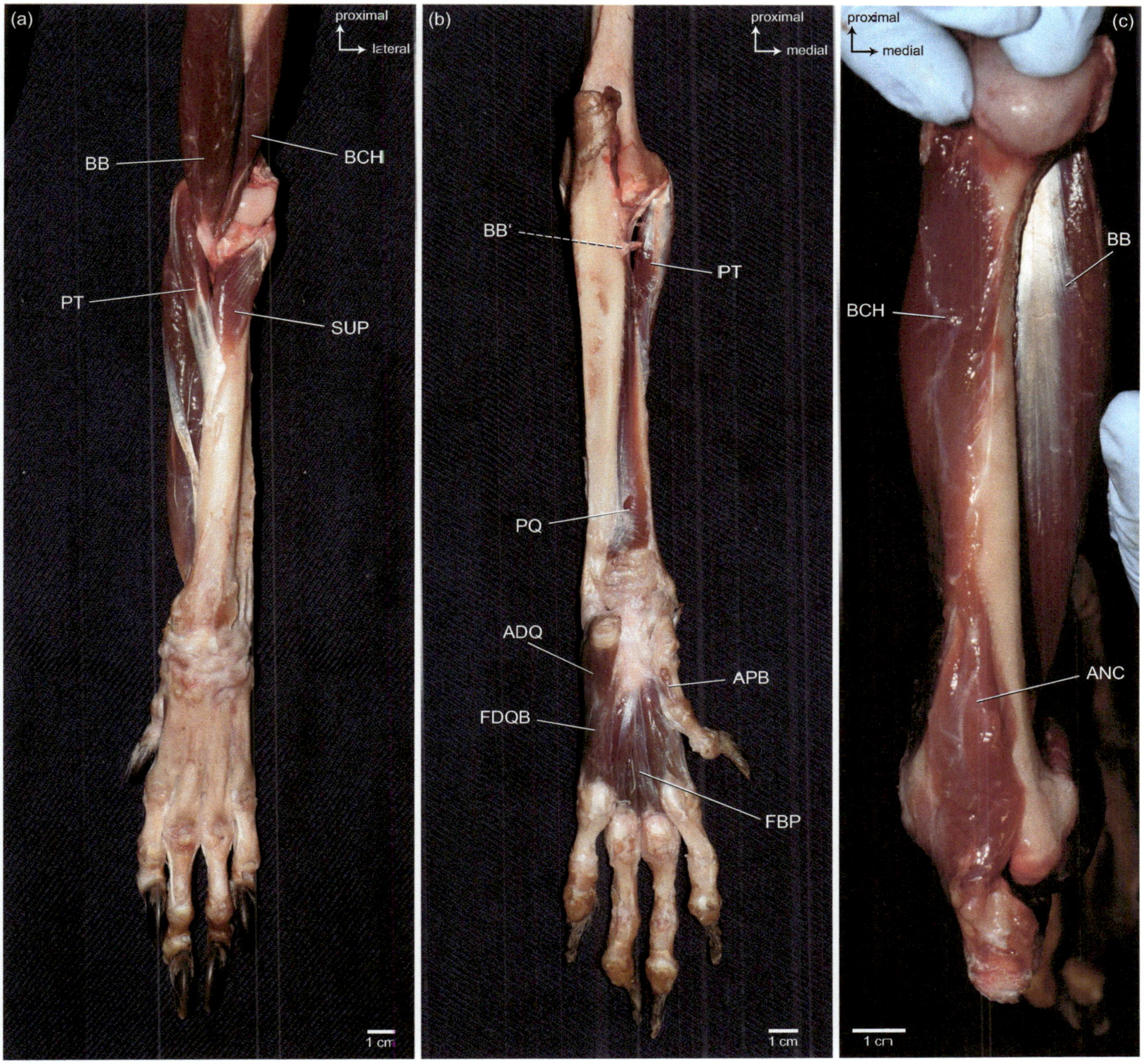
(a)
proximal
lateral
BB
BCH
PT
SUP
1 cm
(b)
proximal
medial
BB'
PT
PQ
ADQ
APB
FDQB
FBP
1 cm
proximal
medial
(c)
BB
BCH
ANC
1 cm

Vulpes vulpes (Red fox)

Classification: Carnivora, Canoidea, Canidae

Mean body mass: 3.0–14.0 kg

Habitat: Forest, grassland, mountains

Locomotor ecology: Cursorial

Peculiarities: Digitigrade

Pectoral muscles. Four pectoral muscles are identified (the PMJ, *M. pectoralis major*; the PAB; *M. pectoantebrachialis*; the PMN, *M. pectoralis minor*; and the XH, *M. xiphihumeralis*).

Deep neck and shoulder muscles. In the neck region, three *Mm. rhomboidei* are present (RCR, *M. rhomboideus cervicis*; RT, *M. rhomboideus thoracis*; and RCP, *M. rhomboideus capitis*). The RPR, *M. rhomboideus profundus* is absent. The RCR, *M. rhomboideus cervicis* does not reach the occiput.

Intrinsic forelimb muscles. The BB, *M. biceps brachii* has one distinct origin. The EDL, *M. extensor digitorum lateralis* inserts with two tendons on the distal phalanges of digits IV through V. The EDC, *M. extensor digitorum communis* inserts with four tendons on the distal phalanges of digits II through V. The EI, *M. extensor digiti I and II* inserts with two tendons on the distal phalanges of digits I and II. The M. *extensor carpi radialis* is identified as one muscle belly, but inserts with one tendon into the base of metacarpal II (ECRL, *M. extensor carpi radialis longus*) and with one tendon into the base of metacarpal III (ECRB, *M. extensor carpi radialis brevis*). The PL, *M. palmaris longus* inserts with four tendons into the distal phalanges of digits II through V. The FDP, *M. flexor digitorum profundus* (four muscle bellies) inserts with four tendons on the distal phalanges of digits II through V.

FIGURE II.6.1 Superficial muscles of forelimb and hind limb in lateral view. (a) Without labels. (b) With labels and muscle outlines. In the neck and shoulder region, the *Mm. trapezii* are clearly divided into three parts (CT, *M. clavotrapezius*; AT, *M. arcomiotrapezius*; and ST, *M. spinotrapezius*). The EPI, *M. epitrochlearis* and the BCR, *M. brachioradialis* are absent. The clavicle is absent. The PAN, *M. panniculus carnosus* is generally very variable between individuals. It is intimately attached to the skin and fascia of most mammals and, thus, often not preserved in skinned specimens. The OEA, *M. obliquus externus abdominis* originates from the thoracic ribs and inserts into the linea alba at the midline and on the pubis near the symphysis. In the hip region, the CAF, *M. caudofemoralis* lies deep to the BIF, *M. biceps femoris*. In the distal hind limb, the four tendons of the EDLO, *M. extensor digitorum longus* insert on the distal phalanges of digits II through V. The PET, *M. peroneus tertius* is absent. Note that the retinacula (ligaments) spanning the wrist and ankle joints have been partially removed.

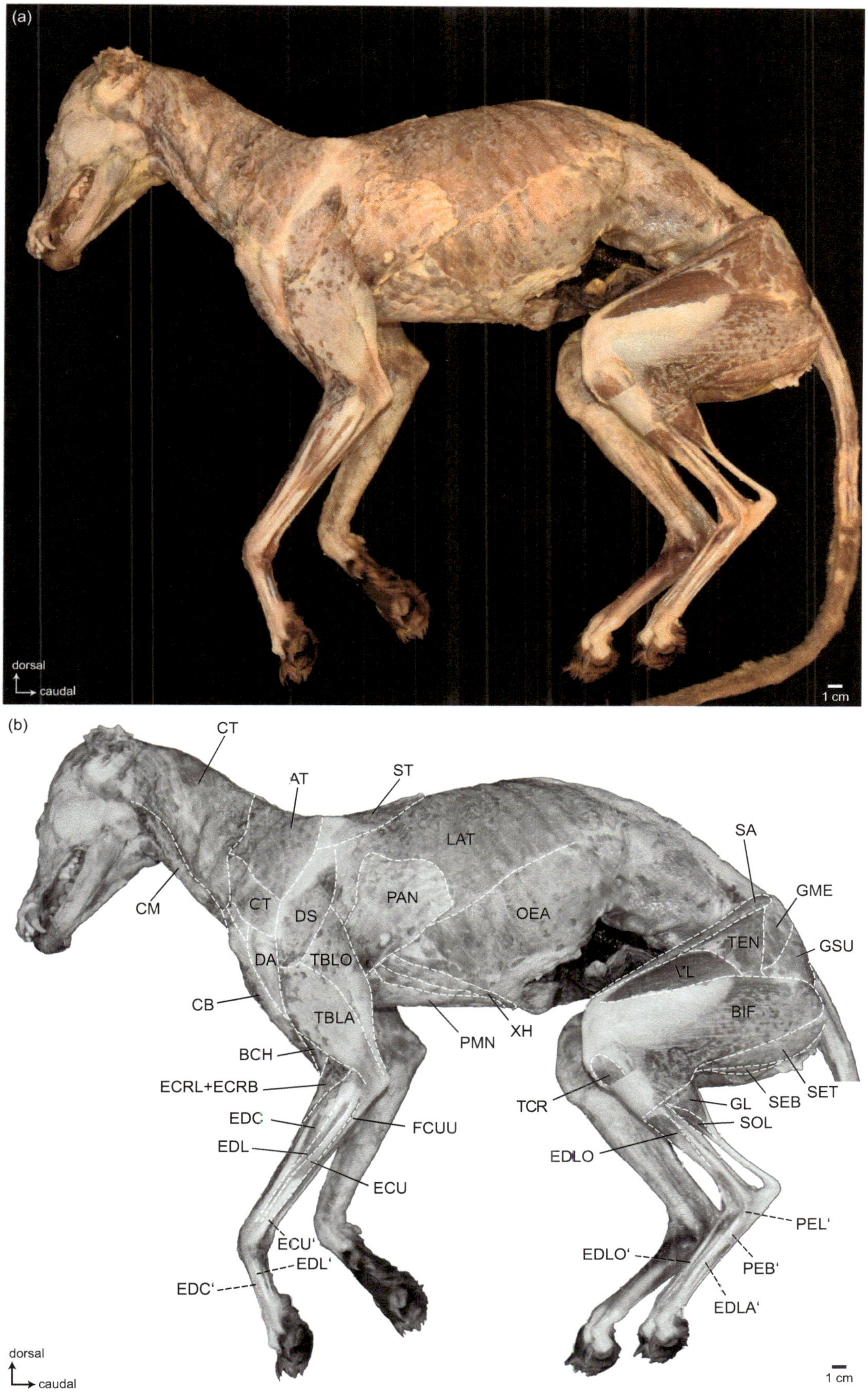
(a)
dorsal
caudal
1 cm
(b)
CT
AT
ST
LAT
SA
GME
GSU
TEN
VL
BIF
CM
CT
DS
PAN
OEA
DA
TBLO
CB
TBLA
BCH
PMN
XH
ECRL+ECRB
EDC
FCUU
EDL
ECU
TCR
GL
SEB
SET
SOL
EDLO
PEL'
ECU'
EDL'
EDC'
EDLO'
PEB'
EDLA'
dorsal
caudal
1 cm

Herpestes auropunctatus (Small Indian mongoose)

Classification: Carnivora, Canoidea, Canidae

Mean body mass: 0.5 kg

Habitat: Forest, scrubland

Locomotor ecology: Ambulatorial

Peculiarities: Digitigrade

Pectoral muscles. Four pectoral muscles are identified (the PMJ, *M. pectoralis major*; the PAB, *M. pectoantebrachialis*; the PMN, *M. pectoralis minor*; and the XH, *M. xiphihumeralis*).

Deep neck and shoulder muscles. In the neck region, three *Mm. rhomboidei* are present (RCR, *M. rhomboideus cervicis*; RT, *M. rhomboideus thoracis*; and RCP, M. rhomboideus capitis). The RPR, *M. rhomboideus profundus* is absent. The RCR, *M. rhomboideus cervicis* does not reach the occiput.

Intrinsic forelimb muscles. The BB, *M. biceps brachii* has one distinct origin. The EDL, *M. extensor digitorum lateralis* inserts with three tendons on the distal phalanges of digits III through V. The EDC, *M. extensor digitorum communis* inserts with four tendons on the distal phalanges of digits II through V. The EI, *M. extensor digiti I and II* covers almost the whole length of the radius and inserts with two tendons on the distal phalanges of digits I and II. The EP, *M. extensor pollicis* inserts with a strong tendon that is distally divided into the base of metacarpal I and a sesamoid proximal to metacarpal I. The ECRL, *M. extensor carpi radialis longus* inserts with one tendon into the base of metacarpal II. The ECRB, *M. extensor carpi radialis brevis* inserts with one tendon into the base of metacarpal III. The PL, *M. palmaris longus* inserts into the palmar aponeurosis. The FDP, *M. flexor digitorum profundus* (four muscle bellies) inserts with five tendons on the distal phalanges of digits I through V.

FIGURE II.7.1 **Superficial muscles of forelimb and hind limb in lateral view. (a) Without labels. (b) With labels and muscle outlines**. In the neck and shoulder region, the *Mm. trapezii* are clearly divided into three parts (CT, *M. clavotrapezius*; AT, *M. arcomiotrapezius*; and ST, *M. spinotrapezius*). The EPI, *M. epitrochlearis* is absent. The BCR, *M. brachioradialis* is present. Note that the CB, *M. clavobrachialis* inserts very distally into the radius and ulna (in relation to BB, *M. biceps brachii* and BCH, *M. brachialis*). The clavicle is absent. In the hip region, the SEB, *M. semimembranosus* lies medially to the SET, *M. semitendinosus*. In the distal hind limb, the four tendons of the EDLO, *M. extensor digitorum longus* insert on the distal phalanges of digits II through V. The EDLA, *M. extensor digitorum lateralis* inserts on the distal phalanges of digits IV and V. The PET, *M. peroneus tertius* is absent. Note that the retinacula (ligaments) spanning the wrist and ankle joints have been partially removed.

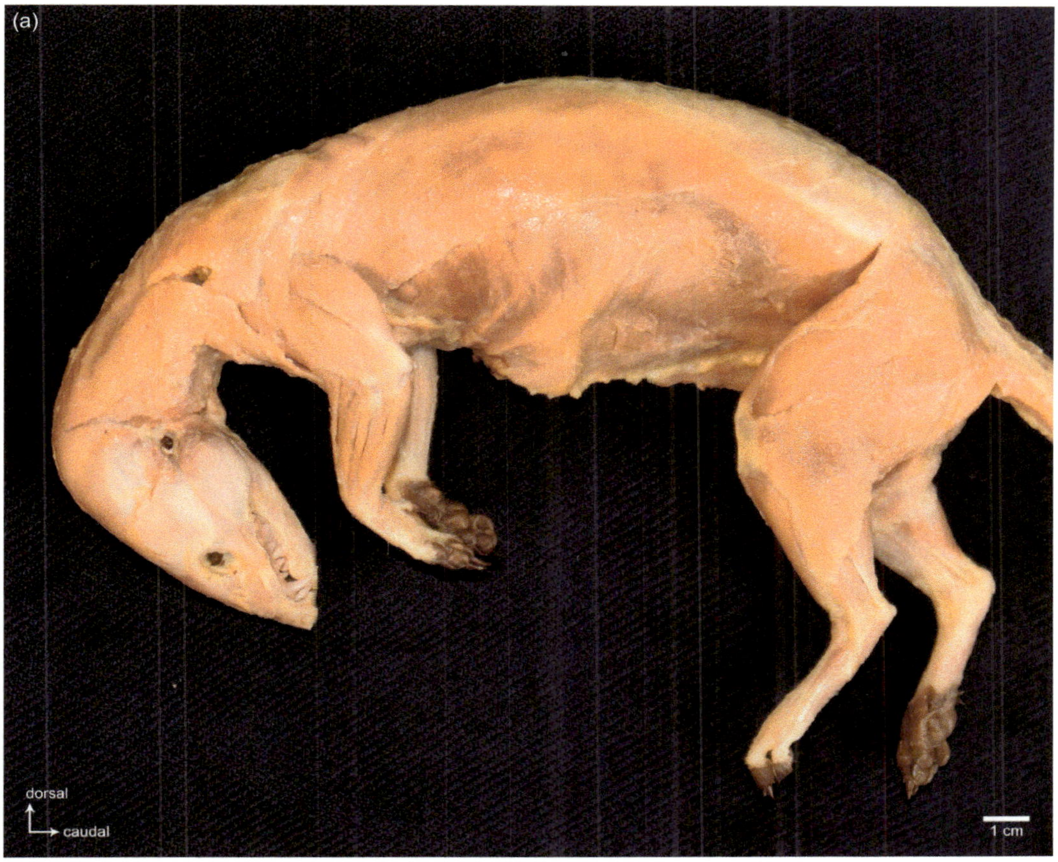
(a)
dorsal
caudal
1 cm

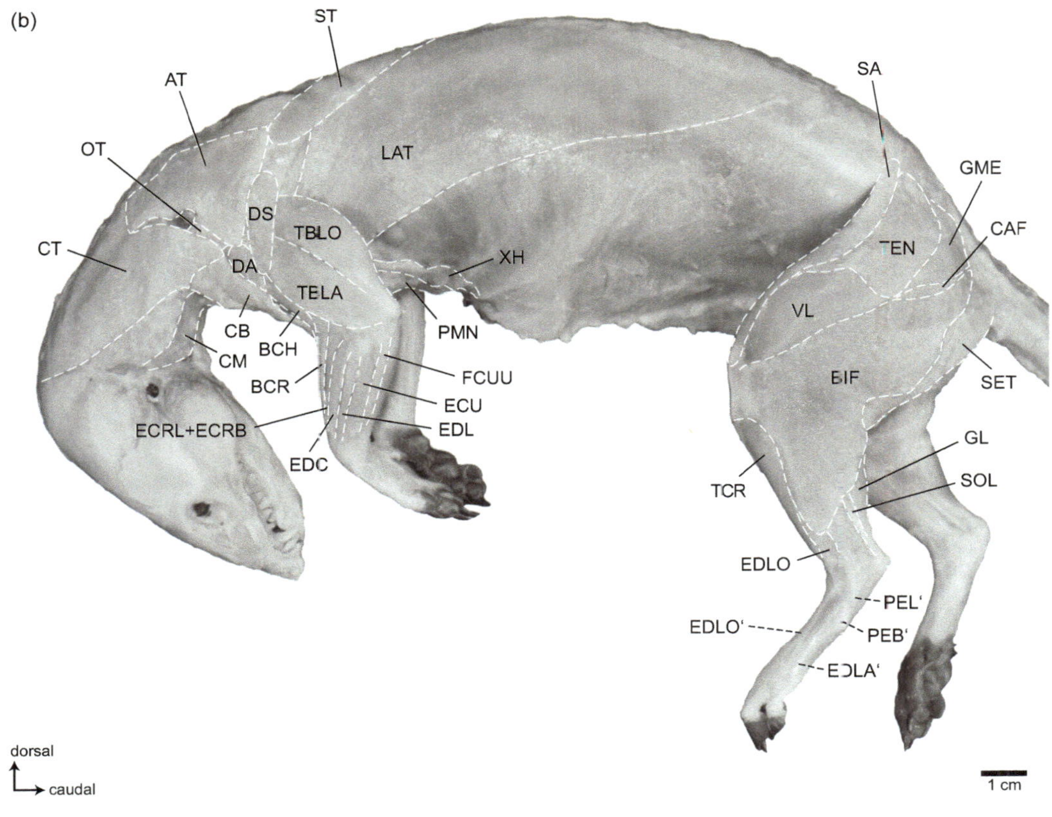
(b)
ST
AT
SA
OT
LAT
GME
DS
TBLO
CAF
CT
DA
XH
TEN
TELA
VL
CB
BCH
PMN
CM
BCR
FCUU
SET
BIF
ECU
ECRL+ECRB
EDL
GL
EDC
TCR
SOL
EDLO
PEL'
EDLO'
PEB'
EDLA'
dorsal
caudal
1 cm

Cryptoprocta ferox (Fossa)

Classification:	Carnivora, Feliformia, Eupleridae
Mean body mass:	9.5 kg
Habitat:	Forest, rainforest (Endemic to Madagascar)
Locomotor ecology:	Scansorial
Peculiarities:	Reversible hind feet; plantigrade; (semi-) retractable claws; capable of head-first descent

Pectoral muscles. Four pectoral muscles are identified (the PMJ, *M. pectoralis major*; the PAB, *M. pectoantebrachialis*; the PMN, *M. pectoralis minor*; and the XH, *M. xiphihumeralis*).

FIGURE II.8.1 **Superficial muscles of forelimb and hind limb in lateral view. (a) Without labels. (b) With labels and muscle outlines**. In the neck and shoulder region, the *Mm. trapezii* are clearly divided into three parts (CT, *M. clavotrapezius*; AT, *M. arcomiotrapezius*; and ST, *M. spinotrapezius*). The AT, *M. arcomiotrapezius* is well developed, and the aponeurosis is relatively small. The clavicle is absent. The EPI, *M. epitrochlearis* is absent. The BCR, *M. brachioradialis* is well developed. In the distal hind limb, the four tendons of the EDLO, *M. extensor digitorum longus* insert on the distal phalanges of digits II through V. The PET, *M. peroneus tertius* is absent. Note that the retinacula (ligaments) spanning the wrist and ankle joints have been partially removed.

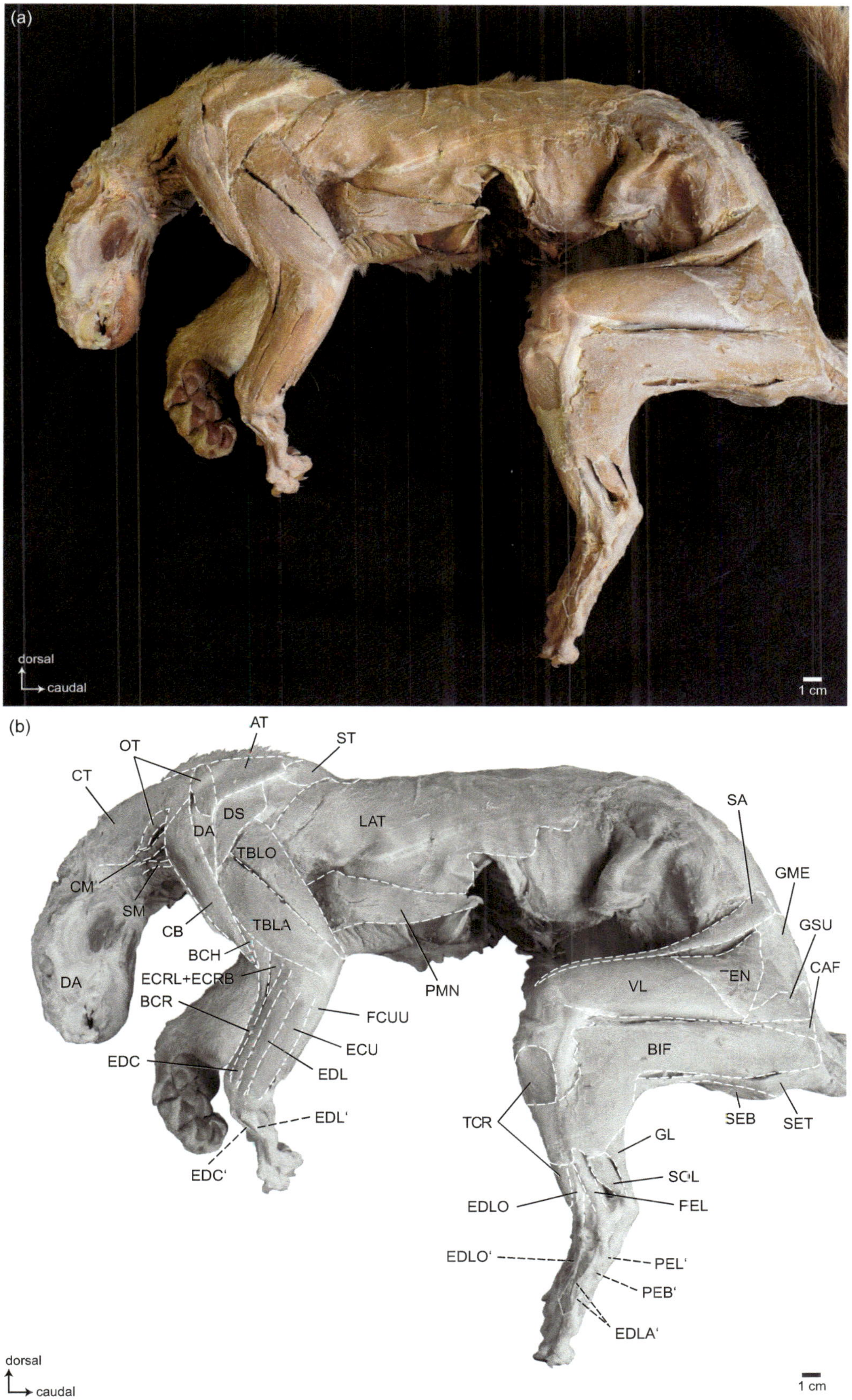
(a)
dorsal
caudal
1 cm
(b)
AT
ST
OT
CT
DS
LAT
DA
TBLO
SA
GME
CM
SM
CB
TBLA
GSU
BCH
CAF
ECRL+ECRB
EN
VL
DA
PMN
BCR
FCUU
ECU
BIF
EDC
EDL
EDL'
TCR
GL
SEB
SET
EDC'
SCL
EDLO
FEL
EDLO'
PEL'
PEB'
EDLA'
dorsal
caudal
1 cm

FIGURE II.8.2 Deep neck and shoulder muscles in dorsal view. In the neck region, three *Mm. rhomboidei* are present (RCR, *M. rhomboideus cervicis*; RT, *M. rhomboideus thoracis*; and RCP, *M. rhomboideus profundus*). The RCR, *M. rhomboideus cervicis* almost reaches the occiput. The RPR, *M. rhomboideus capitis* is absent.

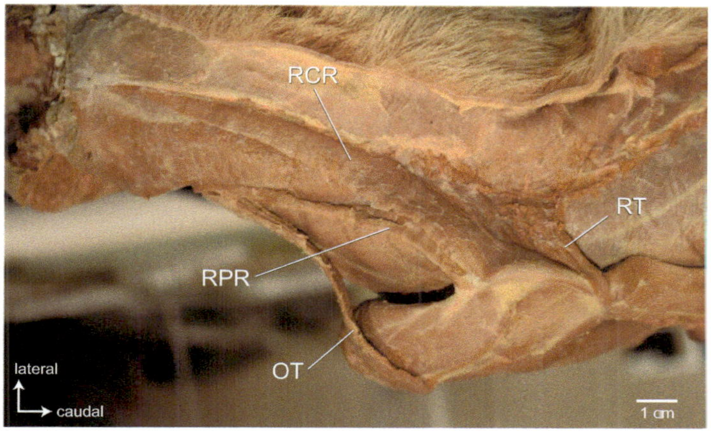
RCR
RT
RPR
OT
lateral
caudal
1 cm

FIGURE II.8.3 Intrinsic forelimb muscles in (a) proximolateral view and (b) proximomedial view. The *Mm. deltoidei* (DS, *M. spinodeltoideus* and DA, *M.acromiodeltoideus)* and the TBMS, *M. triceps brachii caput mediale, short portion* have been removed. Note the very small attachment area for the *M. serratus ventralis* medial on the scapula. The CB, *M. clavobrachialis* inserts distally into the radius and ulna. Note the length of the AH, *M. articularis humeri*. The BB, *M. biceps brachii* has one distinct origin. **Intrinsic forelimb muscles in (c) lateral view and (d) medial view.** The EDL, *M. extensor digitorum lateralis* inserts with three tendons on the distal phalanges of digits III through V. The AH, *M. articularis humeri* has been removed. Note that the PT, *M. pronator teres* inserts distally into the end of the radius. The PL, *M. palmaris longus* inserts with four tendons on the distal phalanges of digits II through V.

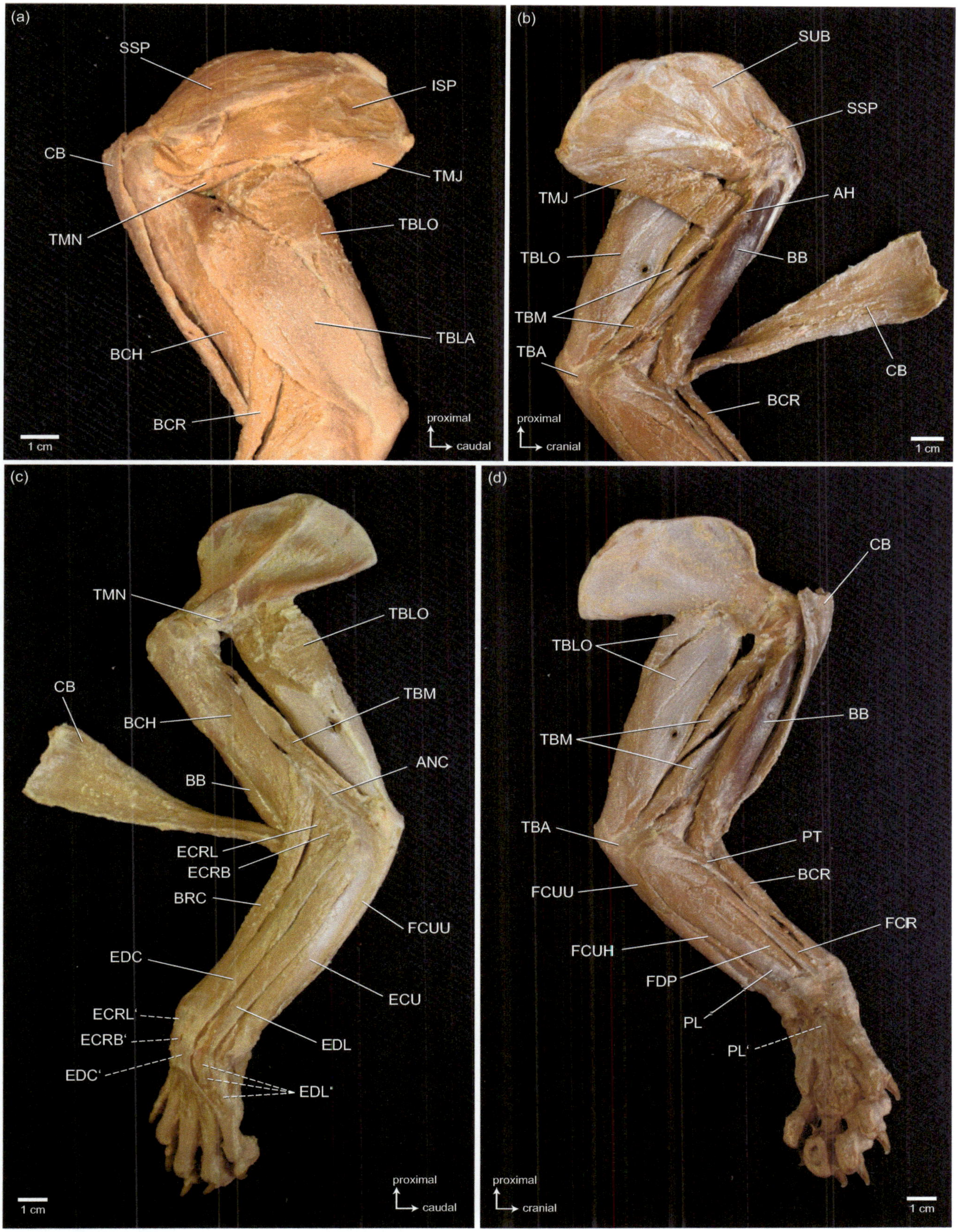
(a)
SSP
ISP
CB
TMJ
TMN
TBLO
TBLA
BCH
BCR
1 cm
proximal
caudal
(b)
SUB
SSP
TMJ
AH
TBLO
BB
TBM
TBA
CB
BCR
proximal
cranial
1 cm
(c)
TMN
TBLO
CB
TBM
BCH
ANC
BB
ECRL
ECRB
BRC
FCUU
EDC
ECU
ECRL'
ECRB'
EDL
EDC'
EDL'
1 cm
proximal
caudal
(d)
CB
TBLO
BB
TBM
TBA
PT
BCR
FCUU
FCR
FCUH
FDP
PL
PL'
proximal
cranial
1 cm

FIGURE II.8.4 **Distal forelimb muscles. (a, b) In dorsal view.** The EDC, *M. extensor digitorum communis* inserts with four tendons on the distal phalanges of digits II through V. Note the proximal origin of the EI, *M. extensor digiti I and II.* It inserts with two tendons on the distal phalanges of digits I and II. **(c) In ventral view.** The FDP, *M. flexor digitorum profundus* (four muscle bellies) inserts with five tendons on the distal phalanges of digits I through V.

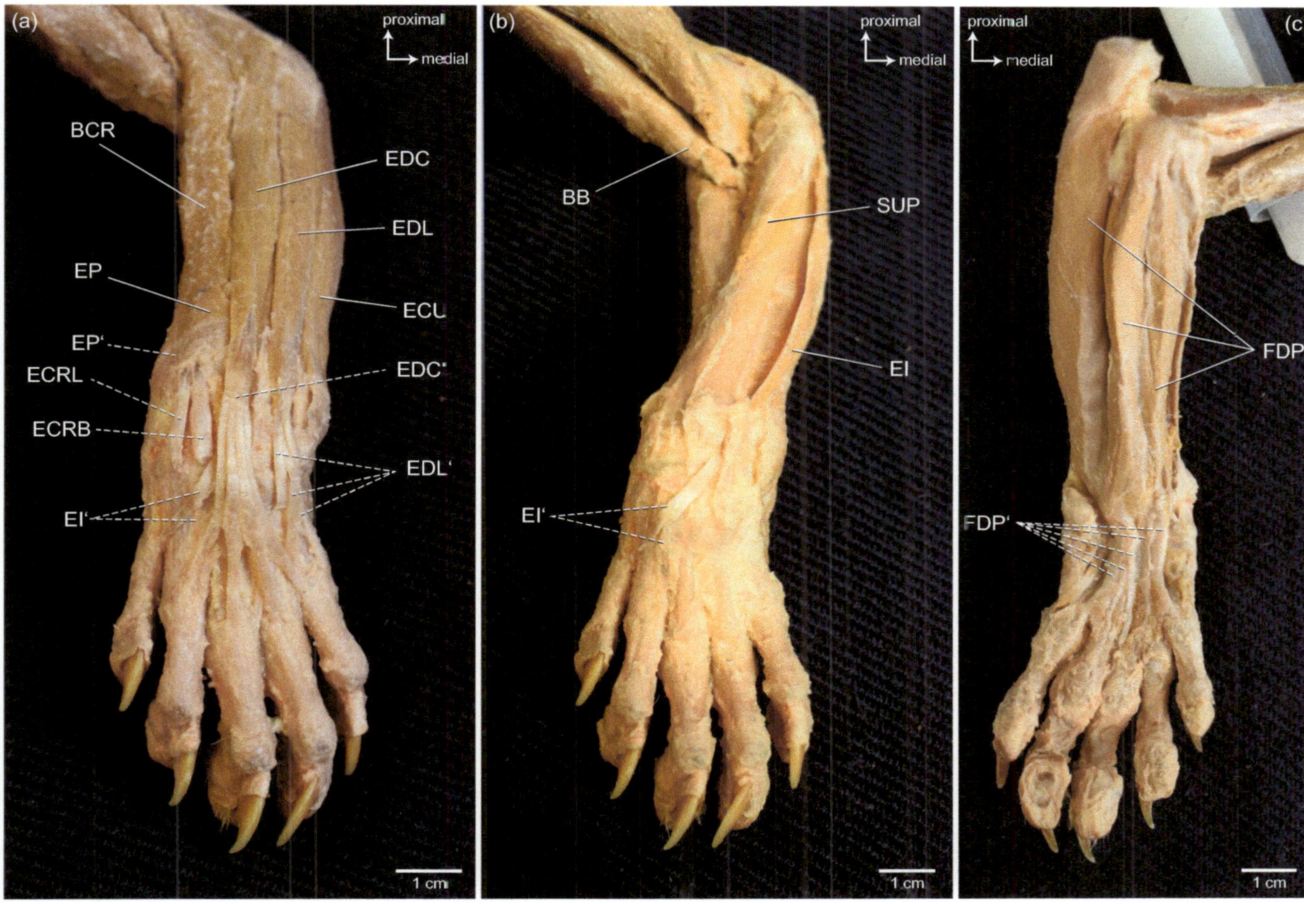
(a)
proximal
medial
BCR
EDC
EDL
EP
ECU
EP'
EDC'
ECRL
ECRB
EDL'
EI'
1 cm
(b)
proximal
medial
BB
SUP
EI
EI'
1 cm
(c)
proximal
medial
FDP
FDP'
1 cm

FIGURE II.8.5 Deep rotator muscles. (a) Distal rotator muscles in dorsal view. (b) Distal rotator muscles in ventral view. (c) Proximal rotator muscles in dorsomedial view. The tendon of the BB, *M. biceps brachii* inserts distally to the tendon of the BCH, *M. brachioradialis* into the radius. The tendon of the BCH, *M. brachioradialis* inserts into the coronoid process into the ulna. **(d) Proximal rotator muscles in ventral view**.

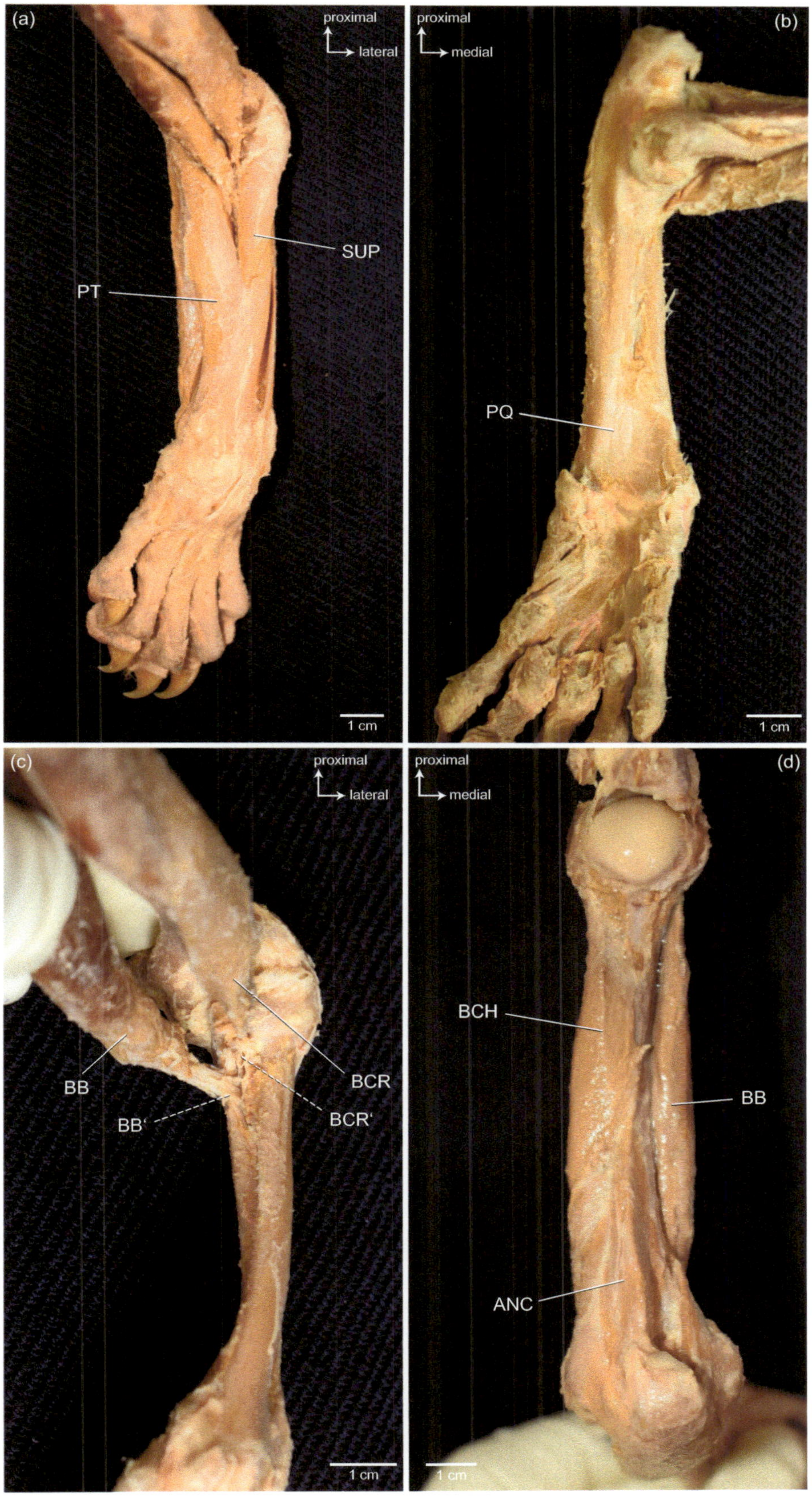
(a)
proximal
lateral
SUP
PT
1 cm
(b)
proximal
medial
PQ
1 cm
(c)
proximal
lateral
BB
BCR
BB'
BCR'
1 cm
(d)
proximal
medial
BCH
BB
ANC
1 cm

Hyaena hyaena (Striped hyaena)

Classification:	Carnivora, Feloidea, Hyaenidae
Mean body mass:	35 kg
Habitat:	Savanna, grassland
Locomotor ecology:	Ambulatorial
Peculiarities:	Relatively short hind limbs; capable of great endurance (lope)

Pectoral muscles. Three pectoral muscles are identified (the PMJ, *M. pectoralis major*; the PAB, *M. pectoantebrachialis*; and the PMN, *M. pectoralis minor*) The XH, *M. xiphihumeralis* is not clearly developed as a separate muscle, but part of the PMN, *M. pectoralis minor.*

Deep neck and shoulder muscles. In the neck region, two *Mm. rhomboidei* are present (RCR, *M. rhomboideus cervicis* and RT, *M. rhomboideus thoracis*) The RCP, *M. rhomboideus capitis* and the RPR, *M. rhomboideus profundus* are absent. The RCR, *M. rhomboideus cervicis* reaches the occiput.

Intrinsic forelimb muscles. The BB, *M. biceps brachii* has one distinct origin. It inserts with a strong tendon into the coronoid process of the ulna (together with the tendon of the CB, *M. clavobrachialis*). The tendon of the BCH, *M. brachioradialis* inserts proximally into the bicipital tuberosity of the radius. The EDL, *M. extensor digitorum lateralis* inserts with two tendons on the distal phalanges of digits IV and V. The EDC, *M. extensor digitorum communis* inserts with four tendons on the distal phalanges of digits II through V. The EI, *M. extensor digiti I and II* inserts with one tendon on the distal phalanx of digit II. The EP, *M. extensor pollicis* inserts with a strong tendon that is distally divided into the base of metacarpal I and a sesamoid proximal to metacarpal I. The ECRL, *M. extensor carpi radialis longus* inserts with one tendon into the base of metacarpal II. The ECRB, *M. extensor carpi radialis brevis* inserts with one tendon into the base of metacarpal III. The PL, *M. palmaris longus* inserts with four tendons on the distal phalanges of digits II through V. The FDP, *M. flexor digitorum profundus* (four muscle bellies) insert with five tendons on the distal phalanges of digits I through V.

FIGURE II.9.1 Superficial muscles of forelimb and hind limb in lateral view. (a) Without labels. (b) With labels and muscle outlines. In the neck and shoulder region, the *Mm. trapezii* are clearly divided into three parts (CT, *M. clavotrapezius*; AT, *M. arcomiotrapezius*; and ST, *M. spinotrapezius*). The EPI, *M. epitrochlearis* and the BCR, *M. brachioradialis* are absent. Note that the CB, *M. clavobrachialis* inserts very distally into the ulna (together with the tendon of the BB, *M. biceps brachii*). A rudimentary clavicle is present (embedded into the raphe between the CM, *M. cleidomastoideus* and the CB, *M. clavobrachialis*). The SC, *M. subclavius* is absent. The TBLO, *M. triceps brachii caput longum* appears inhomogenous, but both parts are strongly fused to each other and form one muscle. The OEA, *M. obliquus externus abdominis* originates from the thoracic ribs and inserts into the linea alba at the midline and on the pubis near the symphysis. The RA, *M. rectus abdominis* originates from the costal cartilages and inserts into the prebubic tendon. In the distal hind limb, the four tendons of the EDLO, *M. extensor digitorum longus* insert on the distal phalanges of digits II through V. The EDLA, *M. extensor digitorum lateralis* inserts on the distal phalanges of digits IV and V. The PET, *M. peroneus tertius* is absent. Note that the retinacula (ligaments) spanning the wrist and ankle joints have been partially removed.

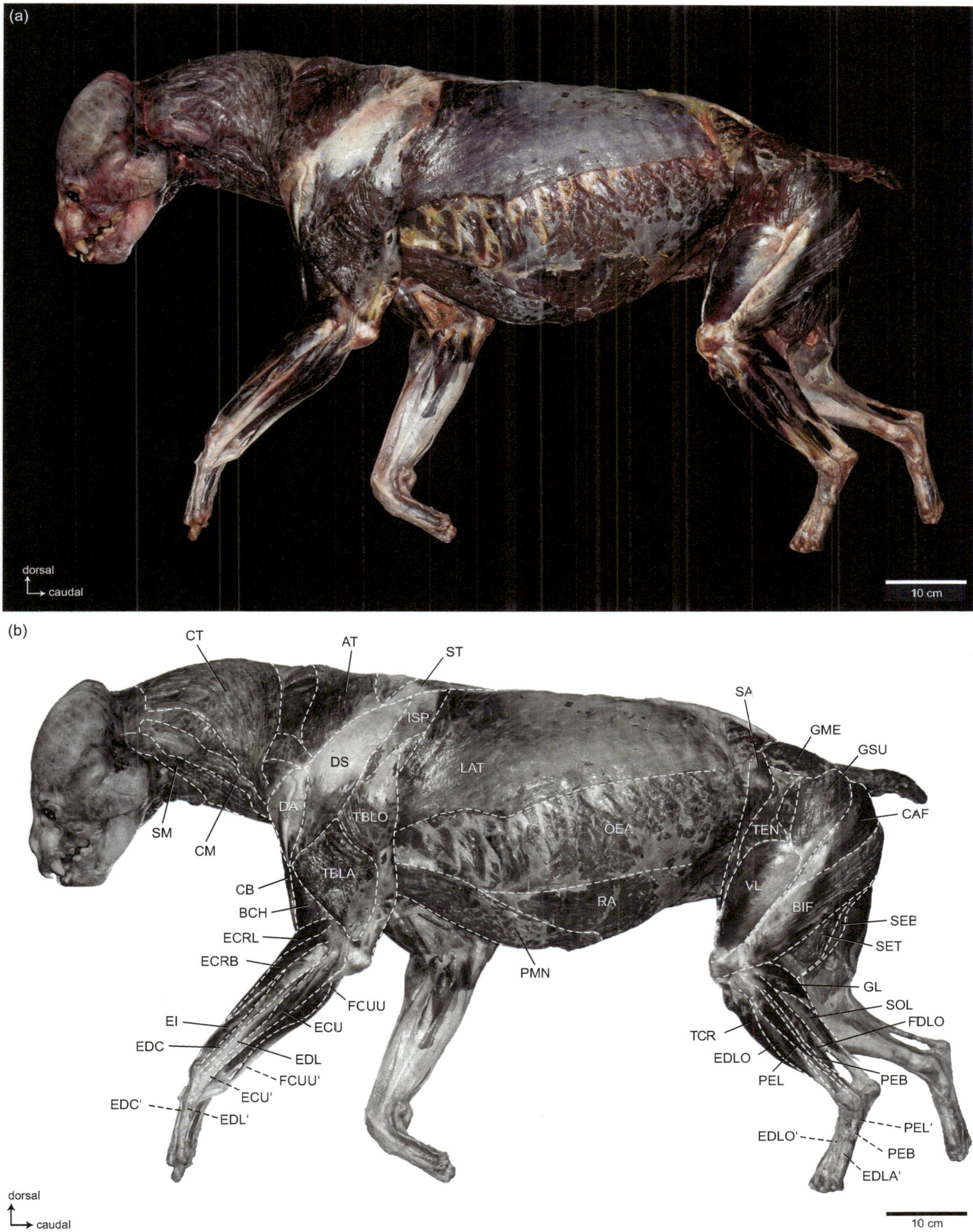
(a)
dorsal
caudal
10 cm
(b)
CT
AT
ST
ISP
DS
LAT
SA
GME
GSU
CAF
DA
TBLO
OEA
TEN
SM
CM
TBLA
CB
BCH
ECRL
ECRB
RA
VL
BIF
SEB
SET
PMN
FCUU
ECU
EI
EDC
EDL
FCUU'
ECU'
EDC'
EDL'
GL
SOL
FDLO
TCR
EDLO
PEL
PEB
PEL'
EDLO'
PEB
EDLA'
dorsal
caudal
10 cm

Acinonyx jubatus (Cheetah)

Classification:	Carnivora, Feliformia, Felidae
Mean body mass:	21.0–72.0 kg
Habitat:	Savanna, grassland
Locomotor ecology:	Cursorial
Peculiarities:	Reduced ability to supinate the forelimb (compared to other felids)

FIGURE II.10.1 Superficial muscles of forelimb and hind limb in lateral view. (a) Without labels. (b) With labels and muscle outlines. In the neck and shoulder region, the *Mm. trapezii* are clearly divided into three parts (CT, *M. clavotrapezius*; AT, *M. arcomio-trapezius*; and ST, *M. spinotrapezius*). The EPI, *M. epitrochlearis* and the BCR, *M. brachio-radialis* are absent. The OEA, *M. obliquus externus abdominis* originates from the thoracic ribs and inserts into the linea alba at the midline and on the pubis near the symphysis. The RA, *M. rectus abdominis* originates from the costal cartilages and inserts into the prebubic tendon. In the distal hind limb, the four tendons of the EDLO, *M. extensor digitorum longus* insert on the distal phalanges of digits II through V. The EHL, *M. hallucis longus* lies deep to the TCR, *M. tibialis cranialis* and the EDLO, *M. extensor digitorum longus*. The PET, *M. peroneus tertius* is absent. Note that the retinacula (ligaments) spanning the wrist and ankle joints have been largely removed.

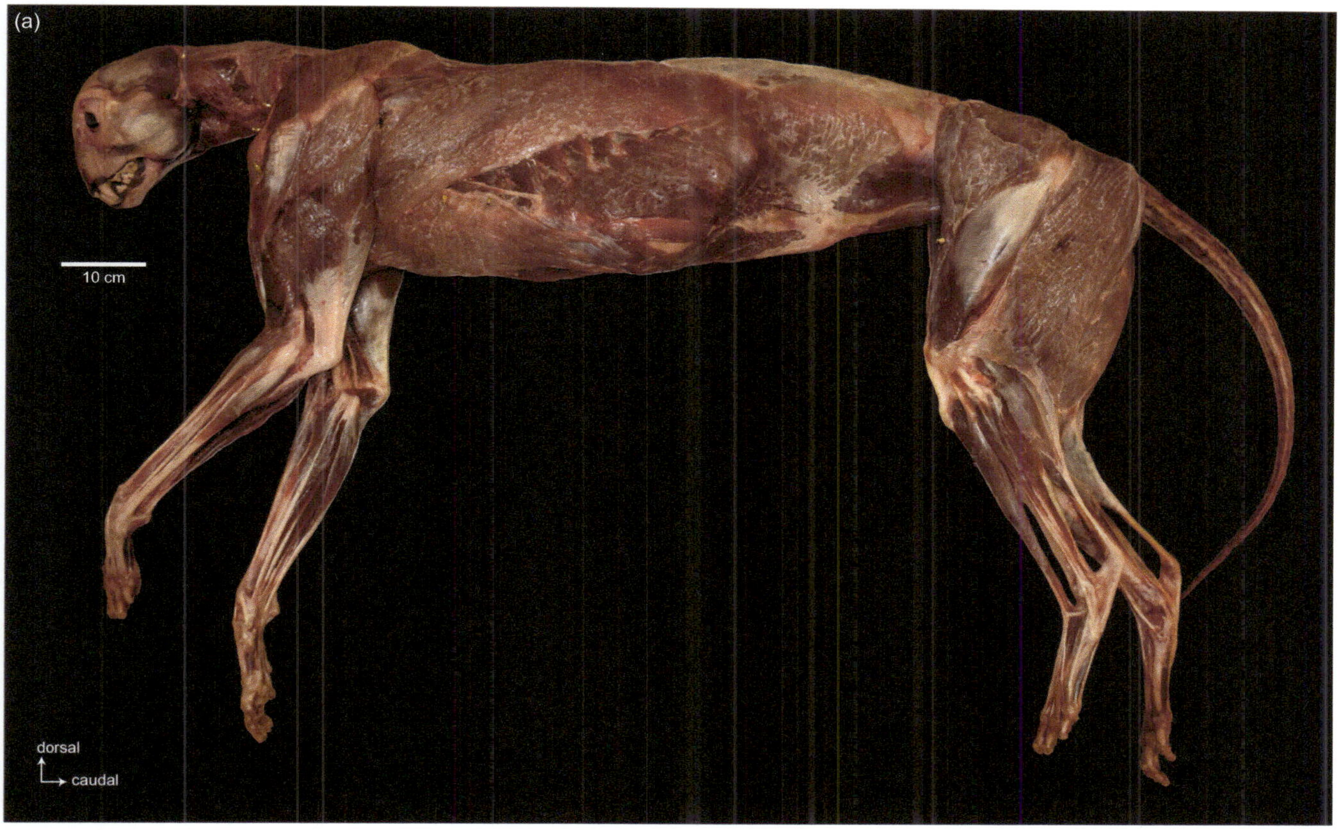
(a)
10 cm
dorsal
caudal

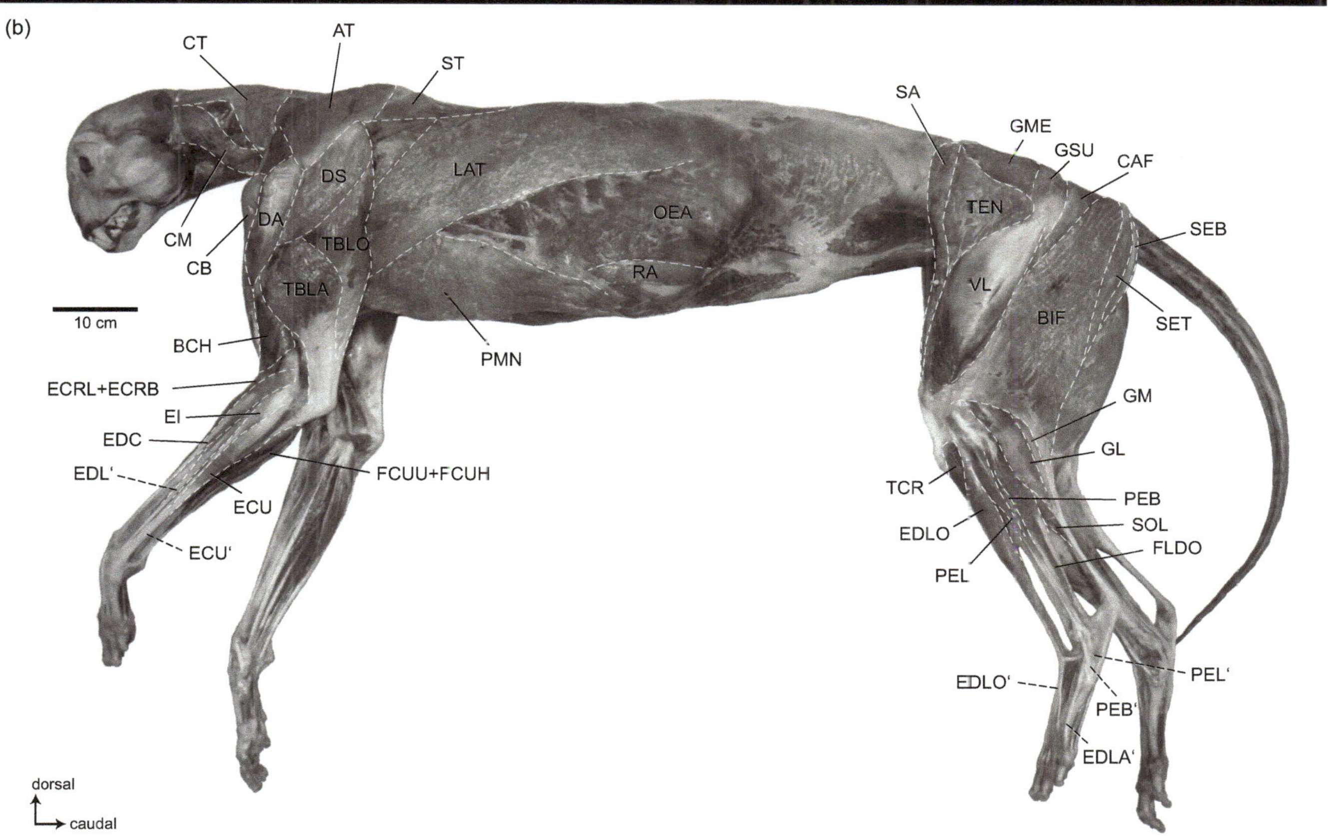
(b)
CT
AT
ST
SA
GME
GSU
CAF
SEB
SET
DS
LAT
OEA
TEN
CM
CB
DA
TBLO
TBLA
RA
VL
BIF
10 cm
BCH
PMN
ECRL+ECRB
EI
EDC
EDL'
FCUU+FCUH
ECU
ECU'
GM
GL
TCR
PEB
SOL
EDLO
FLDO
PEL
EDLO'
PEL'
PEB'
EDLA'
dorsal
caudal

FIGURE II.10.2 **Pectoral muscles in ventral view. (a, b) Superficial view.** A rudimentary clavicle is present in this specimen (embedded into the raphe between the CM, *M. cleidomastoideus* and the CB, *M. clavobrachialis*). The SC, *M. subclavius* is absent. The PAB, *M. pectoantebrachialis* is clearly separated from the PMJ, *M. pectoralis major*, but relatively thin. The XH, *M. xiphihumeralis* is not clearly developed as a separate muscle, but part of the PMN, *M. pectoralis minor*.

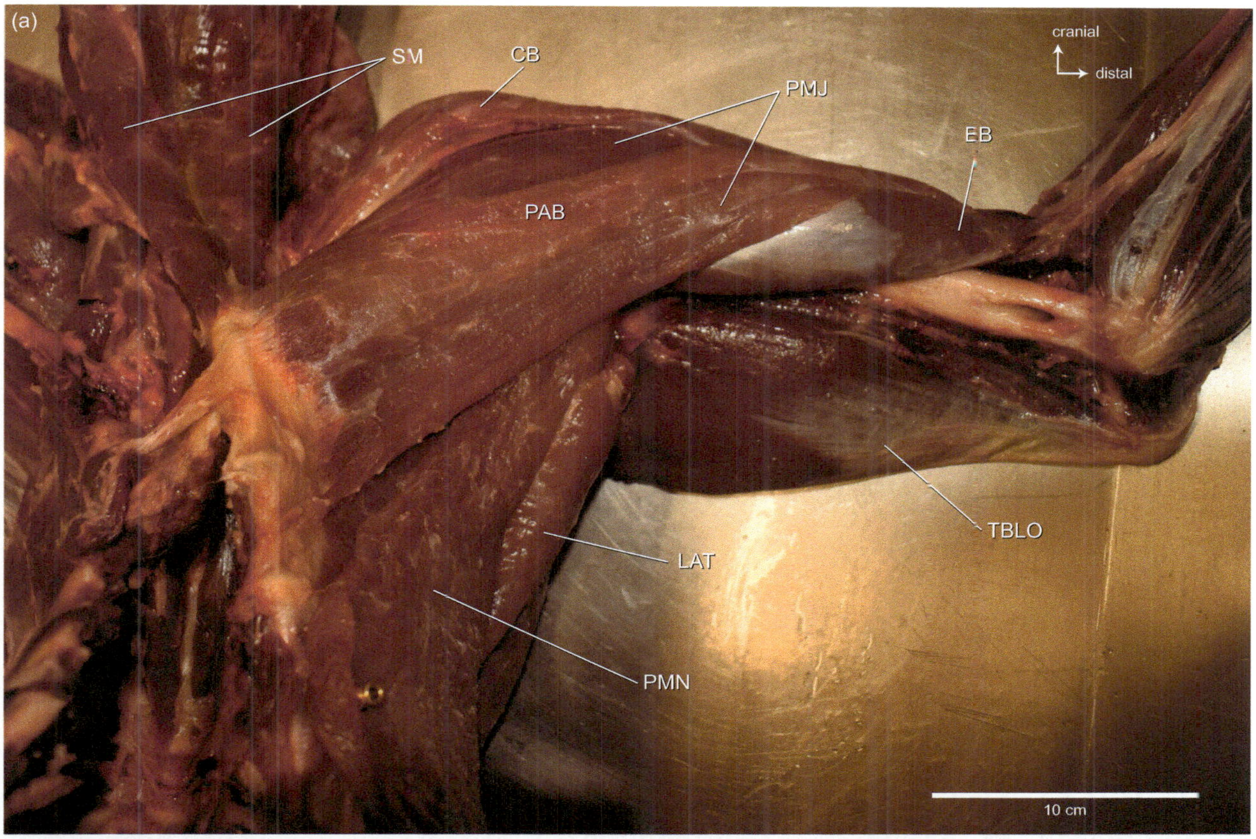
(a)
SM
CB
PMJ
cranial
distal
EB
PAB
TBLO
LAT
PMN
10 cm

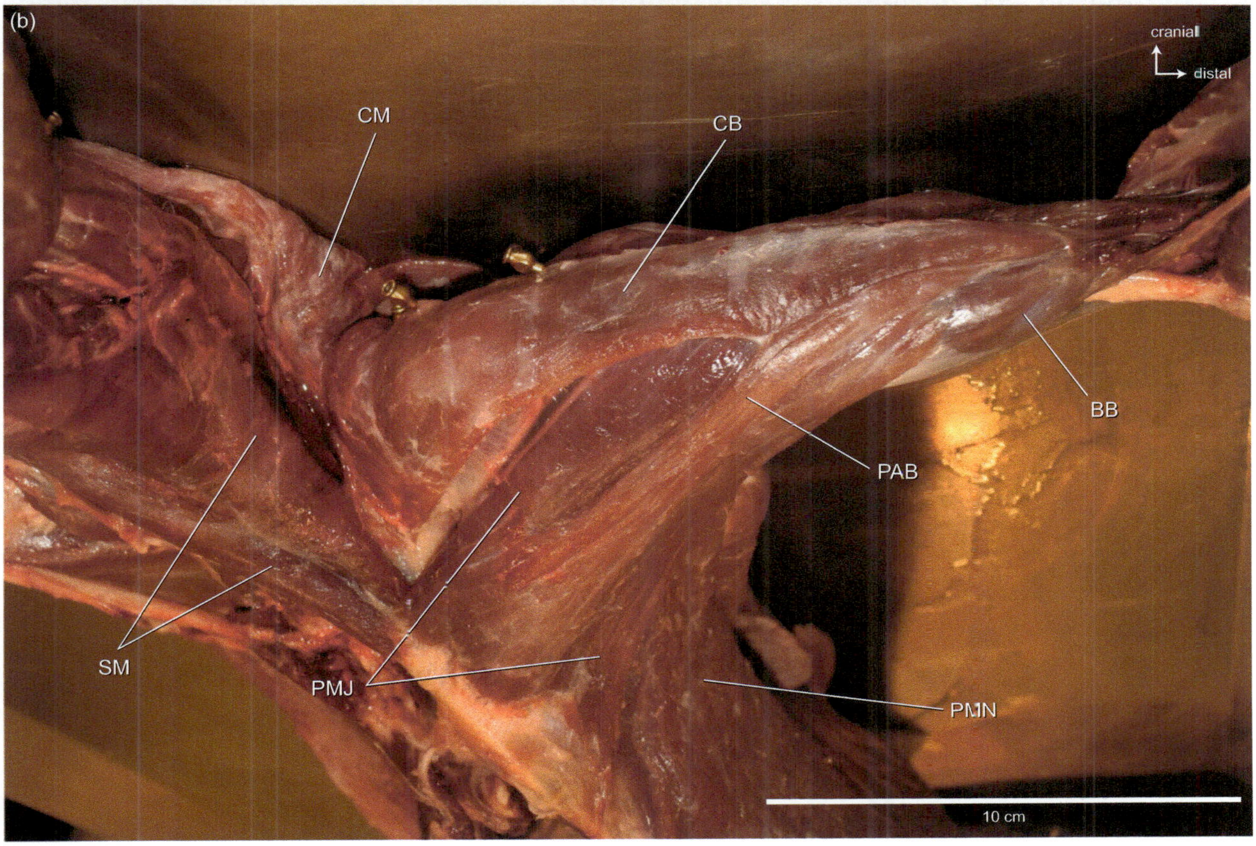
(b)
cranial
distal
CM
CB
BB
PAB
SM
PMJ
PMN
10 cm

FIGURE II.10.3 **Neck and shoulder muscles. In dorsal view.** Note that the soft tissue separating the AT, M. *acromiotrapezius* and the ST, *M. spinotrapezius* has been removed. This exposes the RCR, *M. rhomboideus cervicis*. In the neck region, two *Mm. rhomboidei* are present (RCR, *M. rhomboideus cervicis* and RCP, *M. rhomboideus capitis*). Not visible in the picture, but the RCR, *M. rhomboideus cervicis* covers about half of the length of the neck. It does not reach the occiput.

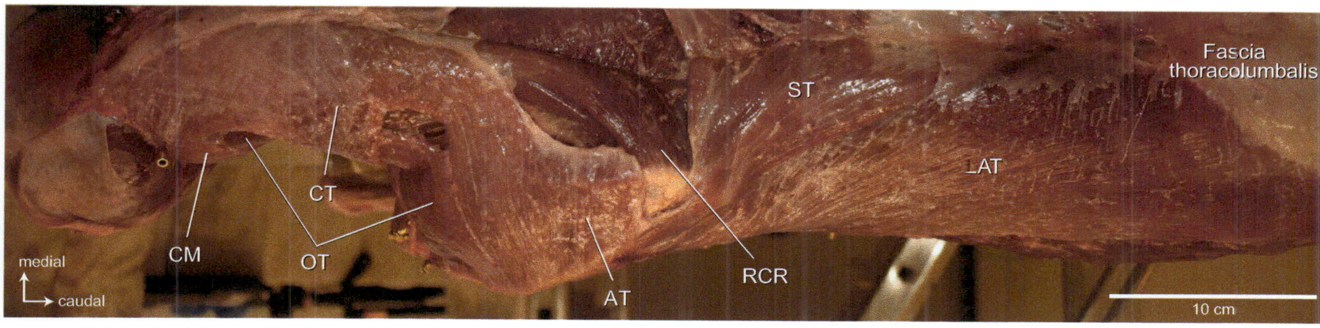
Fascia thoracolumbalis
ST
LAT
CT
CM
OT
AT
RCR
medial
caudal
10 cm

FIGURE II.10.4 **Intrinsic forelimb muscles in (a) proximolateral view and (b) proximomedial view.** The *Mm. deltoideii* have been removed. The TBLO, *M. triceps brachii caput longum* appears inhomogenous, but both parts are strongly fused to each other and form one muscle. **Intrinsic muscles in (c) lateral view and (d) medial view.** The BB, *M. biceps brachii* has one distinct origin. The ECRL, *M. extensor carpi radialis longus* and the ECRB, *M. extensor carpi radialis brevis* are strongly fused to each other. The FCUU, *M. flexor carpi ulnaris caput ulnare* and the FCUH, *M. flexor carpi ulnaris caput humerale* are also strongly fused to each other. The PL, *M. palmaris longus* inserts with four tendons on the distal phalanges of digits II through V. **Distal intrinsic forelimb muscles in (e) lateral view** The EI, *M. extensor digiti I and II* and the ECU, *M. extensor carpi ulnaris* has been removed. **(f) In medial view.** The ECRL, *M. extensor carpi radialis longus* and the ECRB, *M. extensor carpi radialis brevis* have been removed.

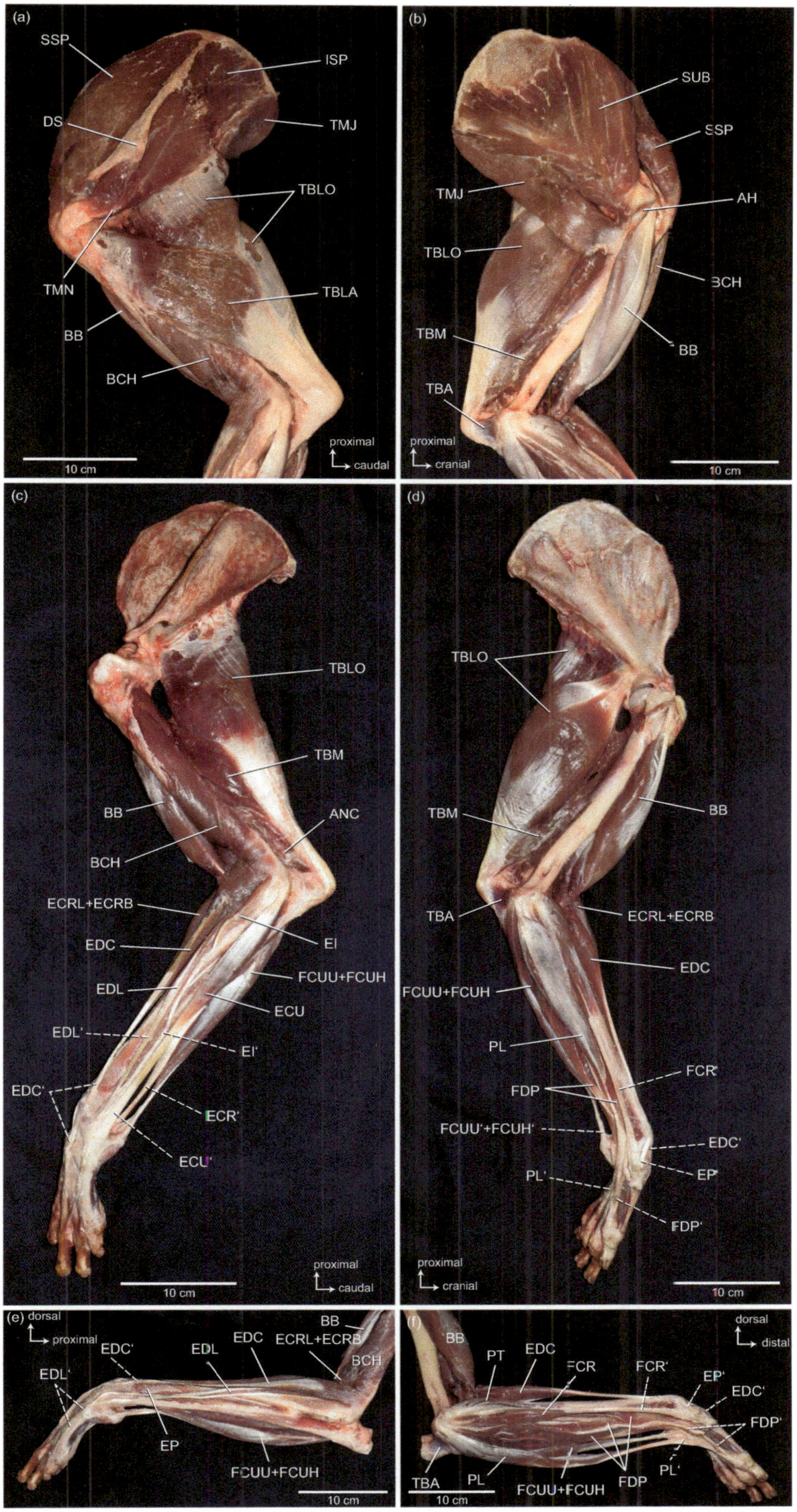
(a)
SSP
ISP
DS
TMJ
TBLO
TMN
TBLA
BB
BCH
proximal
caudal
10 cm
(b)
SUB
SSP
TMJ
AH
TBLO
BCH
TBM
BB
TBA
proximal
cranial
10 cm
(c)
TBLO
TBM
BB
ANC
BCH
ECRL+ECRB
EI
EDC
FCUU+FCUH
EDL
ECU
EDL'
EI'
EDC'
ECR'
ECU'
proximal
caudal
10 cm
(d)
TBLO
TBM
BB
TBA
ECRL+ECRB
EDC
FCUU+FCUH
PL
FCR'
FDP
FCUU'+FCUH'
EDC'
EP'
PL'
FDP'
proximal
cranial
10 cm
(e) dorsal
proximal
BB
EDC'
EDL
EDC
ECRL+ECRB
BCH
EDL'
EP
FCUU+FCUH
10 cm
(f)
BB
PT
EDC
FCR
FCR'
dorsal
distal
EP'
EDC'
FDP'
PL'
TBA
PL
FCUU+FCUH
FDP
10 cm

FIGURE II.10.5 **Distal forelimb muscles. (a) In dorsal view.** The EDL, *M. extensor digitorum lateralis* inserts with three tendons on the distal phalanges of digits III through V. Note that the tendon of digit V has accidently been ruptured during preparation of the specimen. **(b) In ventral view.** The FDP, *M. flexor digitorum profundus* (four muscle bellies) inserts with four tendons on the distal phalanges of digits II through V.

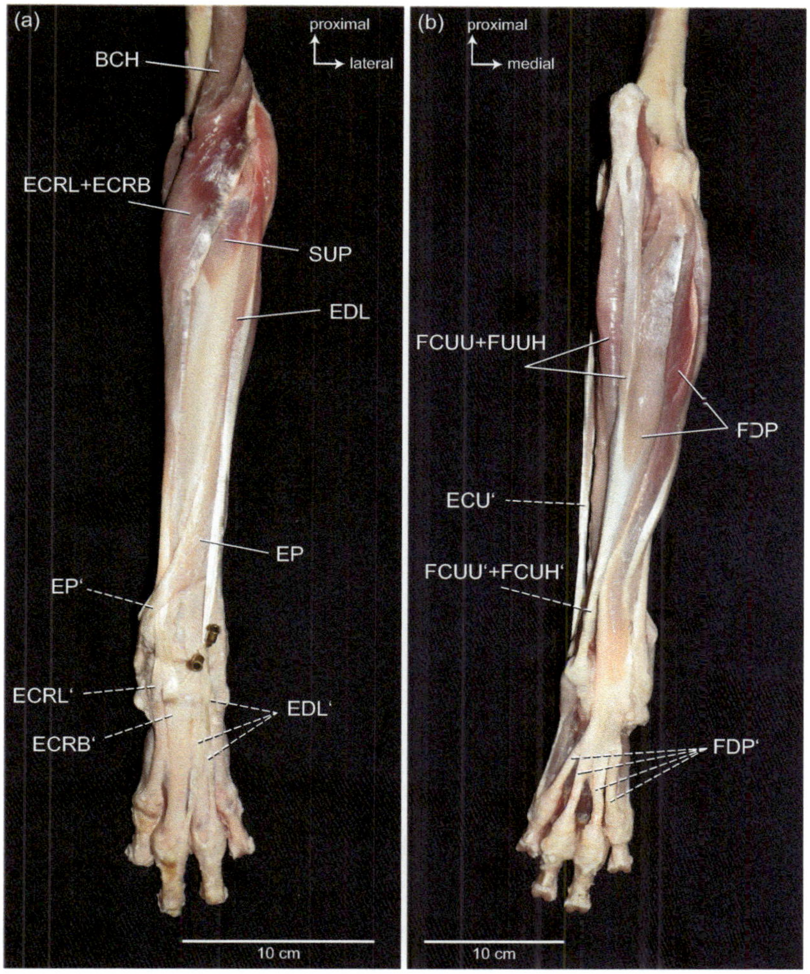
(a)
proximal
lateral
BCH
ECRL+ECRB
SUP
EDL
EP
EP'
ECRL'
ECRB'
EDL'
10 cm
(b)
proximal
medial
FCUU+FUUH
FDP
ECU'
FCUU'+FCUH'
FDP'
10 cm

FIGURE II.10.6 **Deep rotator muscles. (a) Distal rotator muscles in dorsal view**. The EI, *M. extensor digiti I and II* inserts with two tendons on the distal phalanges of digits I and II. **(b) Distal rotator muscles in ventral view. (c) Proximal rotator muscles in ventral view.** The BB, *M. biceps brachii* inserts with two tendons into the bicipital tuberosity of the radius and the coronoid process of the ulna. The tendon of the BCH, *M. brachioradialis* inserts proximally into the bicipital tuberosity of the radius.

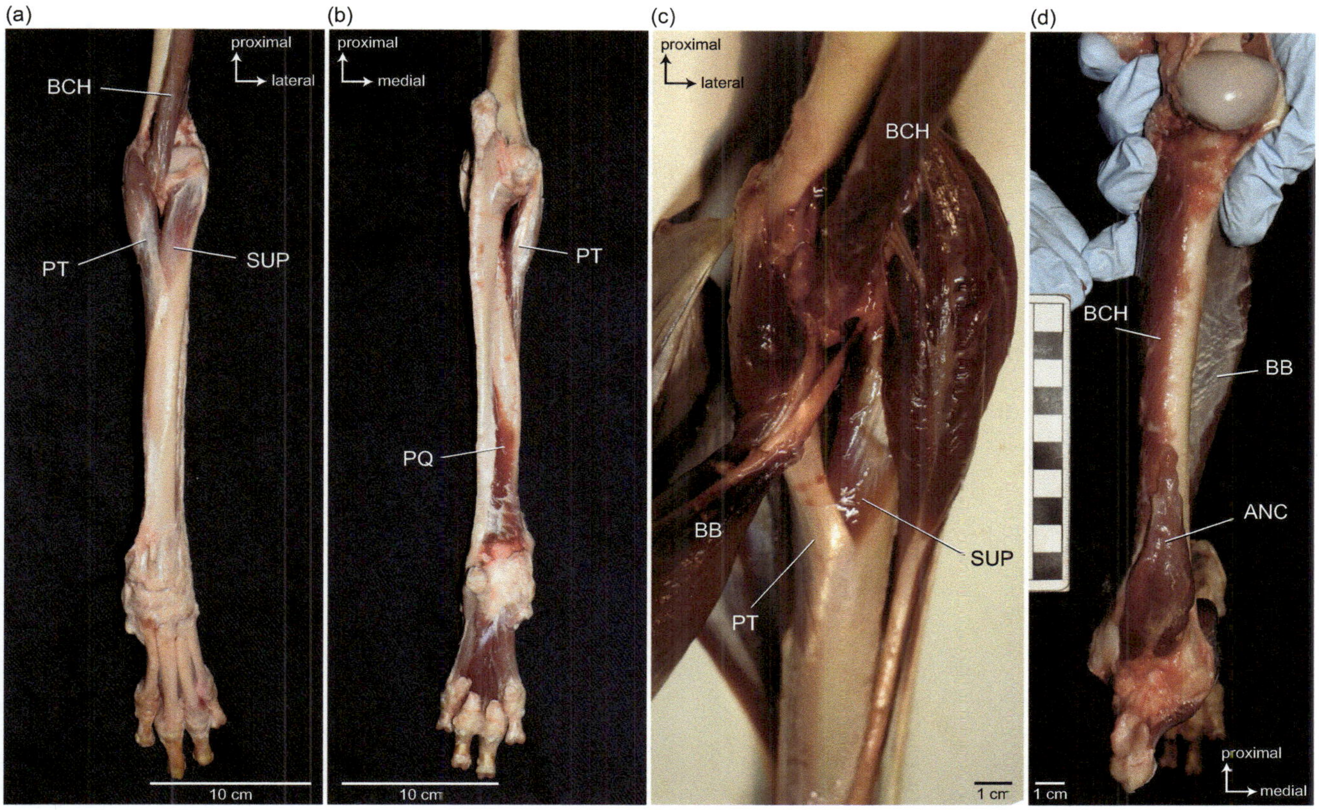
(a)
proximal
lateral
BCH
PT
SUP
10 cm
(b)
proximal
medial
PT
PQ
10 cm
(c)
proximal
lateral
BCH
BB
SUP
PT
1 cm
(d)
BCH
BB
ANC
proximal
medial
1 cm

FIGURE II.10.7 **Hip muscles. (a) In ventral view.** Typically in carnivores, the PSMJ, *M. psoas major* is fused with the ILC, *M. iliacus* forming the *M. iliopsoas.* It originates from the ventral aspect of the last thoracic vertebrae (10–13) and all lumbar vertebrae. The PSMJ, *M. psoas major* and the PSMN, *M. psoas minor* join to insert together into the lesser trochanter of the femur and medial aspect of the neck of the femur. **(b) In medial view.** The OE, *M. obturator externus* lies lateral to the OI, *M. obturator internus.*

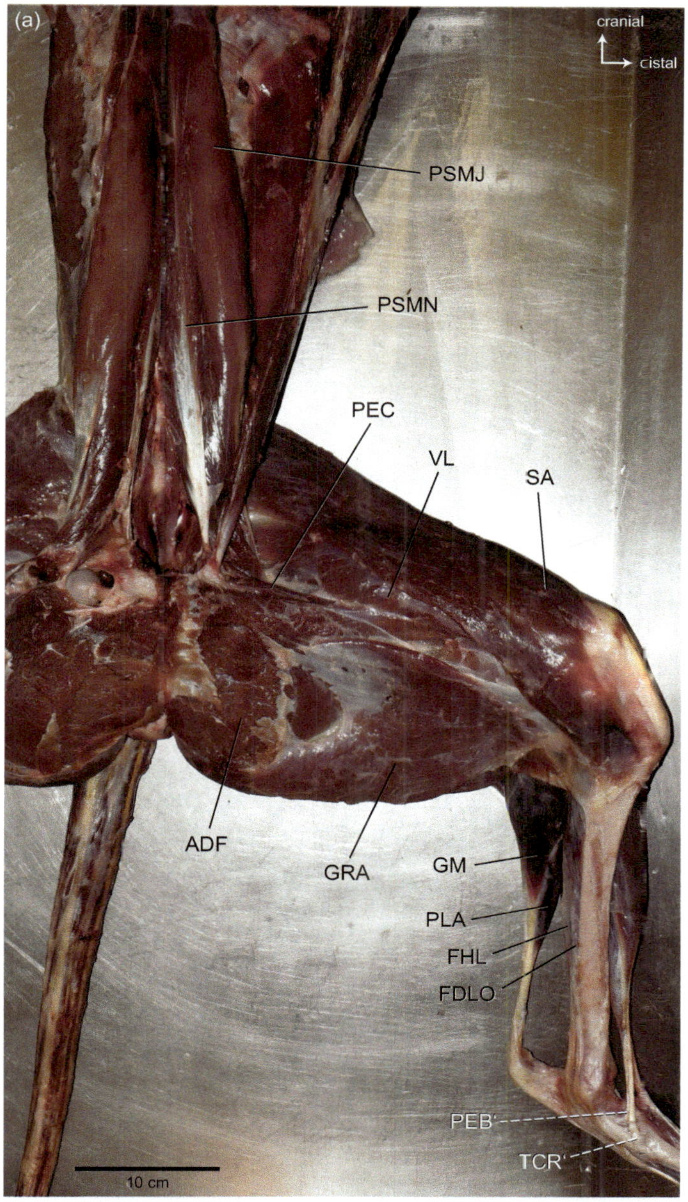
(a)
cranial
cistal
PSMJ
PSMN
PEC
VL
SA
ADF
GRA
GM
PLA
FHL
FDLO
PEB
TCR
10 cm

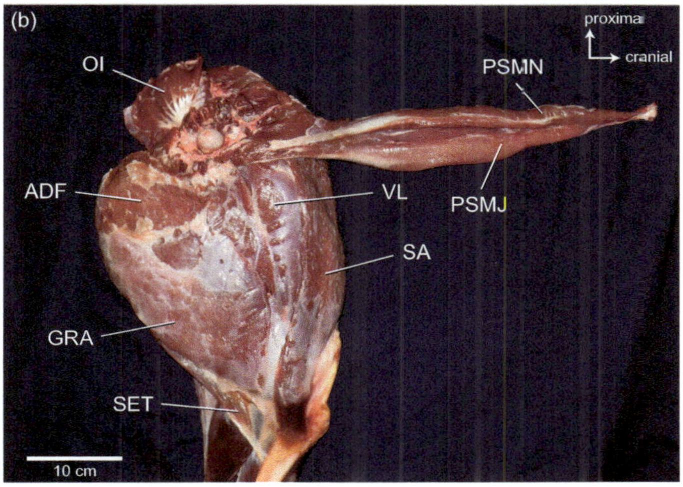
(b)
proximal
cranial
OI
PSMN
PSMJ
ADF
VL
SA
GRA
SET
10 cm

FIGURE II.10.8 **Intrinsic hind limb muscles in (a) lateral view and (b) medial view.** The QF, *M. quadratus femoris*; the GRA, *M. gracilis*; and the SA, *M. sartorius* have been removed. **Distal intrinsic forelimb muscles in (c) lateral view and (d) medial view.** The EDLA, *M. extensor digitorum lateralis* inserts on the distal phalanx of digit V. The EDLO, *M. extensor digitorum longus* inserts into the distal phalanges of digits II–V. The PEB, *M. peroneus brevis* has been removed.

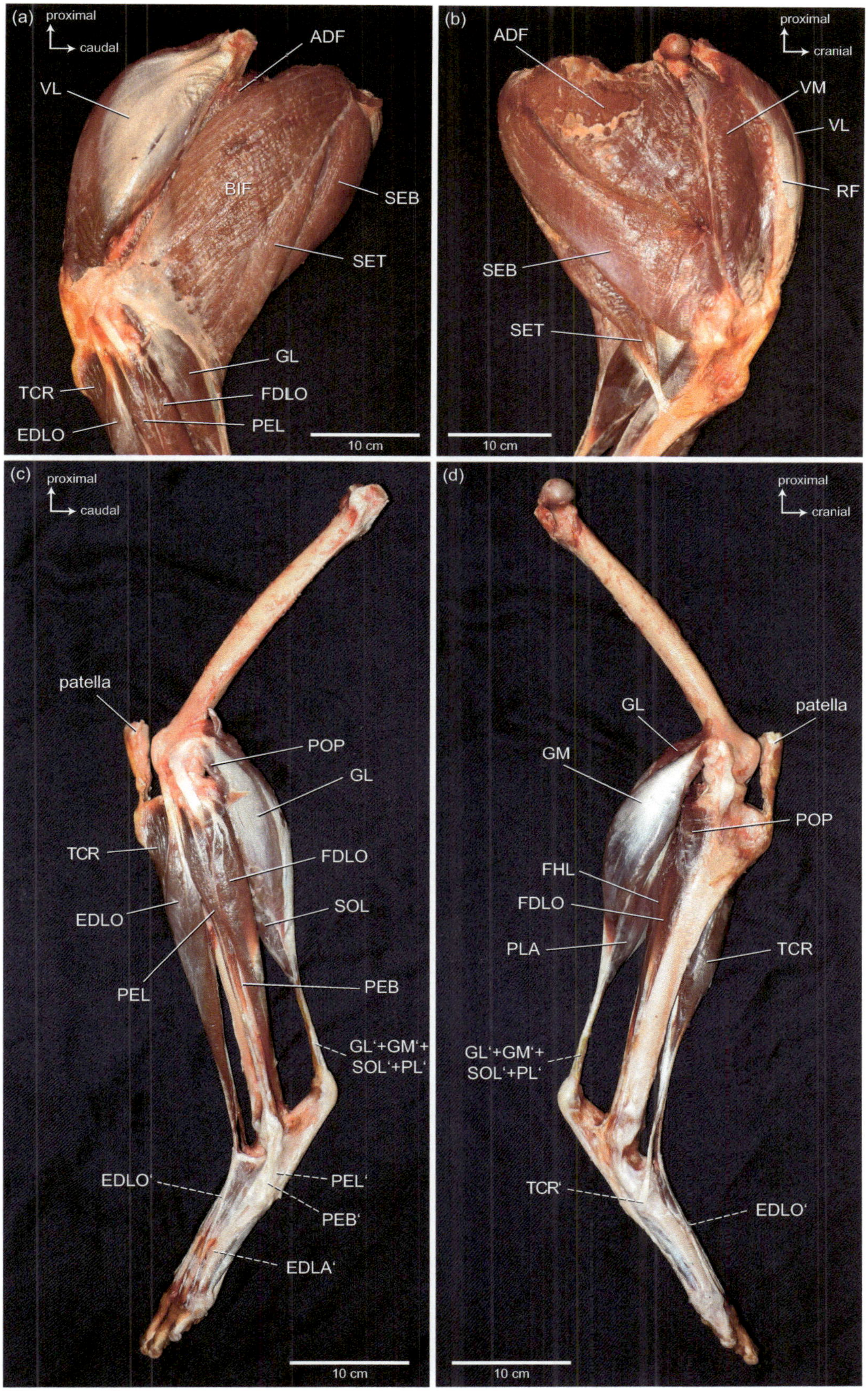
(a)
proximal
caudal
ADF
VL
BIF
SEB
SET
GL
TCR
FDLO
EDLO
PEL
10 cm
(b)
proximal
cranial
ADF
VM
VL
RF
SEB
SET
10 cm
(c)
proximal
caudal
patella
POP
GL
TCR
FDLO
EDLO
SOL
PEL
PEB
GL'+GM'+
SOL'+PL'
EDLO'
PEL'
PEB'
EDLA'
10 cm
(d)
proximal
cranial
GL
patella
GM
POP
FHL
FDLO
PLA
TCR
GL'+GM'+
SOL'+PL'
TCR'
EDLO'
10 cm

Panthera leo (Lion)

Classification: Carnivora, Feloidea, Felidae
Mean body mass: 200 kg
Habitat: Savanna, grassland
Locomotor ecology: Ambulatorial – scansorial
Peculiarities: Cursorial ambusher; uses forelimbs to handle prey

Pectoral muscles. Four pectoral muscles are identified (the PMJ, *M. pectoralis major*; the PAB, *M. pectoantebrachialis*; the PMN, *M. pectoralis minor*; and the XH, *M. xiphihumeralis*).

Deep neck and shoulder muscles. In the neck region, three *Mm. rhomboidei* are present (RCR, *M. rhomboideus cervicis*; RT, *M. rhomboideus thoracis*; and the RCP, *M. rhomboideus capitis*). The RPR, *M. rhomboideus profundus* is absent. The RCR, *M. rhomboideus cervicis* does not reach the occiput.

Intrinsic forelimb muscles. The BB, *M. biceps brachii* has one distinct origin. It inserts with a strong tendon into the bicipital tuberosity of the radius. The tendon of the BCH, *M. brachioradialis* inserts into the coronoid process of the ulna. The EDL, *M. extensor digitorum lateralis* inserts with three tendons on the distal phalanges of digits III and V. The EDC, *M. extensor digitorum communis* inserts with four tendons on the distal phalanges of digits II through V. The EI, *M. extensor digiti I and II* inserts with two tendons on the distal phalanges of digits I and II. The ECRL, *M. extensor carpi radialis longus* inserts with one tendon into the base of metacarpal II. The ECRB, *M. extensor carpi radialis brevis* inserts with one tendon into the base of metacarpal III. The PL, *M. palmaris longus* inserts with four tendons on the distal phalanges of digits II through V. The FDP, *M. flexor digitorum profundus* (four muscle bellies) inserts with four strong tendons on the distal phalanges of digits II through V.

FIGURE II.11.1 Superficial muscles of forelimb and hind limb in lateral view. (a) Without labels. (b) With labels and muscle outlines. In the neck and shoulder region, the *Mm. trapezii* are clearly divided into three parts (CT, *M. clavotrapezius*; AT, *M. arcomiotrapezius*; and ST, *M. spinotrapezius*). The EPI, *M. epitrochlearis* is present. The BCR, *M. brachioradialis* are absent. A rudimentary clavicle is present (embedded into the raphe between the CM, *M. cleidomastoideus* and the CB, *M. clavobrachialis*). The SC, *M. subclavius* is absent. The OEA, *M. obliquus externus abdominis* originates from the thoracic ribs and inserts into the linea alba at the midline and on the pubis near the symphysis. In the distal hind limb, the four tendons of the EDLO, *M. extensor digitorum longus* insert on the distal phalanges of digits II through V. The EDLA, *M. extensor digitorum lateralis* inserts on the distal phalanges of digits IV and V. The PET, *M. peroneus tertius* is absent. Note that the retinacula (ligaments) spanning the wrist and ankle joints have been partially removed.

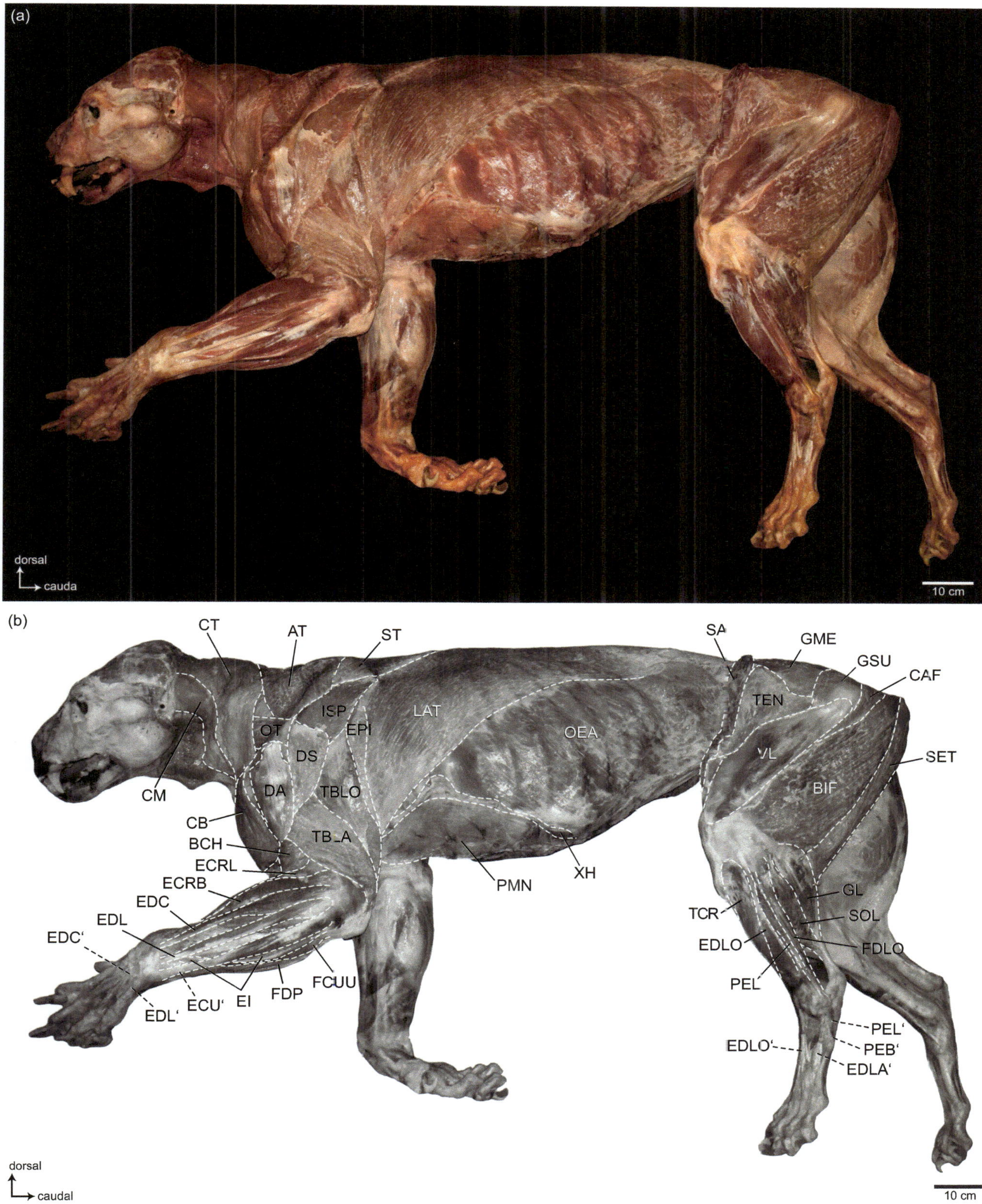
(a)
dorsal
cauda
10 cm
(b)
CT
AT
ST
SA
GME
GSU
CAF
TEN
ISP
EPI
LAT
OT
OEA
DS
VL
SET
CM
DA
TBLO
BIF
CB
BCH
TBLA
ECRL
ECRB
PMN
XH
EDC
GL
TCR
SOL
EDL
EDC‘
EDLO
FDLO
FCUU
PEL
EDL‘
ECU‘
EI
FDP
PEL‘
EDLO‘
PEB‘
EDLA‘
dorsal
caudal
10 cm

Manis tricuspis (Tree Pangolin)

Classification:	Pholidota, Manidae
Mean body mass:	4.5–14.0 kg
Habitat:	Forest, rainforest
Locomotor ecology:	Scansorial
Peculiarities:	Facultative bipedal locomotion on the ground; quadrupedal locomotion in the trees; prehensile tail; strong, curved claws

FIGURE II.12.1 Superficial muscles of forelimb and hind limb in lateral view. (a) Without labels. (b) With labels and muscle outlines. In the neck and shoulder region, the *Mm. trapezii* are divided into two parts (CT, *M. clavotrapezius* together with AT, *M. arcomiotrapezius*; ST, *M. spinotrapezius*). The DA, *M. acromiodeltoideus* is absent. The EPI, *M. epitrochlearis* is present. The BCR, *M. brachioradialis* is absent. In the hip region, the TEN, *M. tensor fasciae latae* covers much of the *Mm. glutei* and the lateral muscles of the *Mm. quadriceps femoris* group. Note the distinct morphology of the BIF, *M. biceps femoris*. In the distal hind limb, the four tendons of the EDLO, *M. extensor digitorum longus* insert on the distal phalanges of digits II through V. Not visible in the picture, but the PET, *M. peroneus tertius* is present. Note that the retinacula (ligaments) spanning the wrist and ankle joints have been partially removed.

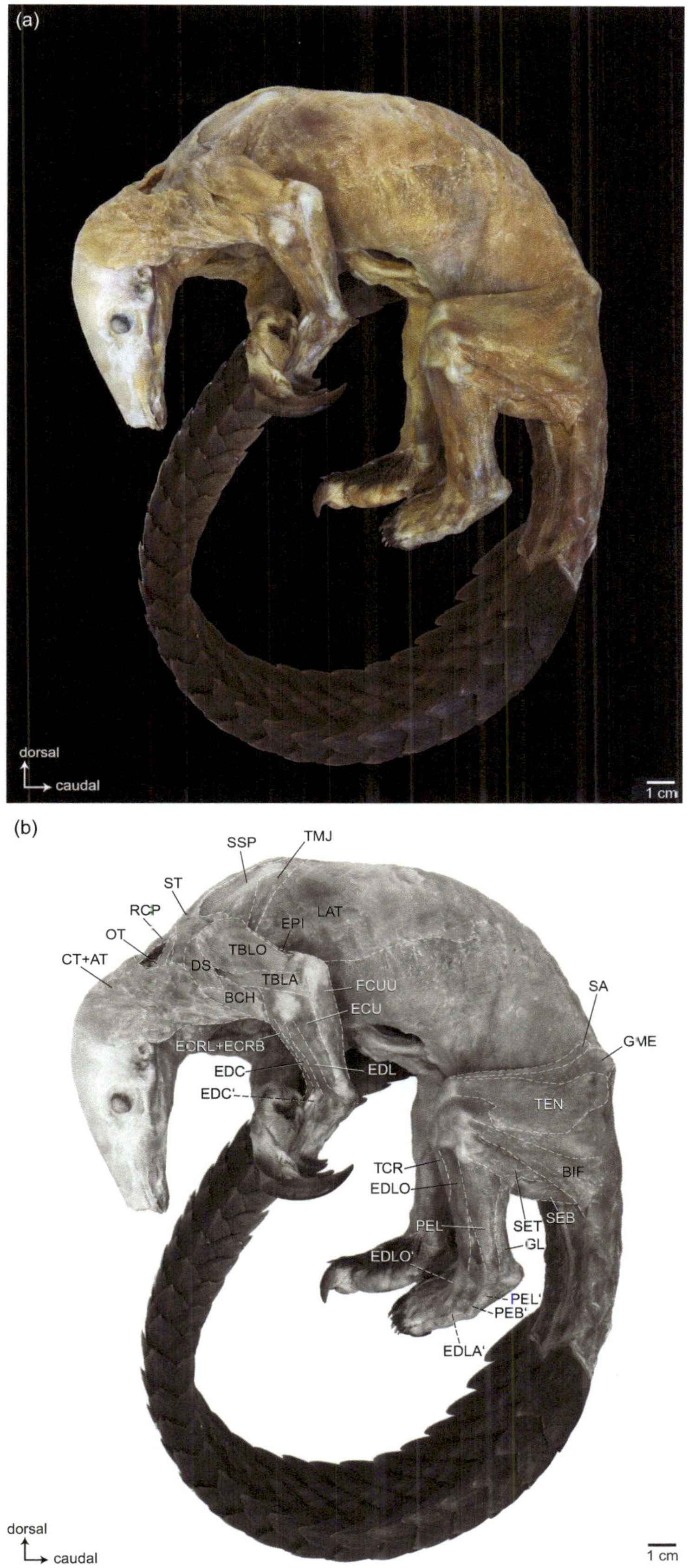
(a)
dorsal
caudal
1 cm
(b)
SSP
TMJ
ST
RCP
OT
CT+AT
LAT
EPI
TBLO
DS
TBLA
FCUU
BCH
ECU
SA
GME
ECRL+ECRB
EDC
EDL
EDC'
TEN
TCR
EDLO
BIF
PEL
SET
SEB
GL
EDLO'
PEL'
PEB'
EDLA'
dorsal
caudal
1 cm

FIGURE II.12.2 **Pectoral muscles in ventral view.** Three pectoral muscles are identified (the PMJ, *M. pectoralis major*; the PAB, *M. pectoantebrachialis*; and the PMN, *M. pectoralis minor*). The XH, *M. xiphihumeralis* is not developed as a separate muscle, but part of the PMN, *M. pectoralis minor.*

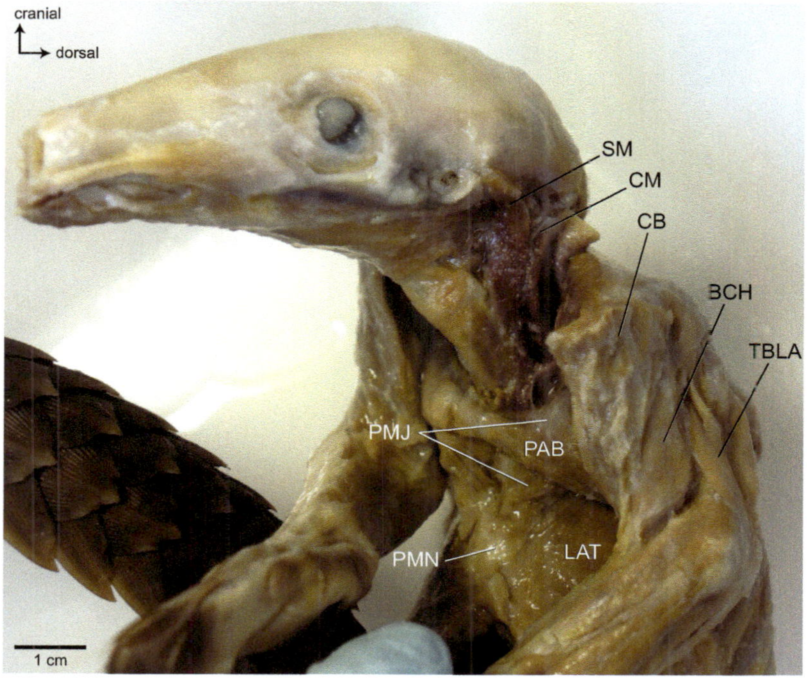
cranial
dorsal
SM
CM
CB
BCH
TBLA
PMJ
PAB
PMN
LAT
1 cm

FIGURE II.12.3 **Deep neck and shoulder muscles in dorsal view.** In the neck region, three *Mm. rhomboidei* are present (RCR, *M. rhomboideus cervicis*; RT, *M. rhomboideus thoracis*; and RCP, *M. rhomboideus capitis*). The RCR, *M. rhomboideus cervicis* does not reach the occiput. The RPR, *M. rhomboideus profundus* is absent.

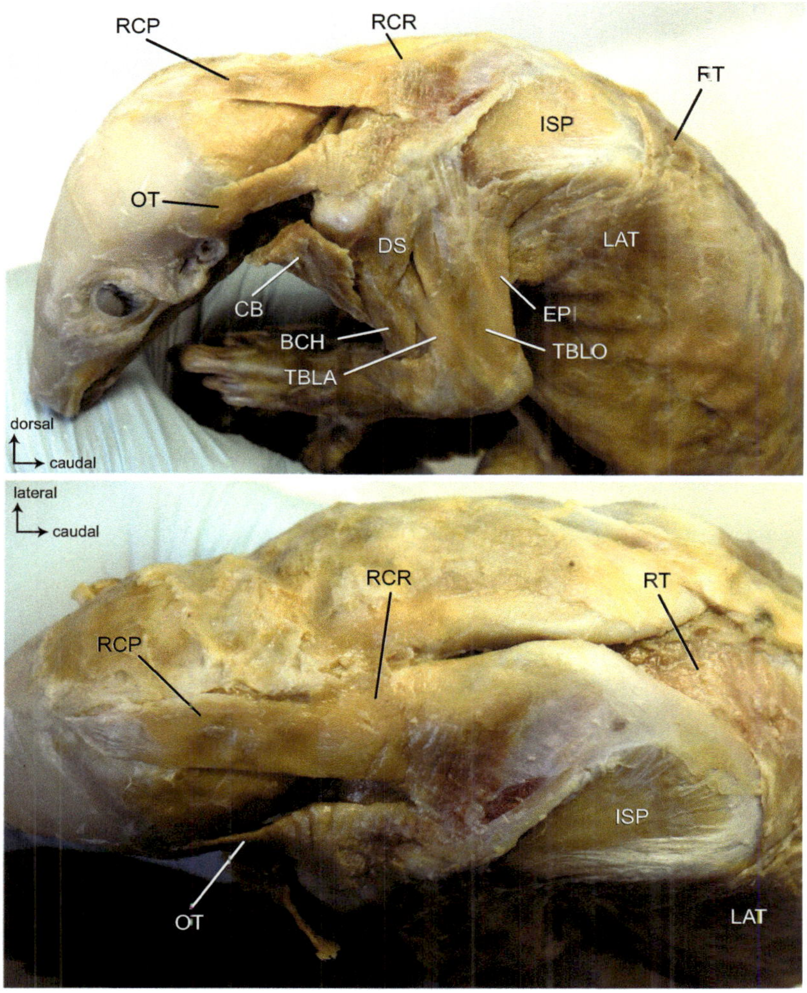
RCP
RCR
RT
ISP
OT
DS
LAT
CB
EPI
BCH
TBLO
TBLA
dorsal
caudal
lateral
caudal
RCR
RT
RCP
ISP
OT
LAT

FIGURE II.12.4 **Intrinsic forelimb muscles in (a) proximolateral view and (b) proximomedial view.** The DA, *M. acromiodeltoideus* is absent. The BB, *M. biceps brachii* has one distinct origin. Note the relatively large attachment area for the *M. serratus ventralis* medial on the scapula. **Intrinsic forelimb muscles in (c) lateral view and (d) medial view.** The CB, *M. clavobrachialis* inserts distally into the radius and ulna. The PL, *M. palmaris longus* inserts into the palmar aponeurosis.

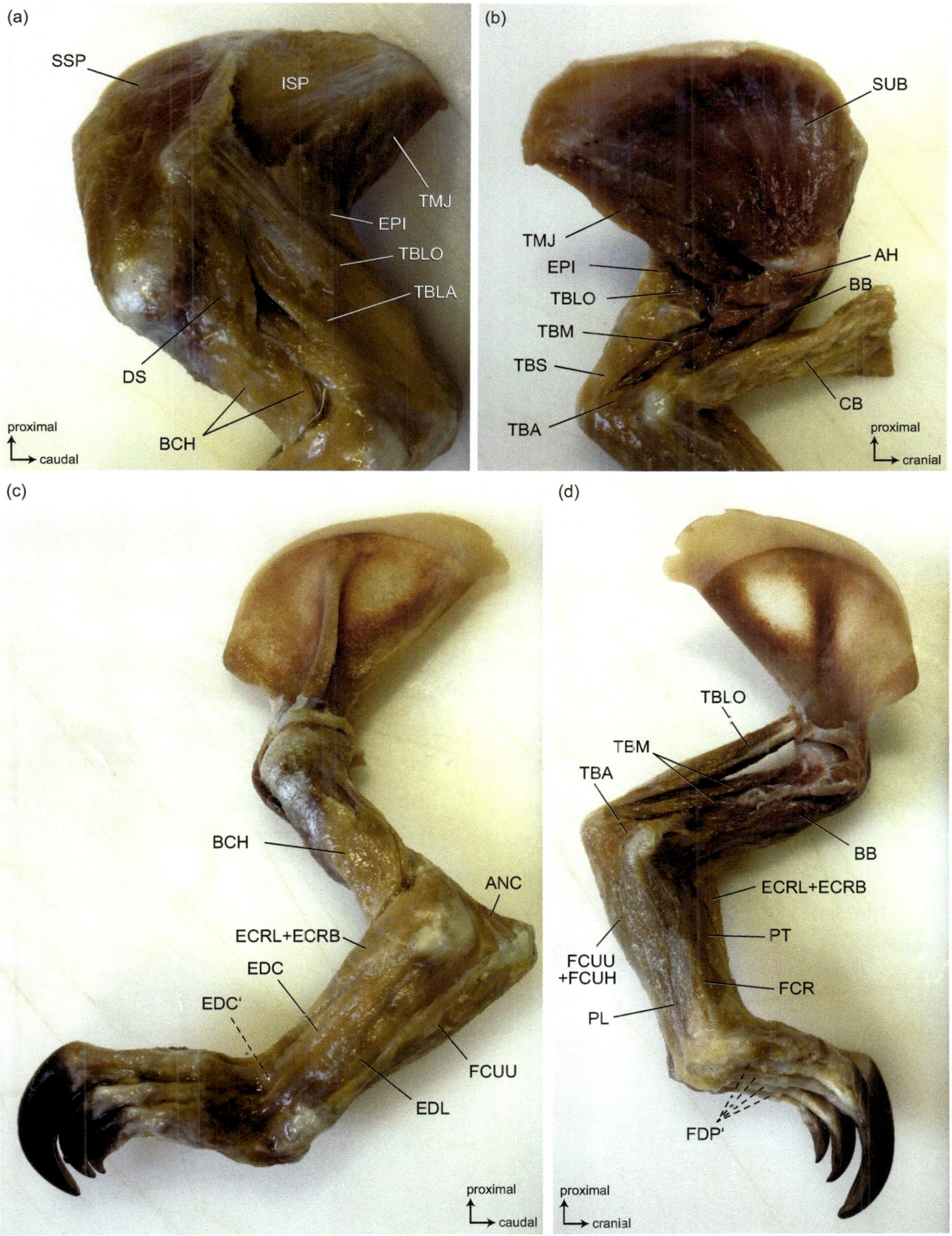
(a)
SSP
ISP
TMJ
EPI
TBLO
TBLA
DS
BCH
proximal
caudal
(b)
SUB
TMJ
EPI
TBLO
TBM
TBS
TBA
AH
BB
CB
proximal
cranial
(c)
BCH
ANC
ECRL+ECRB
EDC
EDC'
FCUU
EDL
proximal
caudal
(d)
TBLO
TBM
TBA
BB
ECRL+ECRB
PT
FCUU
+FCUH
FCR
PL
FDP'
proximal
cranial

FIGURE II.12.5 **Distal forelimb muscles. (a) In dorsal view.** The ECRL, *M. extensor carpi radialis longus* (inserts with one tendon into the base of metacarpal II) and the ECRB, *M. extensor carpi radialis brevis* (inserts with one tendon into the base of metacarpal III) have been removed. The EDC, *M. extensor digitorum communis* inserts with three tendons on the distal phalanges of digits III through V. The EDL, *M. extensor digitorum lateralis* inserts with one strong tendon into the proximal phalanx of digit V. **(b) In ventral view.** The FDP, *M. flexor digitorum profundus* (four muscle bellies) inserts with four tendons on the distal phalanges of digits II through V.

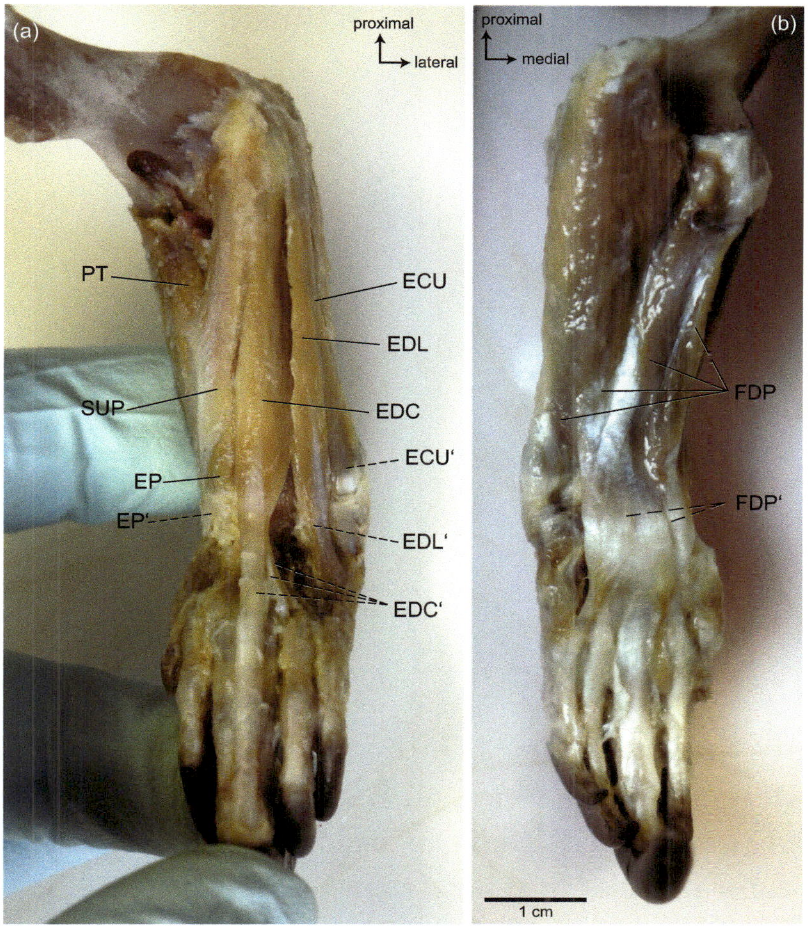

(a)
proximal
lateral
PT
ECU
EDL
SUP
EDC
ECU'
EP
EP'
EDL'
EDC'
(b)
proximal
medial
FDP
FDP'
1 cm

FIGURE II.12.6 **Deep rotator muscles. (a) Distal rotator muscles in dorsal view.** The EI, *M. extensor digiti I and II* inserts with two tendons on the distal phalanges of digits I and II. **(b) Proximal rotator muscles in ventral view.**

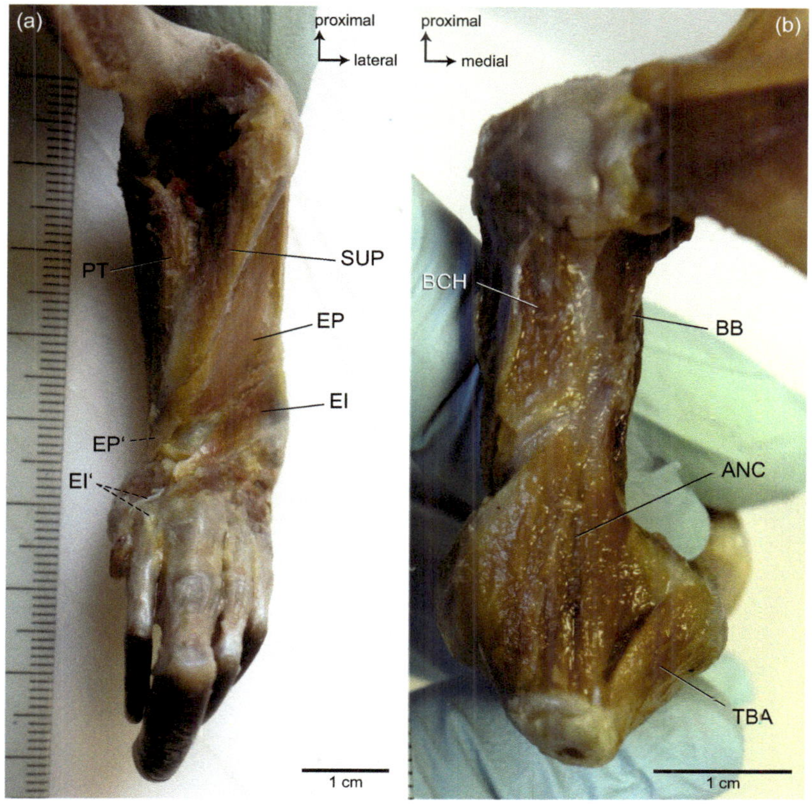
(a)
proximal
lateral
PT
SUP
EP
EI
EP'
EI'
1 cm
(b)
proximal
medial
BCH
BB
ANC
TBA
1 cm

Sus scrofa (Wild boar)

Classification:	Artiodactyla, Suina, Suidae
Mean body mass:	170 kg
Habitat:	Forest, grassland
Locomotor ecology:	Cursorial
Peculiarities:	Four digits; digitigrade

Pectoral muscles. Three pectoral muscles are identified (the PMJ, *M. pectoralis major*; the PAB, *M. pectoantebrachialis*; and the PMN, *M. pectoralis minor*). The XH, *M. xiphihumeralis* is not developed as a separate muscle, but part of the PMN, *M. pectoralis minor.*

Deep neck and shoulder muscles. In the neck region, three *Mm. rhomboidei* are present (RCR, *M. rhomboideus cervicis*; RT, *M. rhomboideus thoracis*; and RCP, *M. rhomboideus capitis*). The RPR, *M. rhomboideus profundus* is absent. The RCR, *M. rhomboideus cervicis* almost reaches the occiput.

Intrinsic forelimb muscles. The BB, *M. biceps brachii* has one distinct origin. The EDL, *M. extensor digitorum lateralis* inserts with two tendons on the distal phalanges of digits IV through V. The EDC, *M. extensor digitorum communis* inserts with four tendons on the distal phalanges of digits II through V. The EI, *M. extensor digiti I and II* inserts with two tendons on the distal phalanges of digits I and II. The M. *extensor carpi radialis* is identified as one muscle belly, but inserts with one tendon into the base of metacarpal II (ECRL, *M. extensor carpi radialis longus*) and with one tendon into the base of metacarpal III (ECRB, *M. extensor carpi radialis brevis*). The PL, *M. palmaris longus* inserts with four tendons into the distal phalanges of digits II through V. The FDP, *M. flexor digitorum profundus* (four muscle bellies) inserts with four tendons on the distal phalanges of digits II through V.

FIGURE II.13.1 Superficial muscles of forelimb and hind limb in lateral view. (a) Without labels. (b) With labels and muscle outlines. In the neck and shoulder region, the *Mm. trapezii* are clearly divided into two parts (CT, *M. clavotrapezius* together with AT, *M. arcomiotrapezius* – already removed in this specimen; ST, *M. spinotrapezius*). The EPI, *M. epitrochlearis* is present. The BCR, *M. brachioradialis* is absent. The clavicle is absent. In the hip region, the CAF, *M. caudofemoralis* is absent. In the distal hind limb, the four tendons of the EDLO, *M. extensor digitorum longus* insert on the distal phalanges of digits II through V. The PEB, *M. peroneus brevis* is absent.

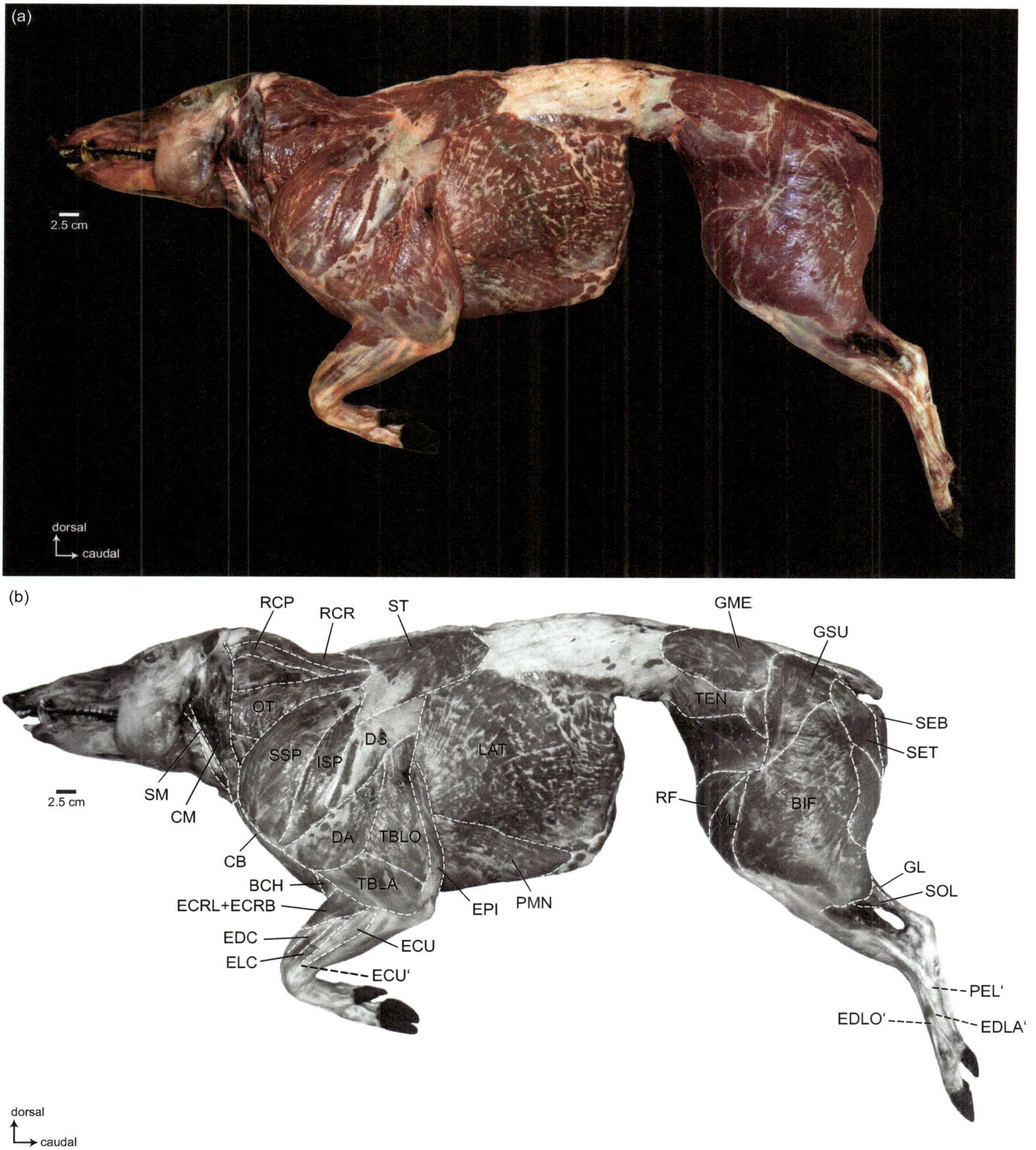
(a)
2.5 cm
dorsal
caudal
(b)
RCP
RCR
ST
GME
GSU
OT
TEN
SEB
SET
DS
SSP
ISP
LAT
2.5 cm
SM
CM
RF
VL
BIF
DA
TBLO
CB
GL
SOL
BCH
TBLA
ECRL+ECRB
EPI
PMN
EDC
ECU
ELC
ECU'
PEL'
EDLO'
EDLA'
dorsal
caudal

Capreolus capreolus (European roe deer)

Classification:	Artiodactyla, Ruminantia, Cervidae
Mean body mass:	22.0–30.0 kg
Habitat:	Forest, forest steppe, high-grass meadows
Locomotor ecology:	Cursorial
Peculiarities:	Four digits; digitigrade; strong ligamentum nuchae

FIGURE II.14.1 Superficial muscles of forelimb and hind limb in lateral view. (a) Without labels. (b) With labels and muscle outlines. In the neck and shoulder region, the *Mm. trapezii* are divided into two parts (CT, *M. clavotrapezius* together with AT, *M. arcomiotrapezius*; ST, *M. spinotrapezius*). The OT, *M. omotransversarius* is fused with the cervical part of the *Mm. trapezii* (CT and AT). The DS, *M. spinodeltoideus* originates mainly from the infraspinatus fascia (connective tissue that covers the ISP, *M. infraspinatus*). The EPI, *M. epitrochlearis* is present, but relatively weakly developed. The BCR, *M. brachioradialis* is absent. The OEA, *M. obliquus externus abdominis* originates from the thoracic ribs and inserts into the linea alba at the midline and on the pubis near the symphysis. The SA, *M. sartorius* is present, but not visible since it lies medially to the TEN, *M. tensor fasciae latae*. The GSU, *M. gluteus superficialis* is fused together with the BIF, *M. biceps femoris* (also named *M. glutaeobiceps* in ungulates). The CAF, *M. caudofemoralis* is absent. In the distal hind limb, the two tendons of the EDLO, *M. extensor digitorum longus* insert on the distal phalanges of digits III and IV. The EHL, *M. hallucis longus* is fused with the TCR, *M. tibialis cranialis*. The PEB, *M. peroneus brevis* is absent. Note that the retinacula (ligaments) spanning the wrist and ankle joints have been largely removed.

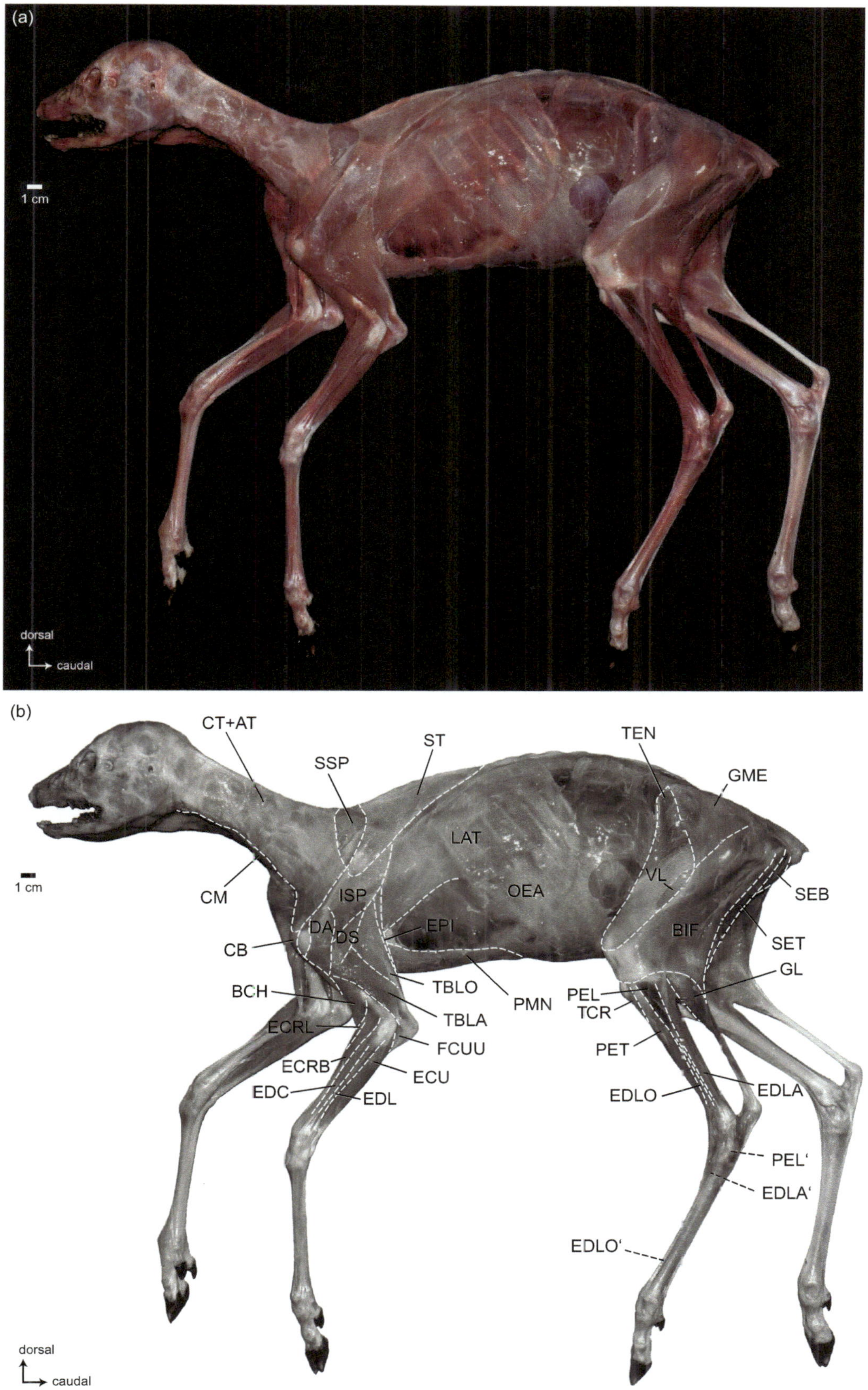
(a)
1 cm
dorsal
caudal
(b)
CT+AT
SSP
ST
TEN
GME
LAT
1 cm
CM
ISP
OEA
VL
SEB
DA
DS
EPI
BIF
CB
SET
GL
TBLO
BCH
PMN
PEL
TCR
TBLA
ECRL
FCUU
PET
ECRB
ECU
EDC
EDL
EDLO
EDLA
PEL'
EDLA'
EDLO'
dorsal
caudal

FIGURE II.14.2 **Pectoral muscles in ventral view. (a) Superficial view.** The clavicle is absent. **(b) Deep view.** The PMJ, *M. pectoralis major* and the PAB, *M. pectoantebrachialis* have been detached at their insertion on the humerus. The XH, *M. xiphihumeralis* is not developed as a separate muscle, but part of the PMN, *M. pectoralis minor.*

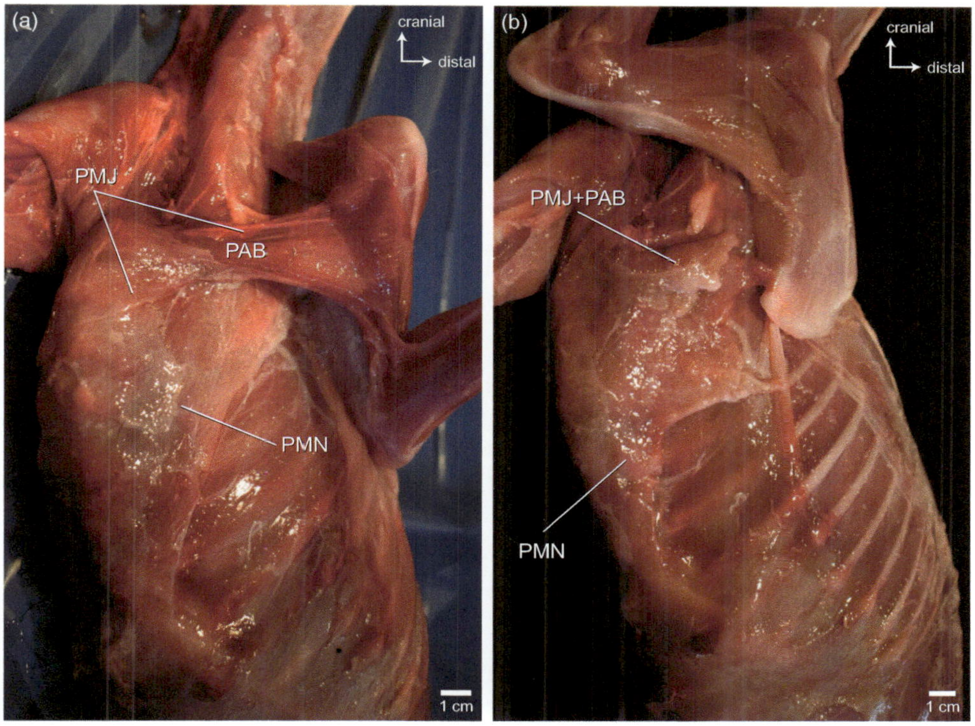
(a)
cranial
distal
PMJ
PAB
PMN
1 cm
(b)
cranial
distal
PMJ+PAB
PMN
1 cm

FIGURE II.14.3 **Deep neck and shoulder muscles. (a) In craniolateral view.** The CT, *M. clavotrapezius*; AT, *M. arcomiotrapezius*; and the OT, *M. omotransversarius* are fused together and in close relation to the CM, *M. cleidomastoideus.* **(b) In dorsolateral view.** In the neck region, two *Mm. rhomboidei* are present (RCR, *M. rhomboideus cervicis* and RT, *M. rhomboideus thoracis*). The RCR, *M. rhomboideus cervicis* does not reach the occiput. The RCP, *M. rhomboideus capitis* and the RPR, *M. rhomboideus profundus* are absent. The clavicle is absent. **(c) In lateral view.** The well-developed ligamentum nuchae connects the occiput to the neural spines of the thoracic vertebrae.

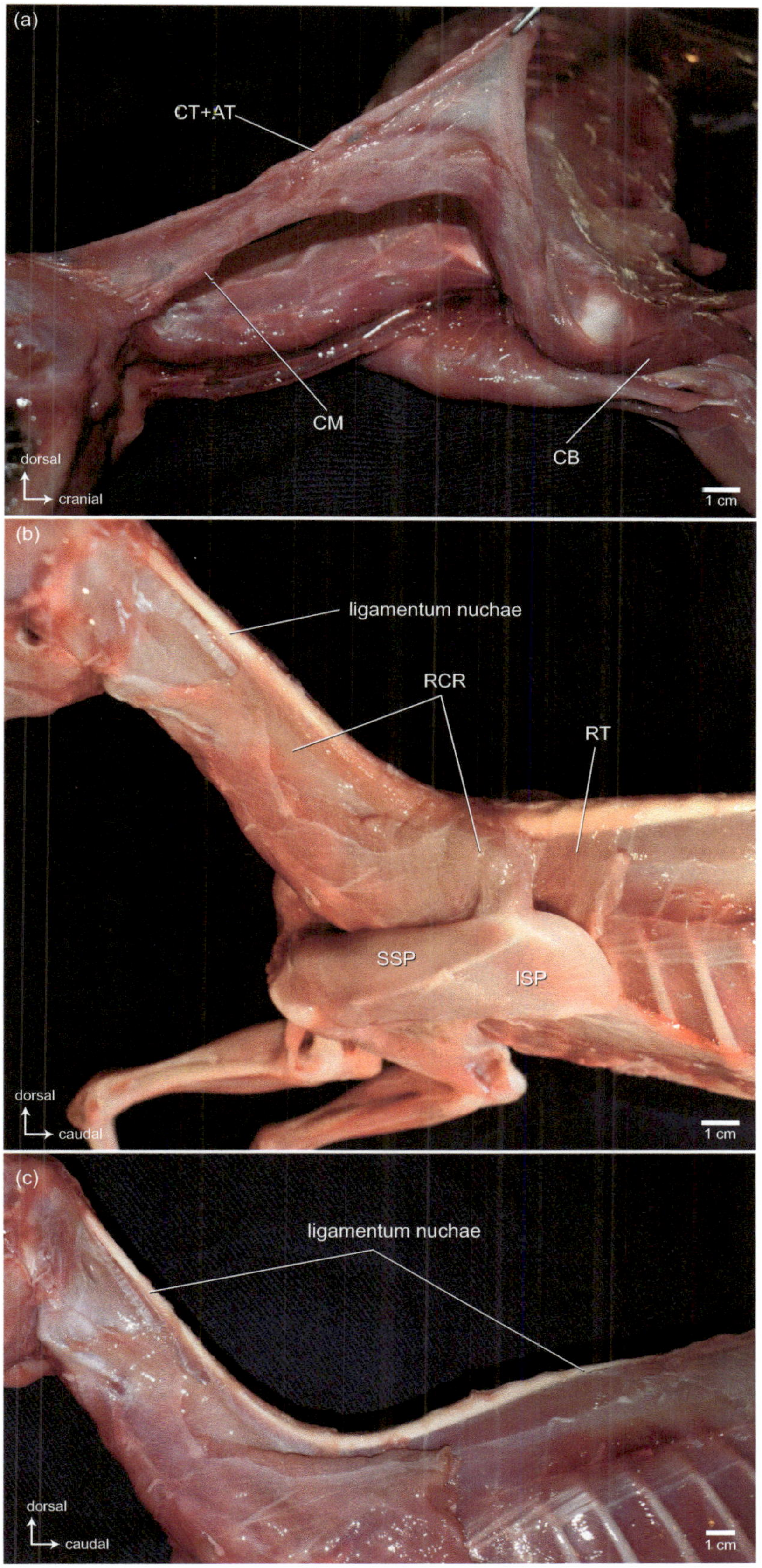
(a)
CT+AT
CM
CB
dorsal
cranial
1 cm
(b)
ligamentum nuchae
RCR
RT
SSP
ISP
dorsal
caudal
1 cm
(c)
ligamentum nuchae
dorsal
caudal
1 cm

FIGURE II.14.4 Intrinsic forelimb muscles in (a) proximolateral view and (b) proximomedial view. The DS, *M. spinodeltoideus* has been removed. The TBA, *M. triceps accessorium* is absent. **Deep proximal muscles in (c) lateral view and (d) medial view.** The SSP, *M. supraspinatus* has been detached from the scapular spine to show the muscle's relation with the supraspinous fossa of the scapula. **Deep proximal muscles in (e) cranial view. Intrinsic forelimb muscles (f) in lateral view and (g) medial view. Distal intrinsic forelimb muscles in (h) lateral view and (i) medial view.** The EI, *M. extensor digiti I and II* fuses distally with the EDC, *M. extensor digitorum communis.*

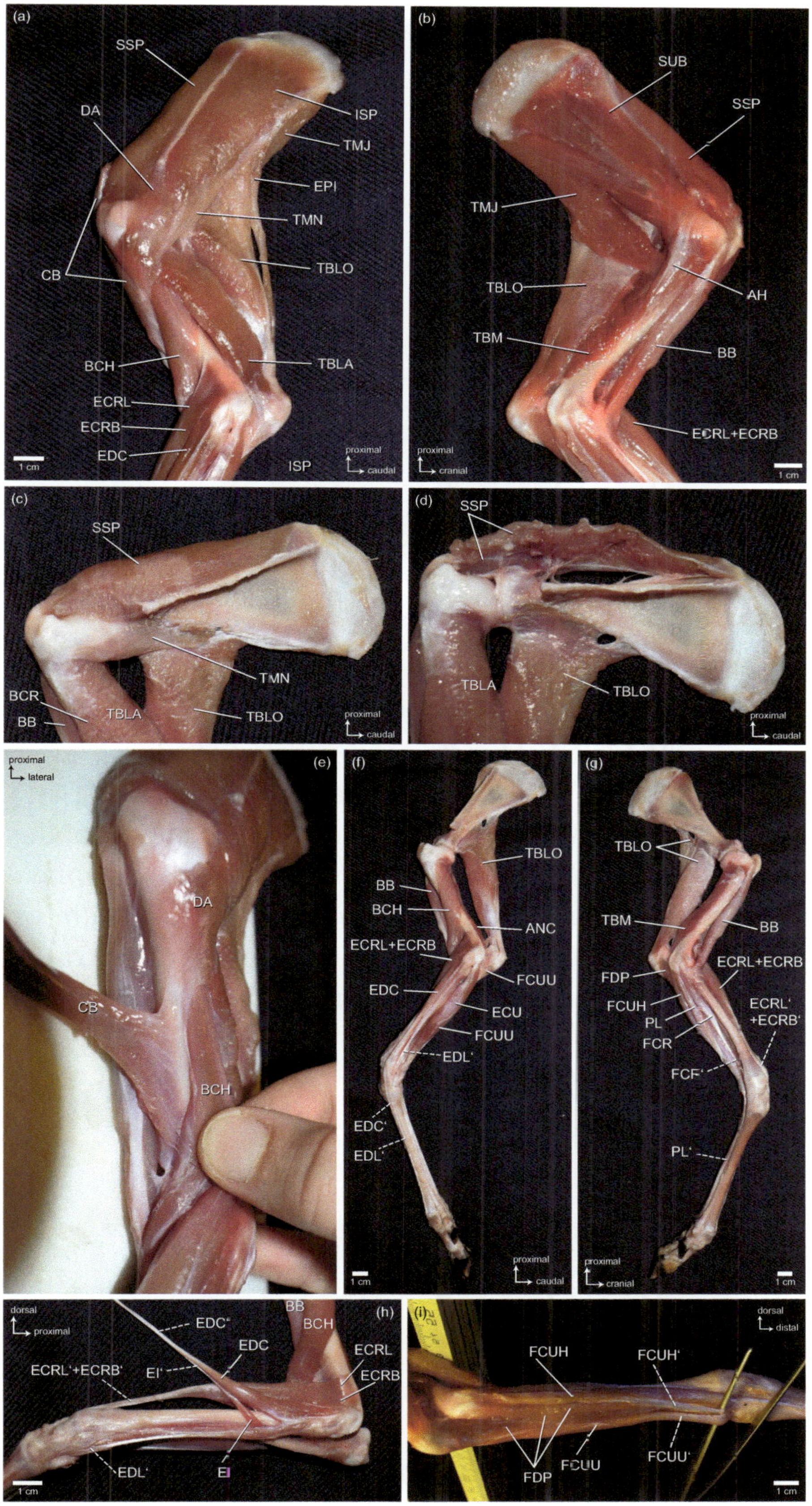
(a)
SSP
DA
ISP
TMJ
EPI
TMN
CB
TBLO
BCH
TBLA
ECRL
ECRB
EDC
ISP
1 cm
proximal
caudal
(b)
SUB
SSP
TMJ
TBLO
AH
TBM
BB
ECRL+ECRB
proximal
cranial
1 cm
(c)
SSP
TMN
BCR
BB
TBLA
TBLO
proximal
caudal
(d)
SSP
TBLA
TBLO
proximal
caudal
proximal
lateral
(e)
DA
CB
BCH
(f)
TBLO
BB
BCH
ANC
ECRL+ECRB
FCUU
EDC
ECU
FCUU
EDL'
EDC'
EDL'
1 cm
proximal
caudal
(g)
TBLO
TBM
BB
ECRL+ECRB
FDP
FCUH
PL
FCR
ECRL'
+ECRB'
FCF'
PL'
proximal
cranial
1 cm
dorsal
proximal
(h)
EDC"
BB
BCH
EDC
ECRL
ECRL'+ECRB'
EI'
ECRB
EDL'
EI
1 cm
(i)
dorsal
distal
FCUH
FCUH'
FDP
FCUU
FCUU'
1 cm

FIGURE II.14.5 **Distal forelimb muscles. (a) In dorsal view.** The ECRL, *M. extensor carpi radialis longus* and the ECRB, *M. extensor carpi radialis brevis* are partially separable, but insert with one tendon to the base of the metacarpal. The EDC, *M. extensor digitorum communis* inserts with two tendons into the distal phalanges of digits III and IV. The EDL, *M. extensor digitorum lateralis* inserts with one tendon into the distal phalanx of digit IV. **(b) In ventral view.** The PL, *M. palmaris longus* inserts with two tendons into the distal phalanges of digits III and IV. **Deep distal muscles in (c) ventral view.** The FDP, *M. flexor digitorum profundus* (one muscle belly) inserts with two tendons into the distal phalanges of digits III and IV.

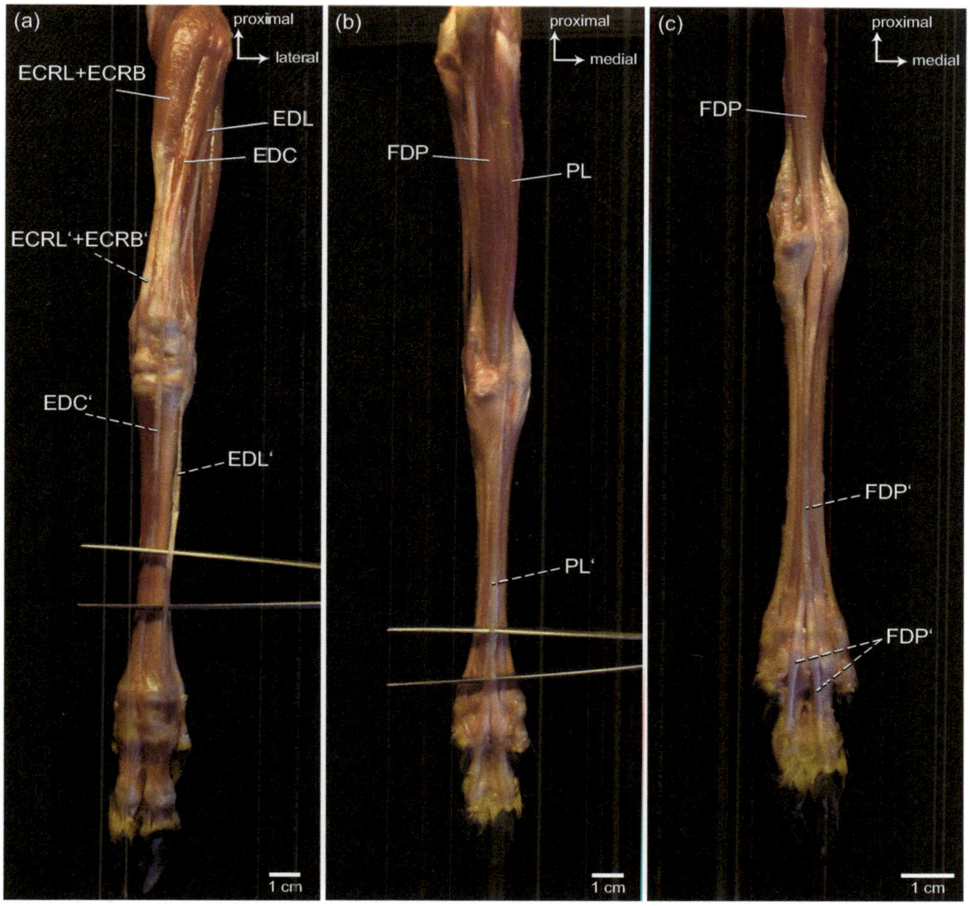
(a)
proximal
lateral
ECRL+ECRB
EDL
EDC
ECRL'+ECRB'
EDC'
EDL'
1 cm
(b)
proximal
medial
FDP
PL
PL'
1 cm
(c)
proximal
medial
FDP
FDP'
FDP'
1 cm

FIGURE II.14.6 Deep rotator muscles. (a) Distal rotator muscles in dorsomedial view. The PT, *M. pronator teres* is very weakly developed. The EI, *M. extensor digiti I and II*; the SUP, *M. supinator*; and the PQ, *M. pronator quadratus* are absent. **(b) Proximal rotator muscles in dorsomedial view.** The tendon of the BB, *M. biceps brachii* inserts into the bicipital tuberosity of the radius. The tendon of the BCH, *M. brachialis* inserts into the coronoid process of the ulna. **(c) Proximal rotator muscles in ventral view.**

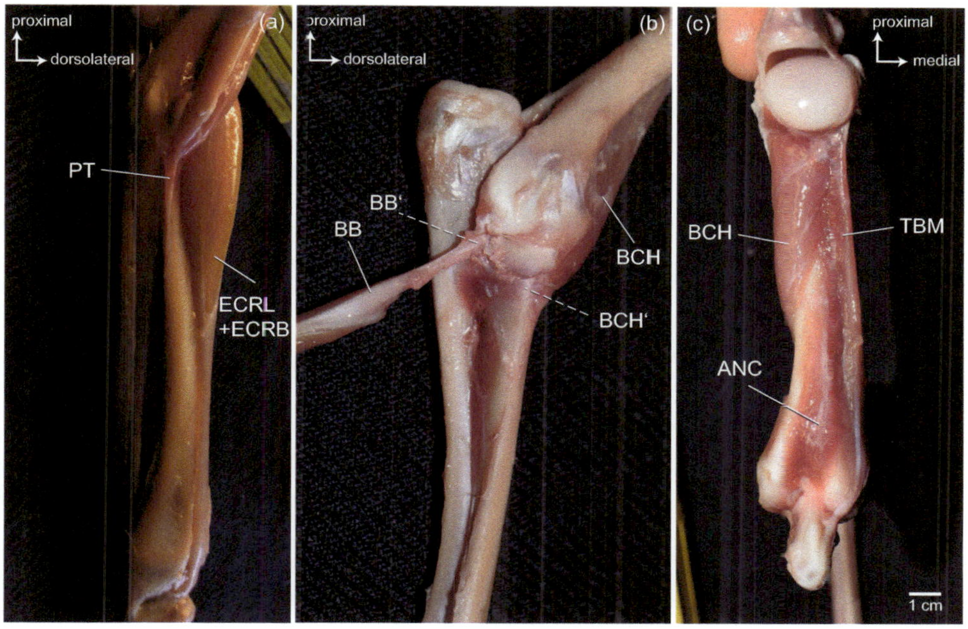

proximal
dorsolateral
(a)
PT
ECRL
+ECRB
proximal
dorsolateral
(b)
BB'
BB
BCH
BCH'
(c)
proximal
medial
BCH
TBM
ANC
1 cm

Erinaceus europaeus (European hedgehog)

Classification:	Eulipotyphla, Erinaceidae, Erinaceinae
Mean body mass:	1.2–0.8 kg
Habitat:	Grassland, forest, field edges, suburban
Locomotor ecology:	Ambulatorial
Peculiarities:	Plantigrade; fossorial

FIGURE II.15.1 Superficial muscles of forelimb and hind limb in lateral view. (a) Without labels. (b) With labels and muscle outlines. In the neck and shoulder region, the *Mm. trapezii* are clearly divided into two parts (CT, *M. clavotrapezius* with AT, *M. arcomiotrapezius*; ST, *M. spinotrapezius*). The EPI, *M. epitrochlearis* is present and originates on the LAT, *M. latissimus dorsi*. The BCR, *M. brachioradialis* is absent. The OEA, *M. obliquus externus abdominis* originates from the thoracic ribs and inserts into the linea alba at the midline and on the pubis near the symphysis. The RA, *M. rectus abdominis* originates from the costal cartilages and inserts into the prebubic tendon. The TEN, *M. tensor fasciae latae* is absent. Not visible in the picture, but the GME, *M. gluteus medius* lies ventral to the GSA, *M. gluteus superficialis*. In the distal hind limb, the four tendons of the EDLO, *M. extensor digitorum longus* insert on the distal phalanges of digits II through V. The EHL, *M. hallucis longus* lies deep to the TCR, *M. tibialis cranialis* and the EDLO, *M. extensor digitorum longus*. Note that the retinacula (ligaments) spanning the wrist and ankle joints have been partially removed.

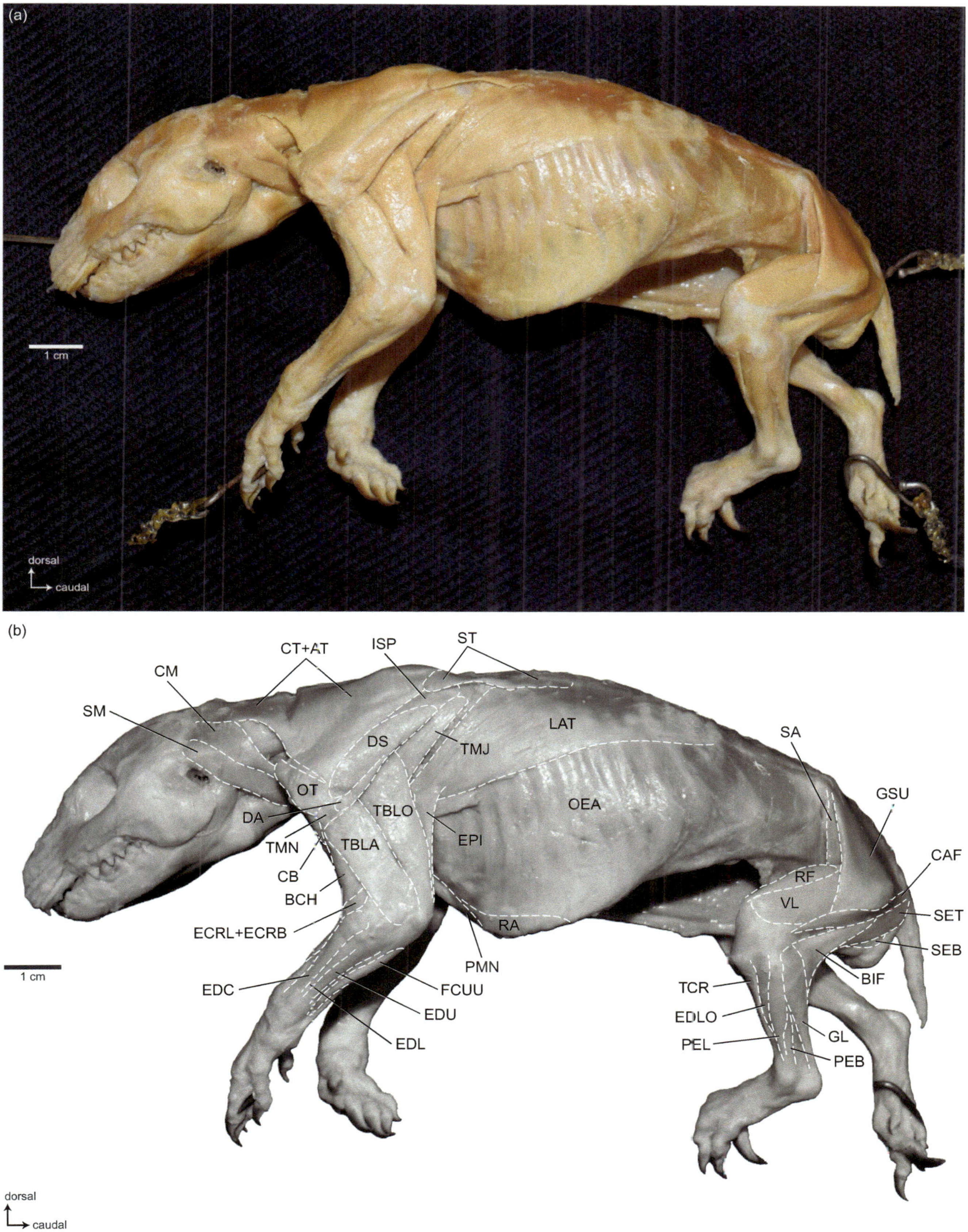
(a)
1 cm
dorsal
caudal
(b)
CT+AT
ISP
ST
CM
SM
DS
LAT
TMJ
SA
OT
DA
TBLO
OEA
GSU
TMN
EPI
TBLA
CB
CAF
RF
BCH
VL
SET
ECRL+ECRB
RA
SEB
PMN
BIF
1 cm
FCUU
TCR
EDC
EDU
EDLO
GL
EDL
PEL
PEB
dorsal
caudal

FIGURE II.15.2 **Pectoral muscles in ventral view. (a) Superficial view.** The PAB, *M. pectoantebrachialis* is not developed as a separate muscle, but part of the PMJ, *M. pectoralis major.* **(b) Deep view.** The SM, *M. sternomastoideus*; the CM, *M. cleidomastoideus*; and the OT, *M. omotransversarius* have been removed. The SC, *M. subclavius* is well developed. The PMJ, *M. pectoralis major* has been removed. The XH, *M. xiphihumeralis* is not developed as a separate muscle, but part of the PMN, *M. pectoralis minor.*

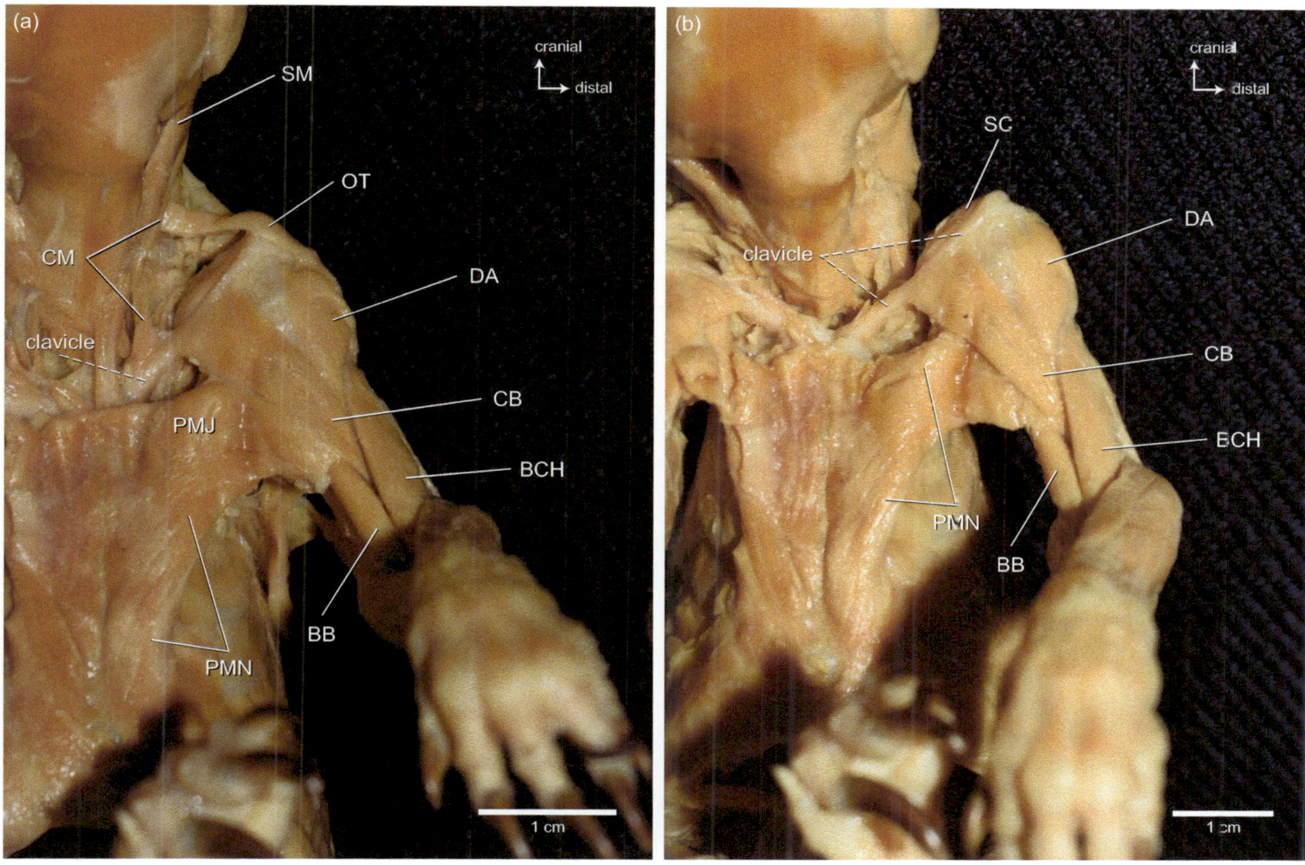
(a)
cranial
distal
SM
OT
CM
DA
clavicle
CB
PMJ
BCH
BB
PMN
1 cm
(b)
cranial
distal
SC
DA
clavicle
CB
BCH
PMN
BB
1 cm

FIGURE II.15.3 **Deep neck and shoulder muscles. (a) In lateral view.** In the neck region, three *Mm. rhomboidei* are present (RCR, *M. rhomboideus cervicis*; RT, M. rhomboideus thoracis; and RCP, *M. rhomboideus capitis*). The RPR, *M. rhomboideus profundus* is absent. The *Mm. deltoidei* have been removed. The EPI, *M. epitrochlearis* originates on the LAT, *M. latissimus dorsi* that has been removed in this picture. **(b) In dorsal view.** The RCR, *M. rhomboideus cervicis* almost reaches the occiput. **(c) In cranial view.**

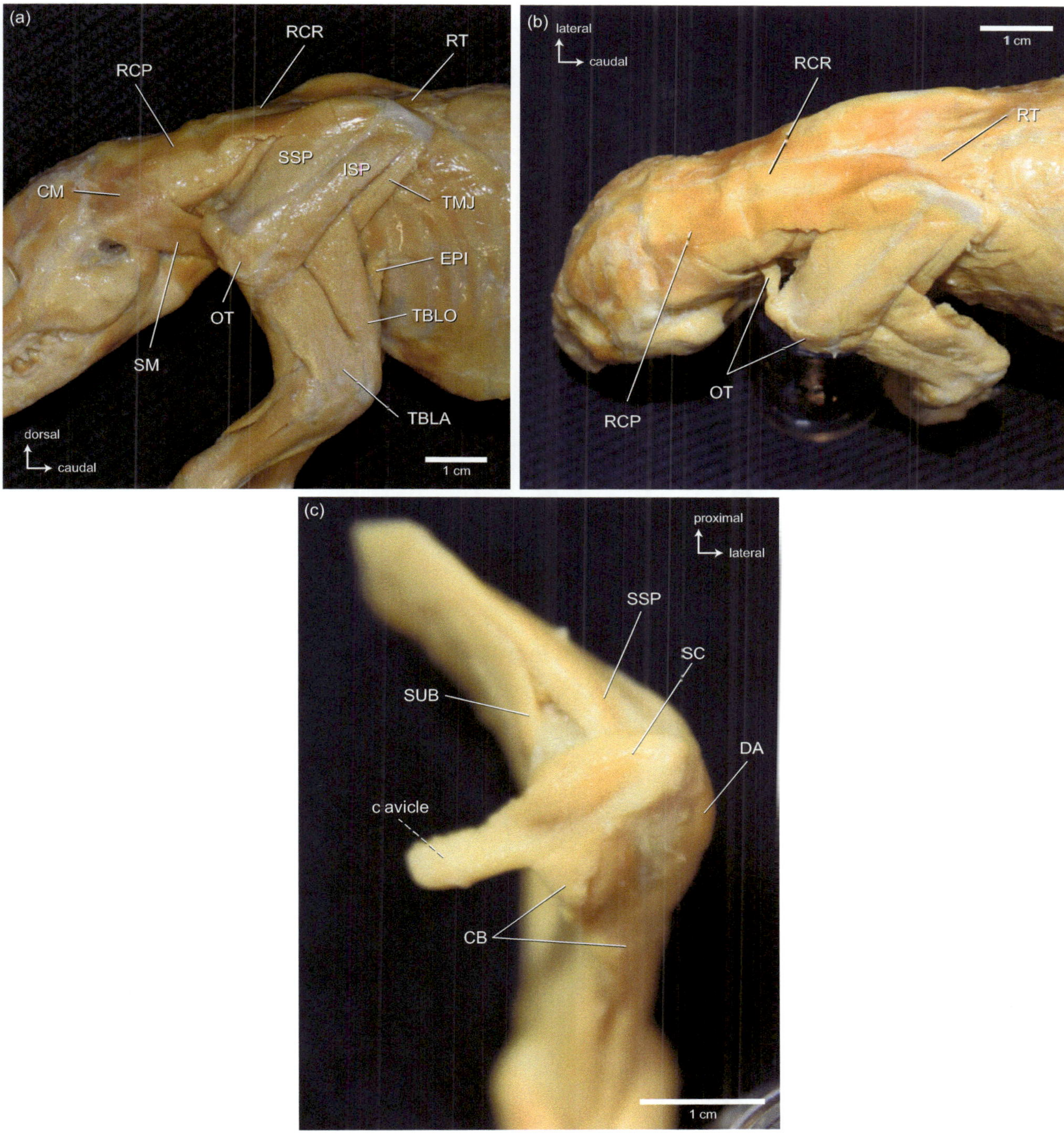
(a)
RCR
RT
RCP
SSP
ISP
CM
TMJ
EPI
OT
TBLO
SM
TBLA
dorsal
caudal
1 cm
(b)
lateral
caudal
1 cm
RCR
RT
OT
RCP
(c)
proximal
lateral
SSP
SC
SUB
DA
c avicle
CB
1 cm

FIGURE II.15.4 **Intrinsic forelimb muscles in (a) lateral view and (b) medial view.** The BB, *M. biceps brachii* has one distinct origin. **(c) Distal intrinsic forelimb muscles in lateral view.**

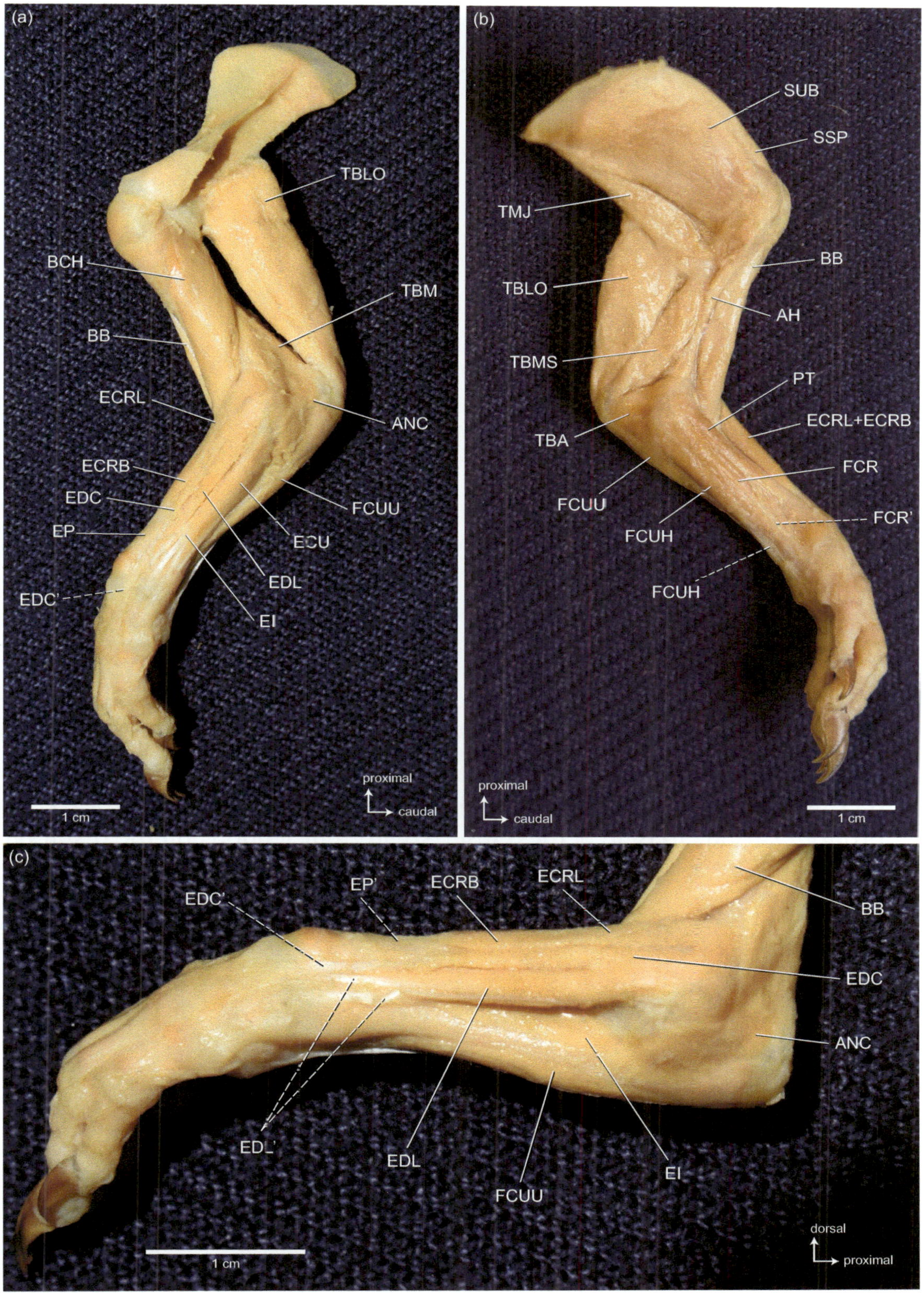

(a)
TBLO
BCH
TBM
BB
ECRL
ANC
ECRB
EDC
FCUU
EP
ECU
EDL
EDC'
EI
proximal
caudal
1 cm
(b)
SUB
SSP
TMJ
BB
TBLO
AH
TBMS
PT
ECRL+ECRB
TBA
FCR
FCUU
FCR'
FCUH
FCUH
proximal
caudal
1 cm
(c)
EDC'
EP'
ECRB
ECRL
BB
EDC
ANC
EDL'
EDL
FCUU
EI
dorsal
proximal
1 cm

FIGURE II.15.5 **Distal forelimb muscles. (a) In dorsal view.** The EDC, *M. extensor digitorum communis* inserts with four tendons on the distal phalanges of digits II through V. The EDL, *M. extensor digitorum lateralis* inserts with two tendons on the distal phalanges of digits IV through V. **(b) In ventral view.** Not visible in the picture, but the PL, *M. palmaris longus* is very weakly developed and inserts into the palmar fascia. The FDP, *M. flexor digitorum profundus* (four muscle bellies) inserts with four tendons on the distal phalanges of digits II through V.

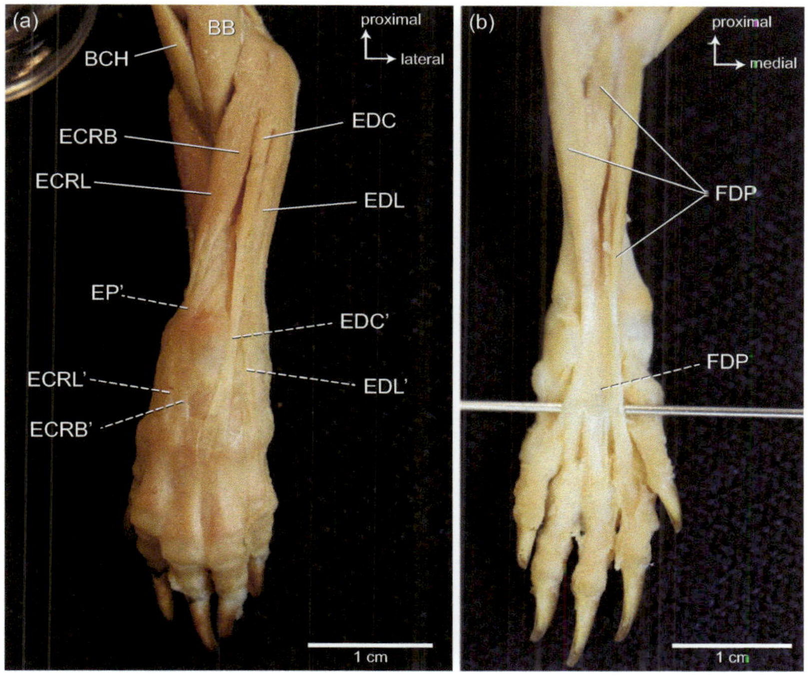
(a)
BB
BCH
proximal
lateral
EDC
ECRB
ECRL
EDL
EP'
EDC'
ECRL'
EDL'
ECRB'
1 cm
(b)
proximal
medial
FDP
FDP
1 cm

FIGURE II.15.6 Deep rotator muscles. (a) Distal rotator muscles in dorsal view. Not visible in the picture, but the tendon of the BB, *M. biceps brachii* inserts into the bicipital tuberosity of the radius. The tendon of the BCH, *M. brachioradialis* inserts into the coronoid process of the ulna. The PQ, *M. pronator quadratus* is absent. **(b) Proximal rotator muscles in ventral view.**

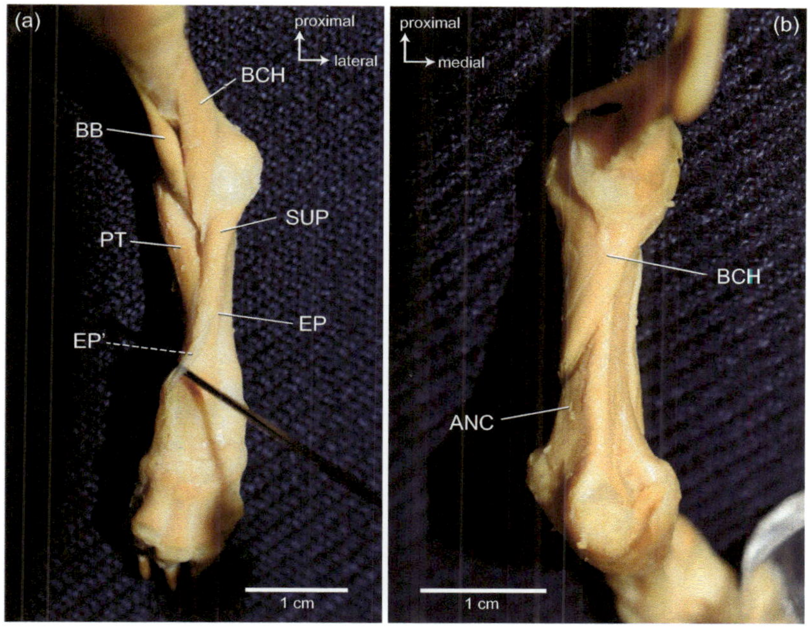
(a)
proximal
lateral
BCH
BB
SUP
PT
EP
EP'
1 cm
(b)
proximal
medial
BCH
ANC
1 cm

Chiropotes satanas (Brown-bearded saki)

Classification:	Primates, Haplorhini, Pitheciinae
Mean body mass:	2.6–3.0 kg
Habitat:	Rainforest
Locomotor ecology:	Scansorial (predominantly quadrupedal)
Peculiarities:	Lacks true opposability of thumb; hind limb suspension (foot tends to deviate laterally; non-prehensile tail)

FIGURE II.16.1 Superficial muscles of forelimb and hind limb in lateral view. (a) Without labels. (b) With labels and muscle outlines. In the neck and shoulder region, the *Mm. trapezii* are clearly divided into two parts (CT, *M. clavotrapezius* with AT, *M. arcomiotrapezius*; ST, *M. spinotrapezius*). The EPI, *M. epitrochlearis* is present, but relatively weakly developed. The BCR, *M. brachioradialis* is present. The OEA, *M. obliquus externus abdominis* originates from the thoracic ribs and inserts into the linea alba at the midline and on the pubis near the symphysis. The RA, *M. rectus abdominis* originates from the costal cartilages and inserts into the prebubic tendon. The TEN, *M. tensor fasciae latae* is present, but has been removed. In the distal hind limb, the four tendons of the EDLO, *M. extensor digitorum longus* insert on the distal phalanges of digits II through V. The EHL, *M. hallucis longus* lies deep to the TCR, *M. tibialis cranialis* and the EDLO, *M. extensor digitorum longus*, and its tendon inserts into the distal phalanx of digit I. Note that the retinacula (ligaments) spanning the wrist and ankle joints have been partially removed.

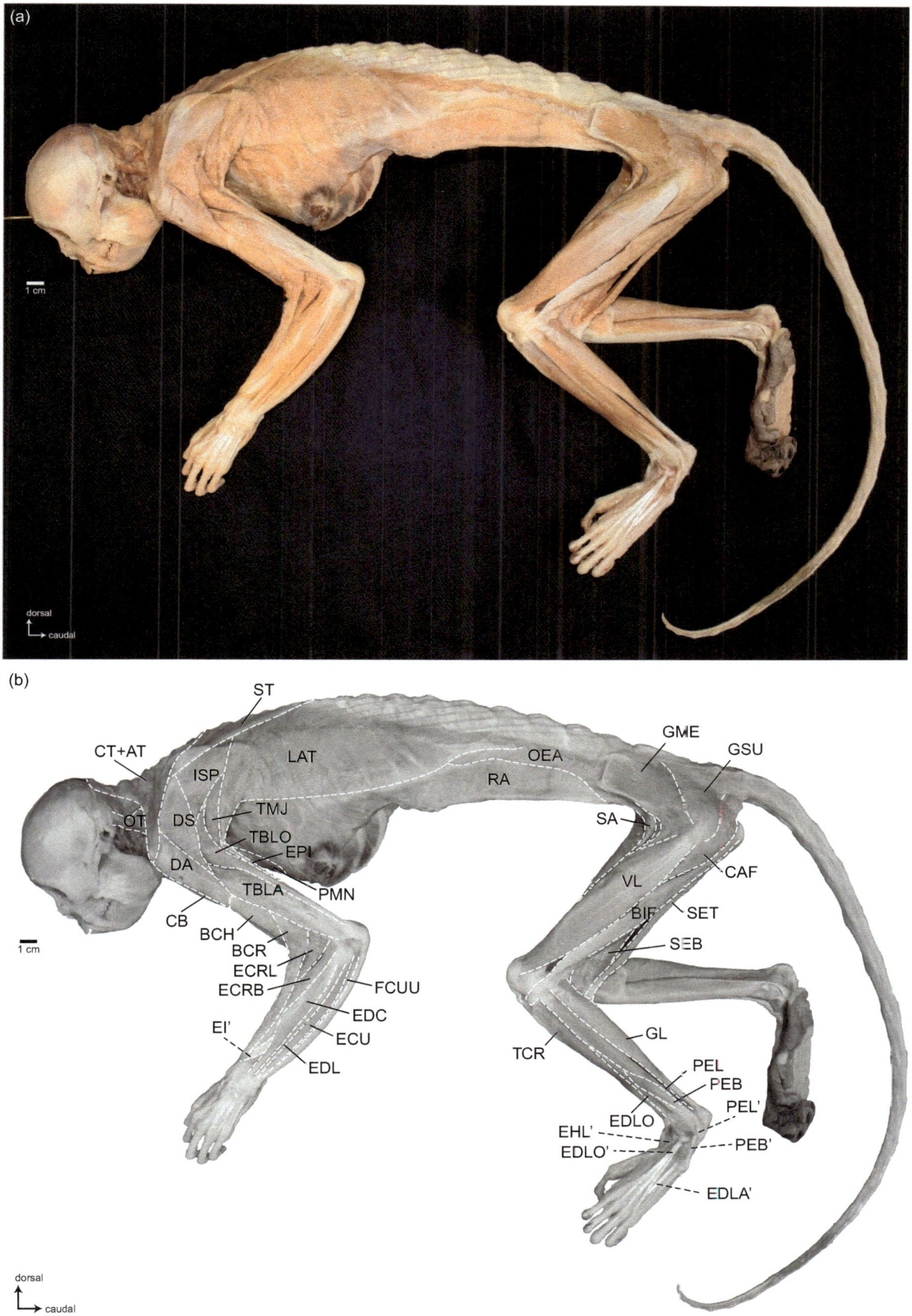
(a)
1 cm
dorsal
caudal
(b)
ST
CT+AT
LAT
ISP
OEA
RA
GME
GSU
OT
DS
TMJ
TBLO
EPI
DA
TBLA
PMN
SA
CAF
VL
SET
BIF
SEB
CB
BCH
BCR
ECRL
ECRB
FCUU
EDC
ECU
EI'
EDL
TCR
GL
PEL
PEB
EDLO
PEL'
EHL'
EDLO'
PEB'
EDLA'
1 cm
dorsal
caudal

FIGURE II.16.2 **Pectoral muscles in ventral view. (a) Superficial view.** The PMN, *M. pectoralis minor* lies deep to the PMJ, *M. pectoralis major.* Not visible in the picture, but the SC, *M. subclavius* is present. **(b) Deep view.** The PMJ, *M. pectoralis major* and the PAB, *M. pectoantebrachialis* have been removed. The XH, *M. xiphihumeralis* is clearly separated from the PMN, *M. pectoralis minor.*

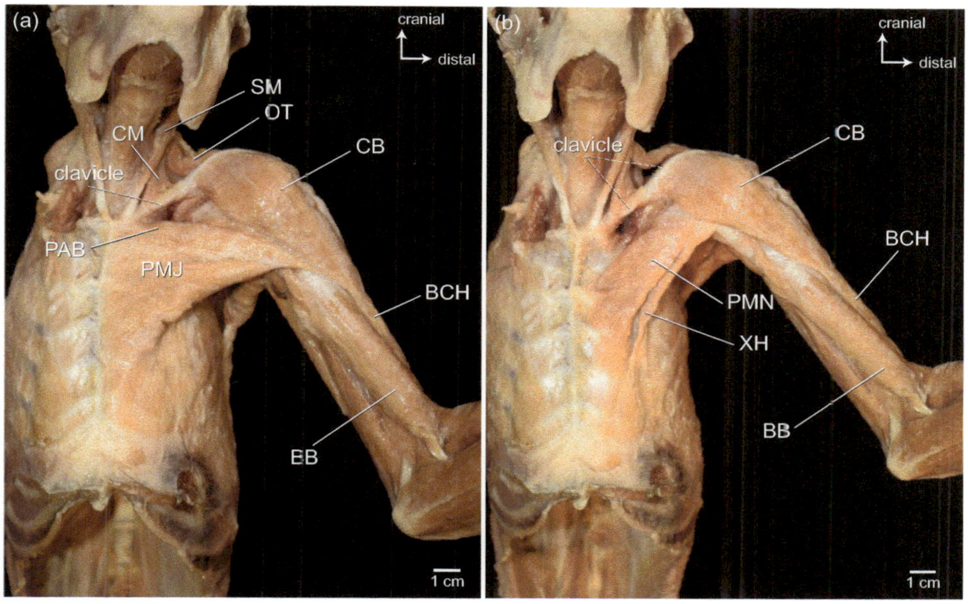
(a)
cranial
distal
SM
OT
CM
CB
clavicle
PAB
PMJ
BCH
BB
1 cm
(b)
cranial
distal
clavicle
CB
BCH
PMN
XH
BB
1 cm

FIGURE II.16.3 **Deep neck and shoulder muscles. (a) In lateral view.** In the neck region, two *Mm. rhomboidei* are present (RCR, *M. rhomboideus cervicis* and RPR, *M. rhomboideus profundus*). **(b) In dorsal view.** The RCR, *M. rhomboideus cervicis* reaches the occiput. **(c) In cranial view.**

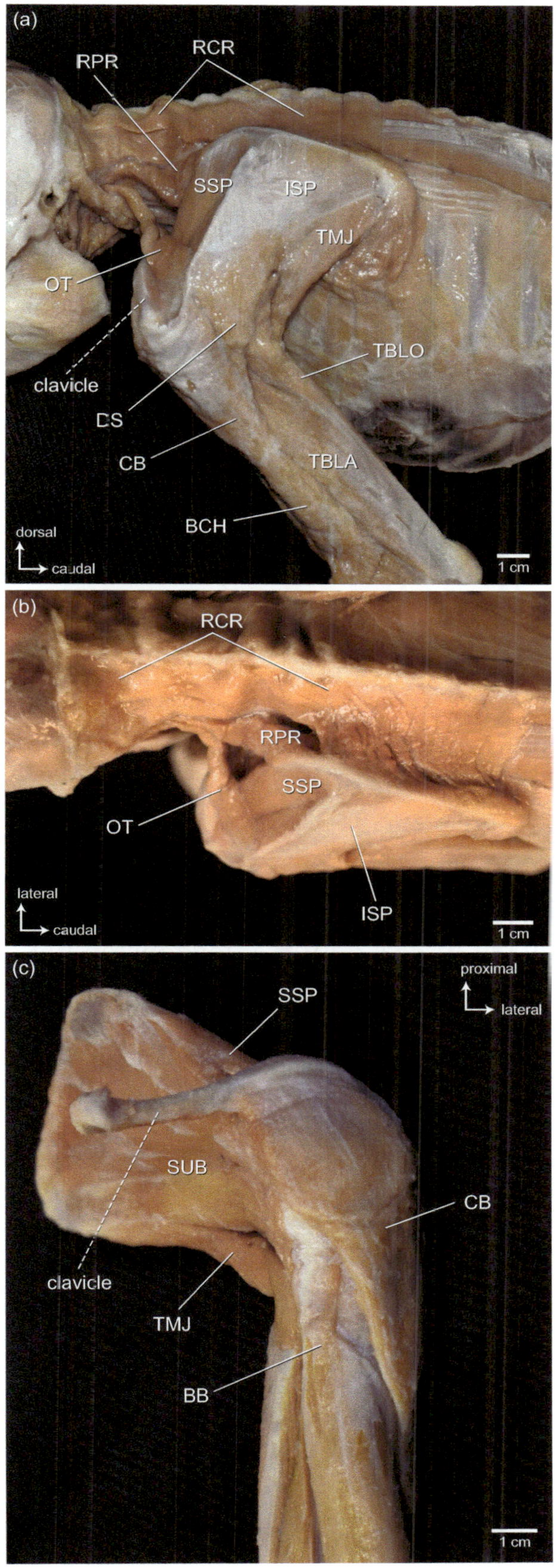
(a)
RPR
RCR
SSP
ISP
TMJ
OT
clavicle
DS
TBLO
CB
TBLA
BCH
dorsal
caudal
1 cm
(b)
RCR
RPR
SSP
OT
ISP
lateral
caudal
1 cm
(c)
proximal
lateral
SSP
SUB
CB
clavicle
TMJ
BB
1 cm

FIGURE II.16.4 **Intrinsic forelimb muscles in (a) lateral view and (b) medial view.** The BB, *M. biceps brachii* has two distinct origins. **Distal intrinsic forelimb muscles in (c) lateral view and (d) medial view.** The EI, *M. extensor digiti I and II* inserts with two tendons on the distal phalanges of digits II and III. The PL, *M. palmaris longus* inserts with four tendons on the distal phalanges of digits II through V. The FDP, *M. flexor digitorum profundus* (four muscle bellies) inserts with five tendons on the distal phalanges of digits I through V.

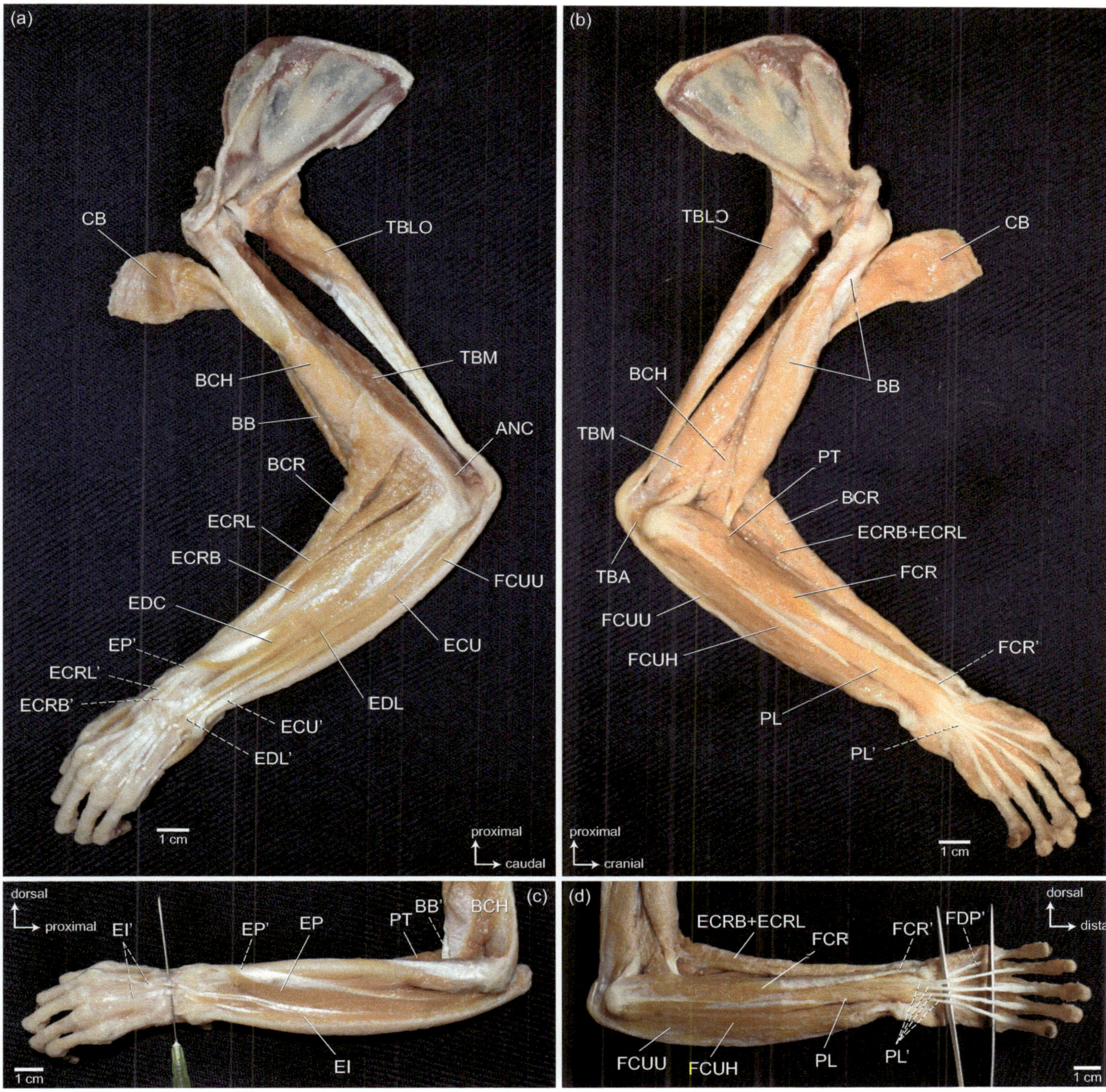

(a)
CB
TBLO
TBM
BCH
BB
ANC
BCR
ECRL
ECRB
FCUU
EDC
ECU
EP'
ECRL'
ECRB'
EDL
ECU'
EDL'
1 cm
proximal
caudal
(b)
TBLO
CB
BCH
BB
TBM
PT
BCR
ECRB+ECRL
TBA
FCR
FCUU
FCUH
FCR'
PL
PL'
proximal
cranial
1 cm
dorsal
proximal
(c)
EI'
EP'
EP
PT
BB'
BCH
EI
1 cm
(d)
ECRB+ECRL
FCR
FCR'
FDP'
dorsal
distal
FCUU
FCUH
PL
PL'
1 cm

FIGURE II.16.5 Distal forelimb muscles. (a) In dorsal view. The EDC, *M. extensor digitorum communis* inserts with four tendons on the distal phalanges of digits II through V. The EDL, *M. extensor digitorum lateralis* inserts with two tendons on the distal phalanges of digits IV through V. **(b) In ventral view.** The FDP, *M. flexor digitorum profundus* (four muscle bellies) inserts with five tendons on the distal phalanges of digits I through V, but the tendon inserting on digit II is lost in this specimen.

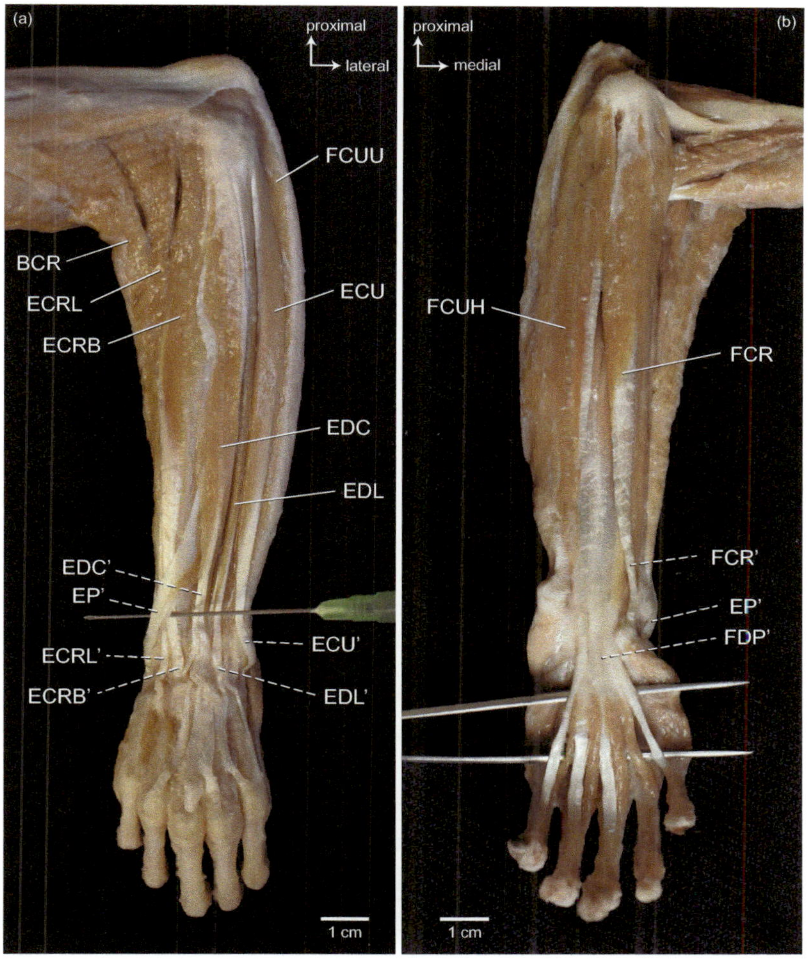
(a)
proximal
lateral
FCUU
BCR
ECRL
ECRB
ECU
EDC
EDL
EDC'
EP'
ECU'
ECRL'
ECRB'
EDL'
1 cm
(b)
proximal
medial
FCUH
FCR
FCR'
EP'
FDP'
1 cm

FIGURE II.16.6 Deep rotator muscles. (a) Distal rotator muscles in dorsal view. (b) Distal rotator muscles in ventral view. Not visible in the picture, but the tendon of the BB, *M. biceps brachii* inserts into the bicipital tuberosity of the radius. The tendon of the BCH, *M. brachioradialis* inserts into the coronoid process of the ulna. **(c) Proximal rotator muscles in ventral view.**

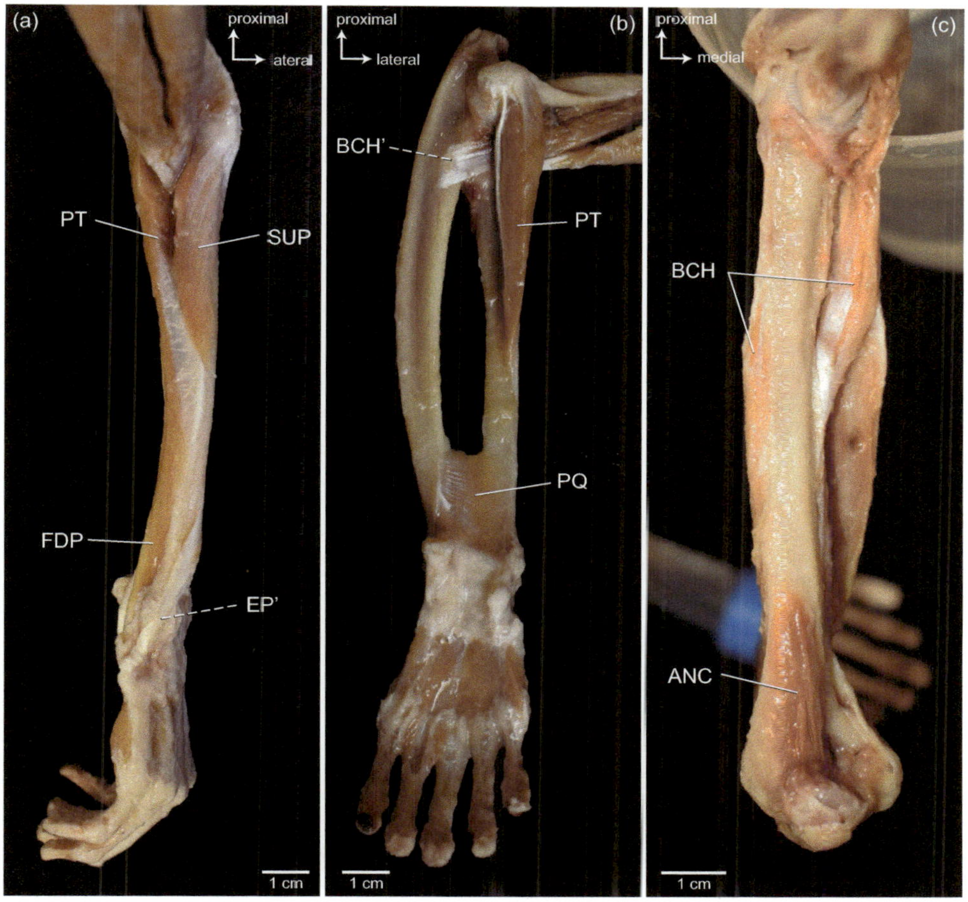
(a)
proximal
lateral
PT
SUP
FDP
EP'
1 cm
(b)
proximal
lateral
BCH'
PT
PQ
1 cm
(c)
proximal
medial
BCH
ANC
1 cm

FIGURE II.16.7 Hip muscles. (a) In ventral view. (b) In lateral view.

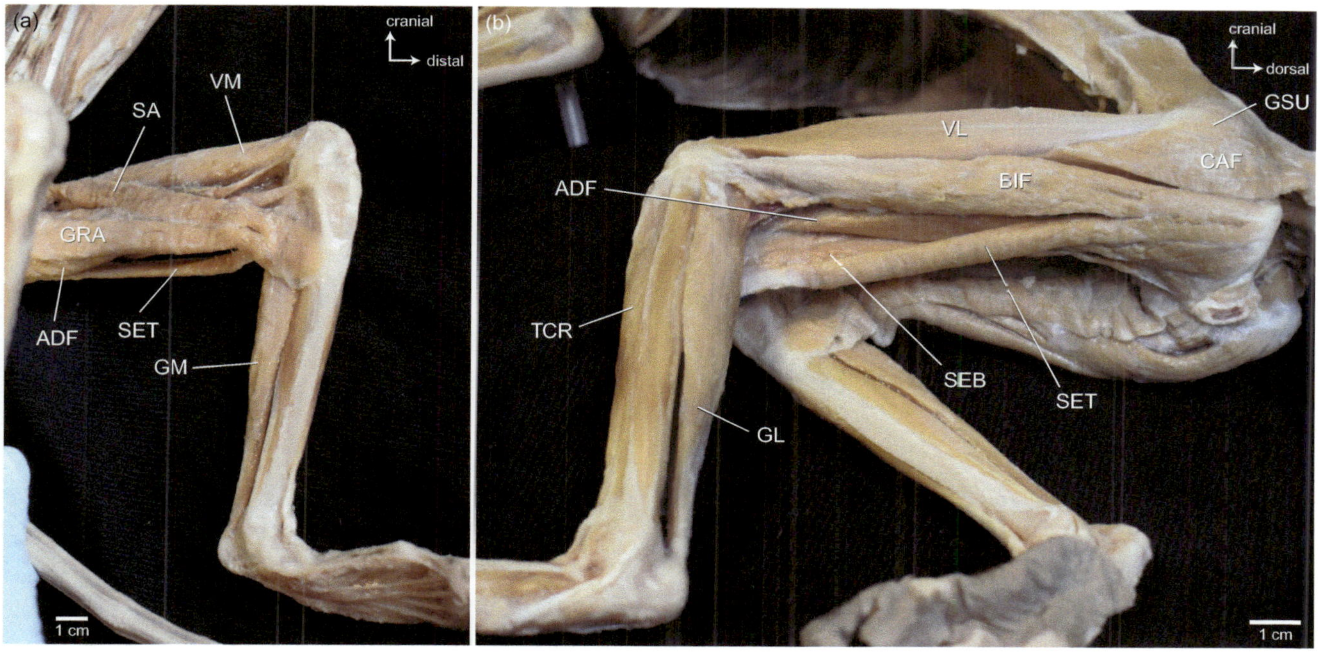
(a)
cranial
distal
VM
SA
GRA
ADF
SET
GM
1 cm
(b)
cranial
dorsal
GSU
VL
CAF
ADF
BIF
TCR
SEB
SET
GL
1 cm

Oryctolagus cuniculus (European rabbit)

Classification: Lagomorpha, Leporidae
Mean body mass: 3.5–6.0 kg
Habitat: Forest, grassland
Locomotor ecology: Saltatorial
Peculiarities: Plantigrade posture of the hind limbs at rest; digitigrade posture during locomotion

Intrinsic forelimb muscles. The BB, *M. biceps brachii* has one distinct origin. The EDL, *M. extensor digitorum lateralis* inserts with two tendons on the distal phalanges of digits IV through V. The EDC, *M. extensor digitorum communis* inserts with four tendons on the distal phalanges of digits II through V. The EI, *M. extensor digiti I and II* inserts with two tendons on the distal phalanges of digits I and II. The tendon of the ECRL, *M. extensor carpi radialis longus* inserts into the base of metacarpal II and tendon of the ECRB, *M. extensor carpi radialis brevis* into the base of metacarpal III. The PL, *M. palmaris longus* inserts with four tendons into the distal phalanges of digits II through V. The FDP, *M. flexor digitorum profundus* (four muscle bellies) inserts with four tendons on the distal phalanges of digits II through V.

FIGURE II.17.1 Superficial muscles of forelimb and hind limb in lateral view. (a) Without labels. (b) With labels and muscle outlines. In the neck and shoulder region, the *Mm. trapezii* are clearly divided into three parts (CT, *M. clavotrapezius*; AT, *M. arcomio-trapezius*; and ST, *M. spinotrapezius*). The EPI, *M. epitrochlearis* and the BCR, *M. brachio-radialis* are absent. In the hip, the well-developed SA, *M. sartorius* lies deep to the TEN, *M. tensor fasciae latae*. The BIF, *M. biceps femoris* is slightly separated into two muscle bellies. The CAF, *M. caudofemoralis* is absent. In the distal hind limb, the four tendons of the EDLO, *M. extensor digitorum longus* insert on the distal phalanges of digits II through V. The EDLA, *M. extensor digitorum lateralis* inserts on the distal phalanges of digits IV and V, joining the EDLO, *M. extensor digitorum longus*. Note that the retinacula (ligaments) spanning the wrist and ankle joints have been partially removed.

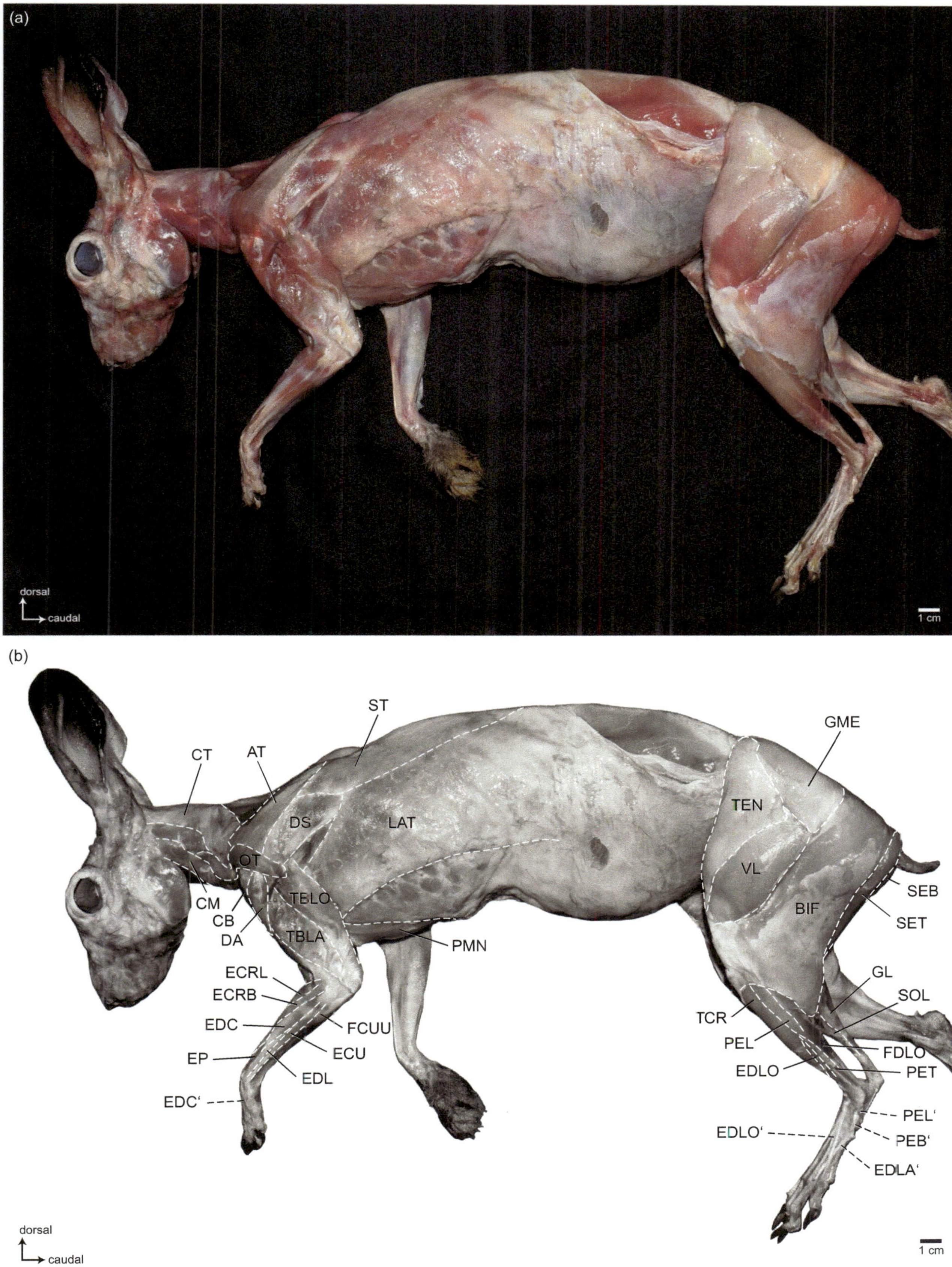
(a)
dorsal
caudal
1 cm
(b)
ST
CT
AT
GME
DS
LAT
TEN
OT
VL
CM
TBLO
SEB
CB
BIF
DA
SET
TBLA
PMN
ECRL
GL
ECRB
SOL
TCR
EDC
FCUU
PEL
FDLO
ECU
EP
EDLO
PET
EDL
EDC'
PEL'
EDLO'
PEB'
EDLA'
dorsal
caudal
1 cm

FIGURE II.17.2 **Pectoral muscles in ventral view. Superficial view.** The clavicle is present. Note that the CM, *M. cleidomastoideus* is slightly separated at its insertion on the clavicle, but not at its origin. The XH, *M. xiphihumeralis* is not clearly developed as a separate muscle, but part of the PMN, *M. pectoralis minor.*

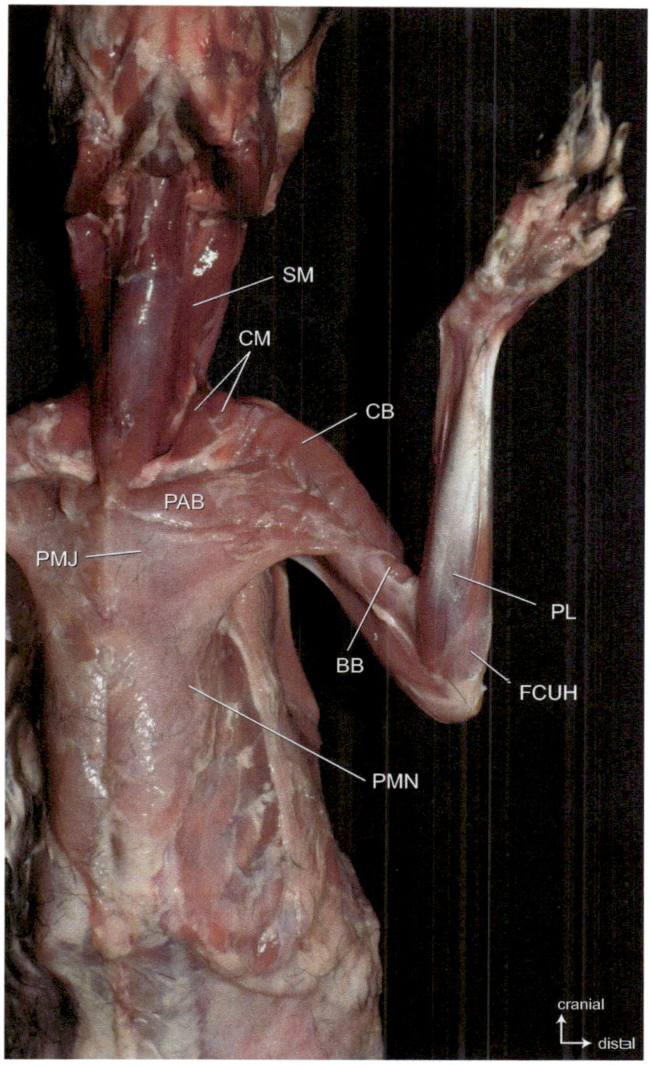
SM
CM
CB
PAB
PMJ
PL
BB
FCUH
PMN
cranial
distal

FIGURE II.17.3 **Deep neck and shoulder muscles. (a, c) In craniolateral view.** Note the distal insertion of the OT, *M. omotransversarius* on the metacromion of the scapula. The SC, *M. subclavius* has a very distinct morphology in the rabbit. It is very large and covers the SUP, *M. supraspinatus*. **(b) In dorsal view.** In the neck region, three *Mm. rhomboidei* are present (RCR, *M. rhomboideus cervicis*; RT, *M. rhomboideus thoracis*; and RCP, *M. rhomboideus capitis*). The RCR, *M. rhomboideus cervicis* does not reach the occiput. The RPR, *M. rhomboideus profundus* is absent.

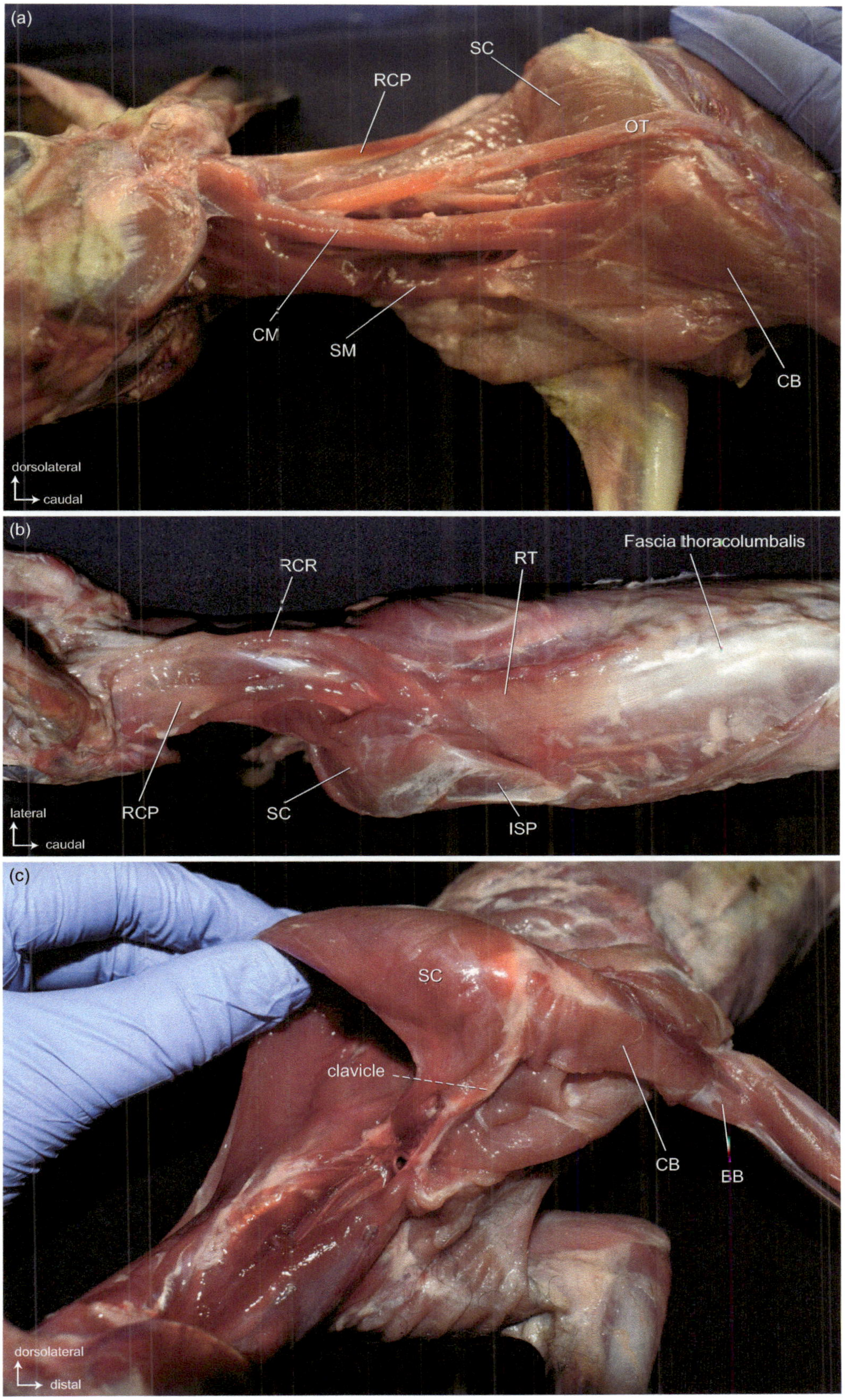
(a)
SC
RCP
OT
CM
SM
CB
dorsolateral
caudal
(b)
RCR
RT
Fascia thoracolumbalis
RCP
SC
ISP
lateral
caudal
(c)
SC
clavicle
CB
BB
dorsolateral
distal

Laonastes aenigmamus (Laotian rock rat or kha-nyou)

Classification:	Rodentia, Hystricomorpha, Diatomyidae
Mean body mass:	0.4 kg
Habitat:	Karst limestone
Locomotor ecology:	Ambulatorial - scansorial
Peculiarities:	Lazarus taxon; rock-dweller

FIGURE II.18.1 **Superficial muscles of forelimb and hind limb in lateral view. (a) Without labels. (b) With labels and muscle outlines**. In the neck and shoulder region, the *Mm. trapezii* are clearly divided into two parts (CT, *M. clavotrapezius* with AT, *M. arcomiotrapezius*; ST, *M. spinotrapezius*). Note the length of the ST, *M. spinotrapezius* and the LAT, *M. latissimus dorsi*. The EPI, *M. epitrochlearis* and the BCR, *M. brachioradialis* are absent. The GSU, *M. gluteus superficialis* is not developed as a separate muscle, but strongly fused with the TEN, *M. tensor fasciae latae*. Not visible in the picture, but the GME, *M. gluteus medius* lies ventral to the TEN, *M. tensor fasciae latae*. The CAF, *M. caudofemoralis* is not developed as a separate muscle, but strongly fused with the BIF, *M. biceps femoris*. In the distal hind limb, the three tendons of the EDLO, *M. extensor digitorum longus* insert on the distal phalanges of digits II through IV. The EHL, *M. hallucis longus* lies deep to the TCR, *M. tibialis cranialis* and the EDLO, *M. extensor digitorum longus*. Note that the retinacula (ligaments) spanning the wrist and ankle joints have been partially removed.

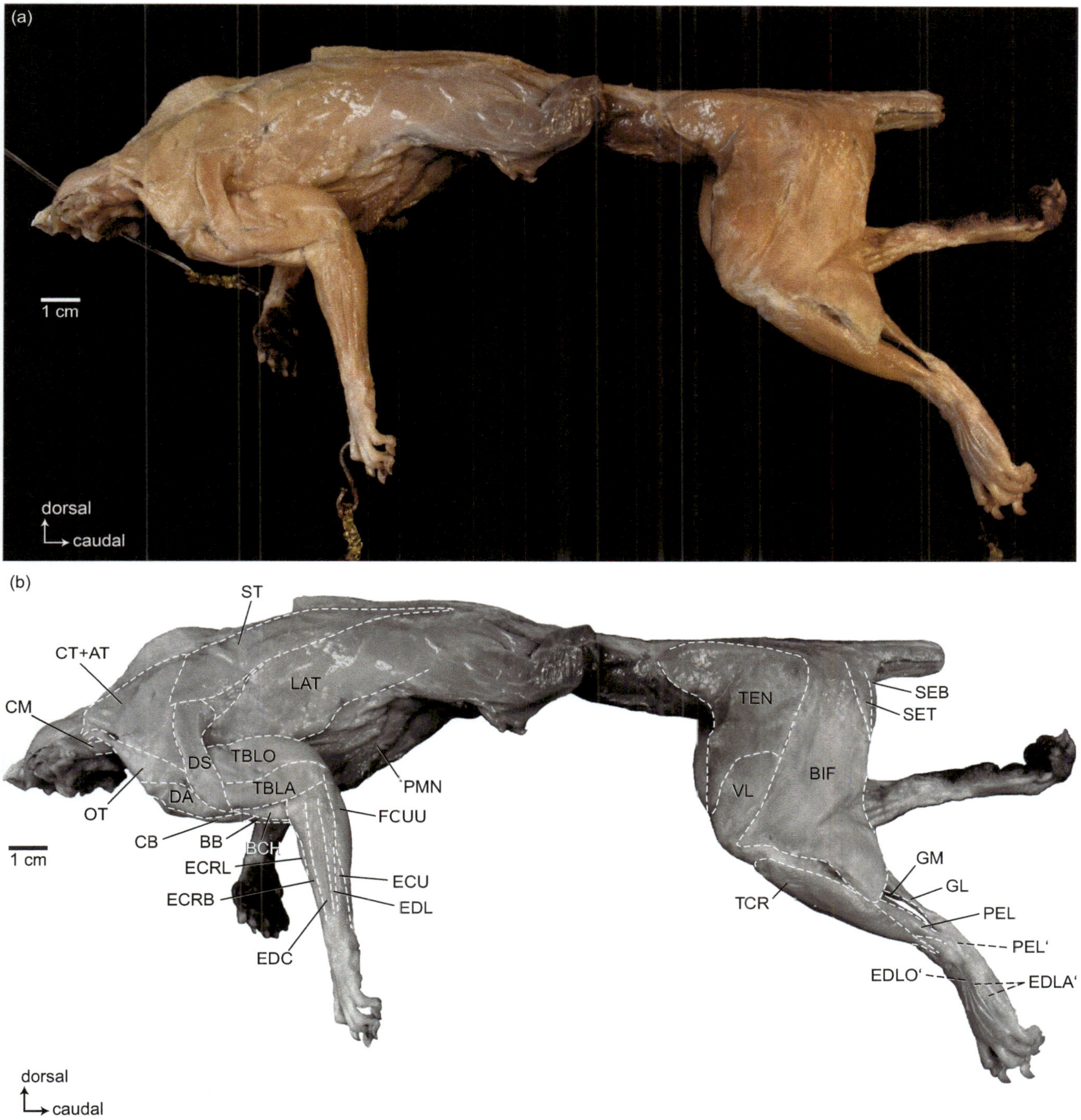
(a)
1 cm
dorsal
caudal
(b)
ST
CT+AT
CM
LAT
DS
TBLO
DA
TBLA
OT
PMN
FCUU
CB
BB
BCH
1 cm
ECRL
ECRB
ECU
EDL
EDC
TEN
SEB
SET
VL
BIF
GM
GL
TCR
PEL
PEL'
EDLO'
EDLA'
dorsal
caudal

FIGURE II.18.2 **Pectoral muscles in ventral view. (a) Superficial view.** The XH, *M. xiphihumeralis* that would distinctly originate from the xiphoid process is not clearly developed as a separate muscle, but part of the PMN *M. pectoralis minor.* **(b) Deep view.** The PMJ, *M. pectoralis major* and the PAB, *M. pectoantebrachialis* have been removed. The cranial and caudal part of the PMN, *M. pectoralis minor* appear as two relatively distinct muscle bellies that are fused at their origin and insertion. The SC, *M. subclavius* is well developed.

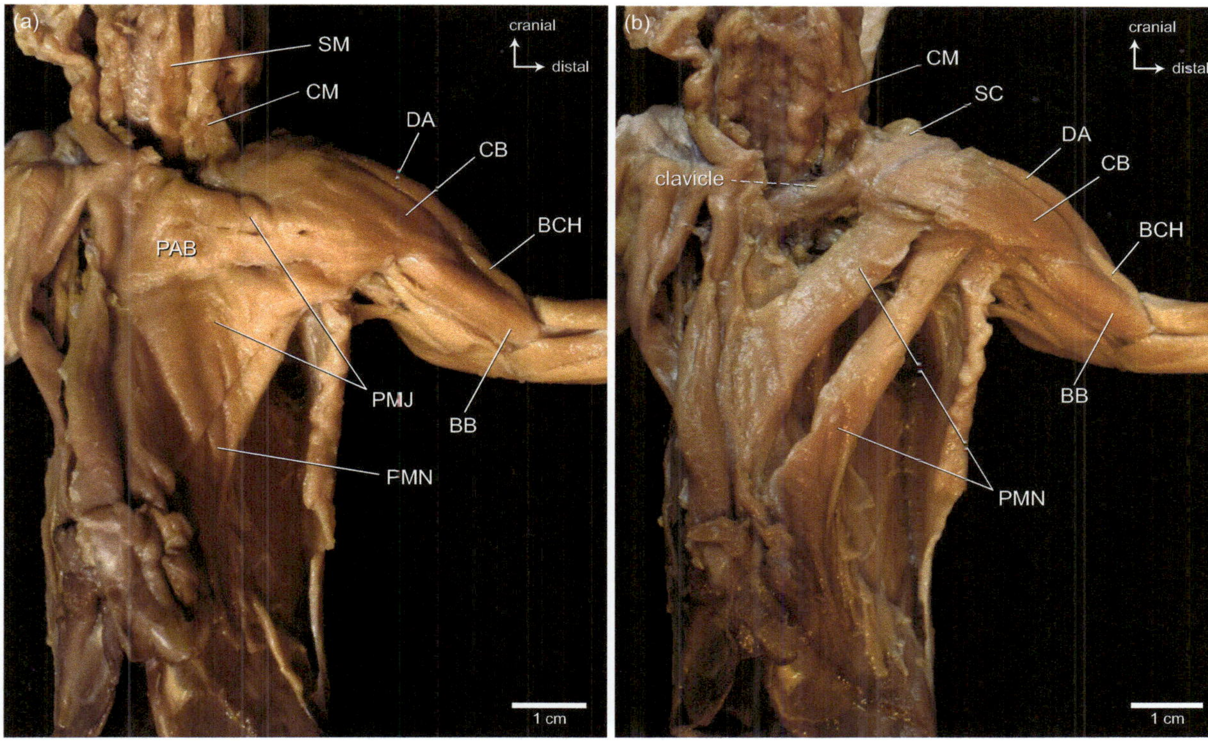
(a)
cranial
distal
SM
CM
DA
CB
BCH
PAB
PMJ
BB
FMN
1 cm
(b)
cranial
distal
CM
SC
DA
CB
clavicle
BCH
BB
PMN
1 cm

FIGURE II.18.3 **Deep neck and shoulder muscles. (a) In dorsolateral view.** The CT, *M. clavotrapezius* and the AT, *M. acromiotrapezius* are fused together. **(b) In craniolateral view and (c) dorsal view.** In the neck region, two *Mm. rhomboidei* are present (RCR, *M. rhomboideus cervicis* and RCP, *M. rhomboideus capitis*). The RCR, *M. rhomboideus cervicis* is slightly damaged at its origin in this specimen, but the muscle almost reaches the occiput.

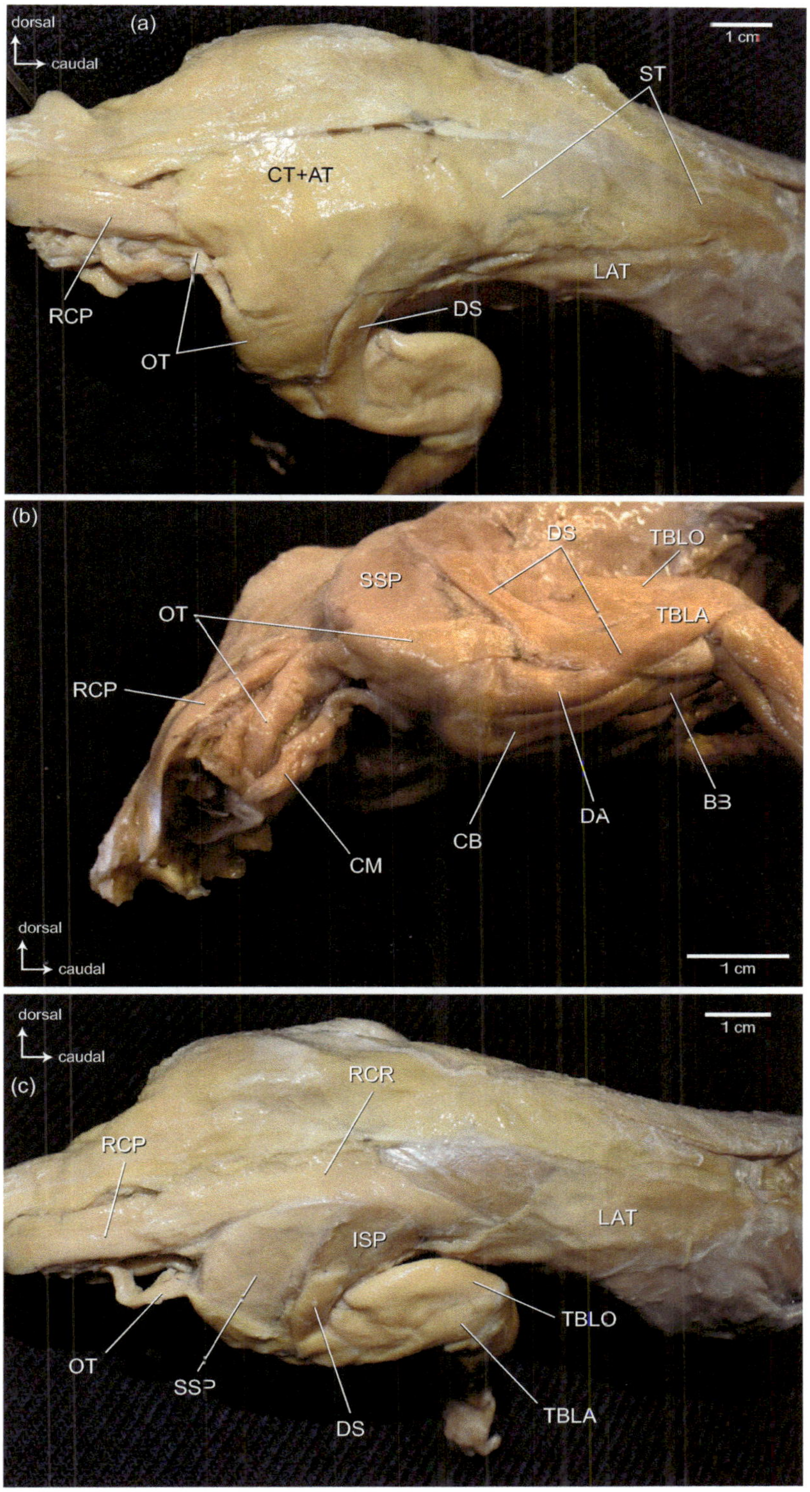
dorsal
caudal
(a)
1 cm
ST
CT+AT
LAT
RCP
OT
DS
(b)
DS
TBLO
SSP
OT
TBLA
RCP
BB
DA
CB
CM
dorsal
caudal
1 cm
dorsal
caudal
1 cm
(c)
RCR
RCP
LAT
ISP
TBLO
OT
SSP
TBLA
DS

FIGURE II.18.4 Intrinsic muscles in (a) proximolateral view and (b) proximomedial view. The DS, *M. spinodeltoideus* has been removed. **Intrinsic forelimb muscles in (c) lateral view and (d) medial view.** The BB, *M. biceps brachii* has two distinct origins. The EDL, *M. extensor digitorum lateralis* inserts with two tendons on the distal phalanges of digits IV and V. **Distal intrinsic forelimb muscles in (e) lateral view and (f) medial view.** The EI, *M. extensor digiti I and II* inserts with two tendons on the distal phalanges of digits I and II.

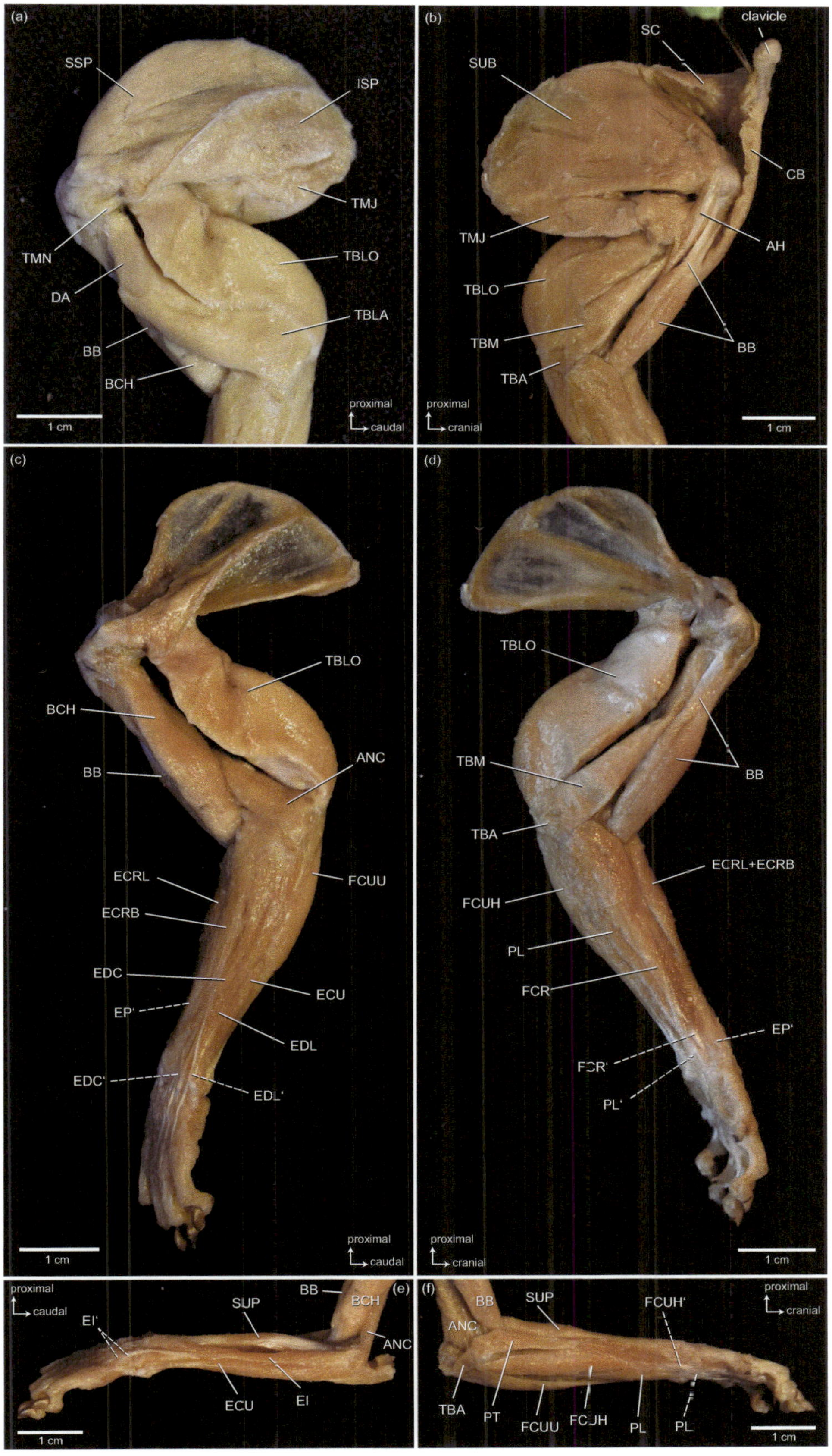
(a)
SSP
ISP
TMJ
TMN
TBLO
DA
TBLA
BB
BCH
1 cm
proximal
caudal
(b)
clavicle
SC
SUB
CB
TMJ
AH
TBLO
TBM
BB
TBA
proximal
cranial
1 cm
(c)
TBLO
BCH
BB
ANC
ECRL
FCUU
ECRB
EDC
ECU
EP'
EDL
EDC'
EDL'
1 cm
proximal
caudal
(d)
TBLO
TBM
BB
TBA
ECRL+ECRB
FCUH
PL
FCR
EP'
PL'
proximal
cranial
1 cm
proximal
caudal
BB
BCH
(e)
SUP
EI'
ANC
ECU
EI
1 cm
(f)
BB
SUP
FCUH'
proximal
cranial
ANC
TBA
PT
FCUU
PL
PL
1 cm

FIGURE II.18.5 **Distal forelimb muscles. (a, b) In dorsal view.** The EDC, *M. extensor digitorum communis* inserts with four tendons on the distal phalanges of digits II through V. The EDL, *M. extensor digitorum lateralis* inserts with two tendons on the distal phalanges of digits IV through V. **(c, d) In ventral view.** The PL, *M. palmaris longus* inserts with three tendons on the distal phalanges of digits II through IV. The FDP, *M. flexor digitorum profundus* (four muscle bellies) inserts with five tendons on the distal phalanges of digits II through V.

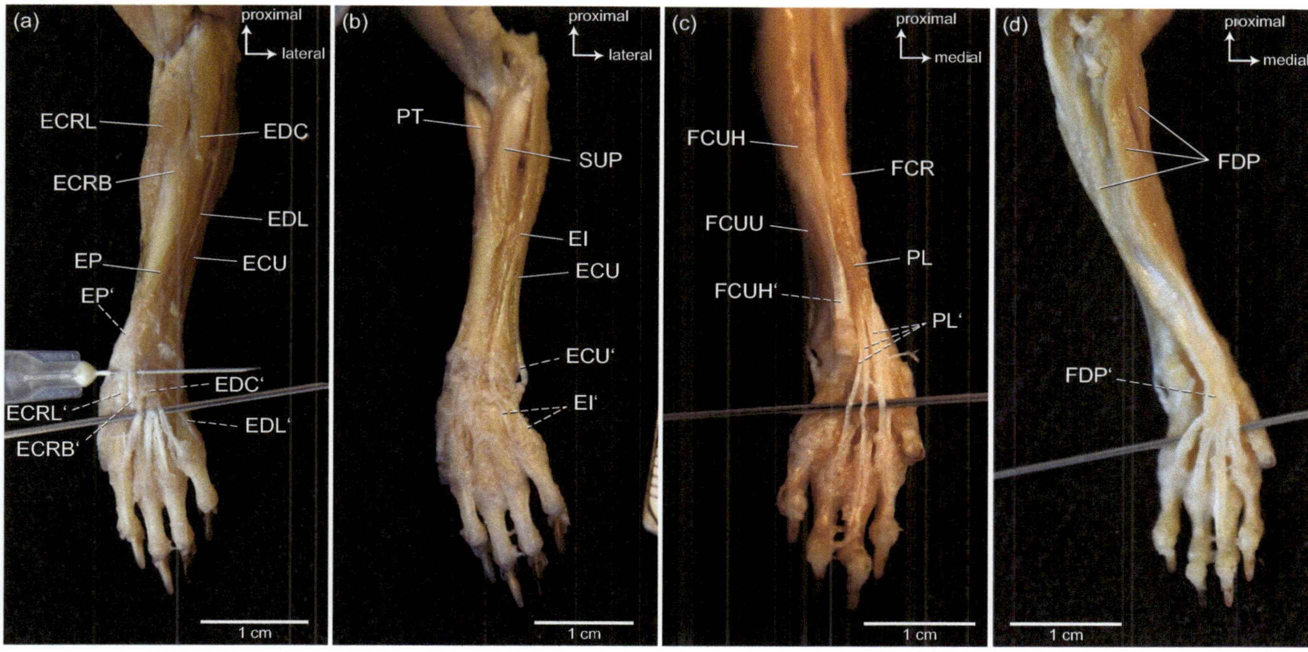
(a)
proximal
lateral
ECRL
EDC
ECRB
EDL
EP
ECU
EP'
EDC'
ECRL'
EDL'
ECRB'
1 cm
(b)
proximal
lateral
PT
SUP
EI
ECU
ECU'
EI'
1 cm
(c)
proximal
medial
FCUH
FCR
FCUU
PL
FCUH'
PL'
1 cm
(d)
proximal
medial
FDP
FDP'
1 cm

FIGURE II.18.6 **Deep rotator muscles. Distal rotator muscles in (a) dorsal view and in (b) ventral view. (c) Proximal rotator muscles in dorsomedial view.** The tendon of the BB, *M. biceps brachii* inserts distally to the tendon of the BCH, *M. brachioradialis* into the coronoid process of the ulna. The tendon of the BCH, *M. brachioradialis* inserts medially into the bicipital tuberosity of the radius. **(d) Proximal rotator muscles in ventral view.**

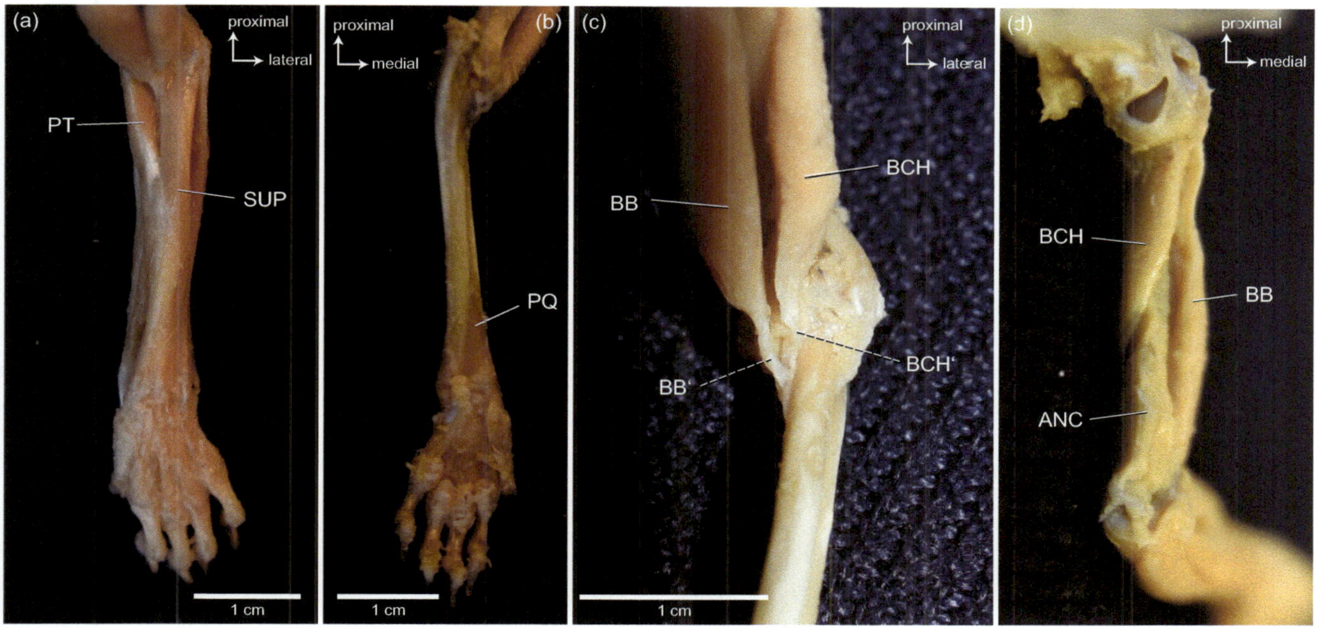
(a)
proximal
lateral
PT
SUP
1 cm
proximal
medial
(b)
PQ
1 cm
(c)
proximal
lateral
BCH
BB
BCH'
BB'
1 cm
proximal
medial
BCH
BB
ANC

Octodon degus (Degu)

Classification:	Rodentia, Hystricomorpha, Octodontidae
Mean body mass:	0.2 kg
Habitat:	Shrubland
Locomotor ecology:	Semi-fossorial - scansorial
Peculiarities:	Dig extensive communal burrow systems

FIGURE II.19.1 Superficial muscles of forelimb and hind limb in lateral view. (a) Without labels. (b) With labels and muscle outlines. In the neck and shoulder region, the *Mm. trapezii* are clearly divided into three parts (CT, *M. clavotrapezius*; AT, *M. arcomiotrapezius*; and ST, *M. spinotrapezius*). Note the length of the ST, *M. spinotrapezius*. The EPI, *M. epitrochlearis* and the BCR, *M. brachioradialis* are absent. The OEA, *M. obliquus externus abdominis* originates from the thoracic ribs and inserts into the linea alba at the midline and on the pubis near the symphysis. The GSU, *M. gluteus superficialis* is not developed as a separate muscle, but strongly fused with the TEN, *M. tensor fasciae latae*. The CAF, *M. caudofemoralis* is not developed as a separate muscle, but strongly fused with the BIF, *M. biceps femoris*. In the distal hind limb, the four tendons of the EDLO, *M. extensor digitorum longus* insert on the distal phalanges of digits II through V. The EHL, *M. hallucis longus* lies deep to the TCR, *M. tibialis cranialis* and the EDLO, *M. extensor digitorum longus*. The EDLA, *M. extensor digitorum lateralis* inserts on the distal phalanx digit V. Note that the retinacula (ligaments) spanning the wrist and ankle joints have been partially removed.

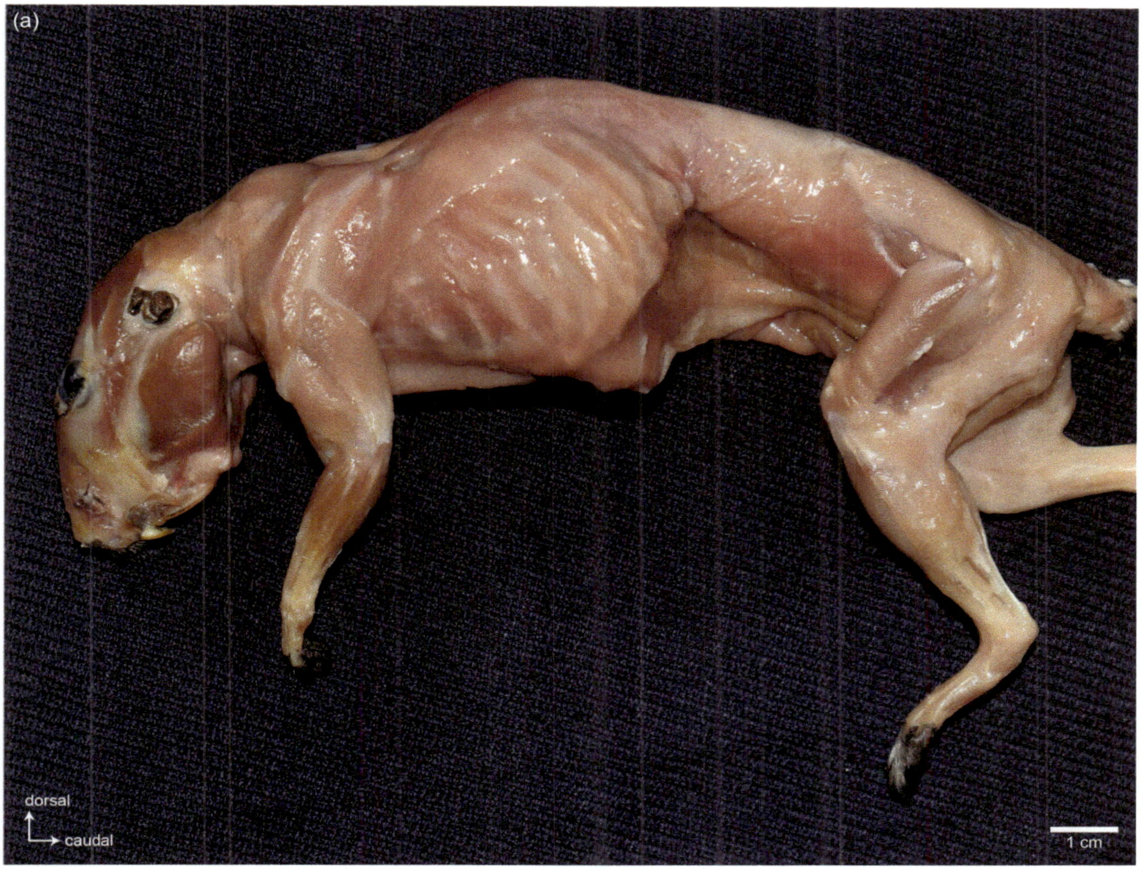
(a)
dorsal
caudal
1 cm

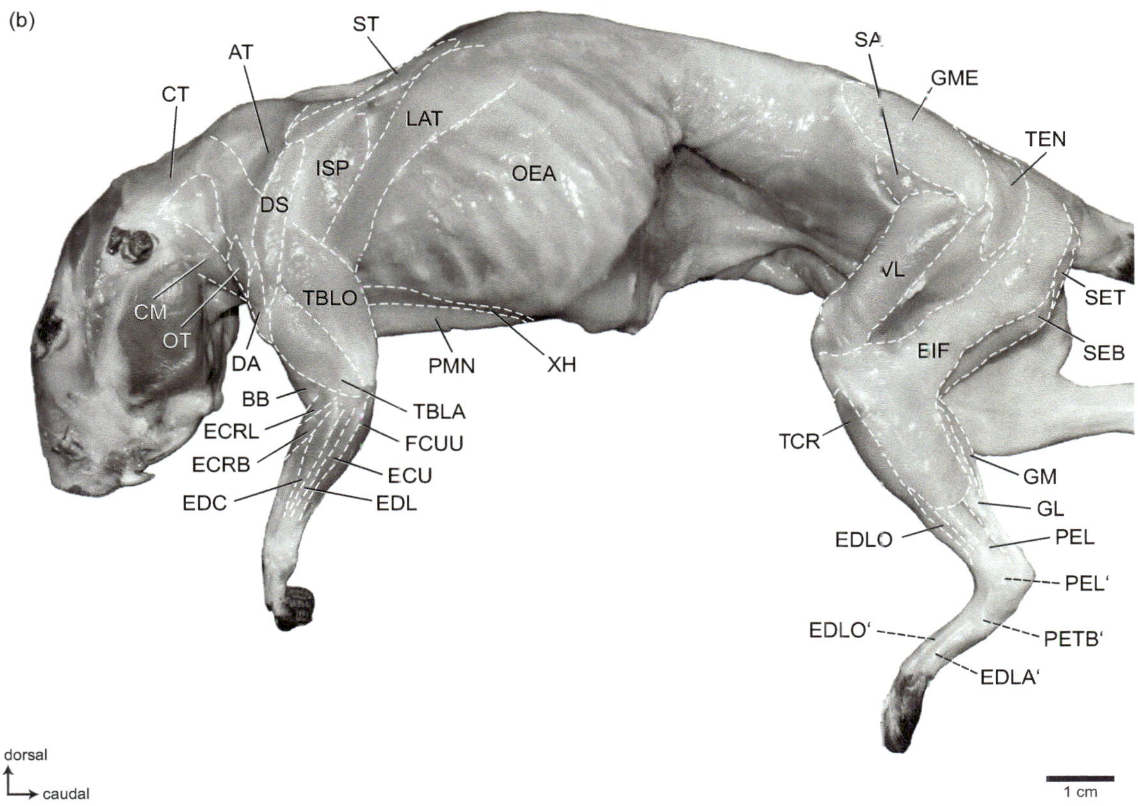
(b)
ST
AT
CT
LAT
ISP
OEA
DS
CM
OT
TBLO
DA
PMN
XH
BB
TBLA
ECRL
FCUU
ECRB
ECU
EDC
EDL
SA
GME
TEN
VL
SET
EIF
SEB
TCR
GM
GL
EDLO
PEL
PEL'
EDLO'
PETB'
EDLA'
dorsal
caudal
1 cm

FIGURE II.19.2 **Pectoral muscles in ventral view. (a) Superficial view.** The clavicle is present. The XH, *M. xiphihumeralis* is clearly separated from the PMN, *M. pectoralis minor.* **(b) Deep view.** The PMJ, *M. pectoralis major*; the PAB, *M. pectoantebrachialis*; and the XH, *M. xiphihumeralis* have been removed.

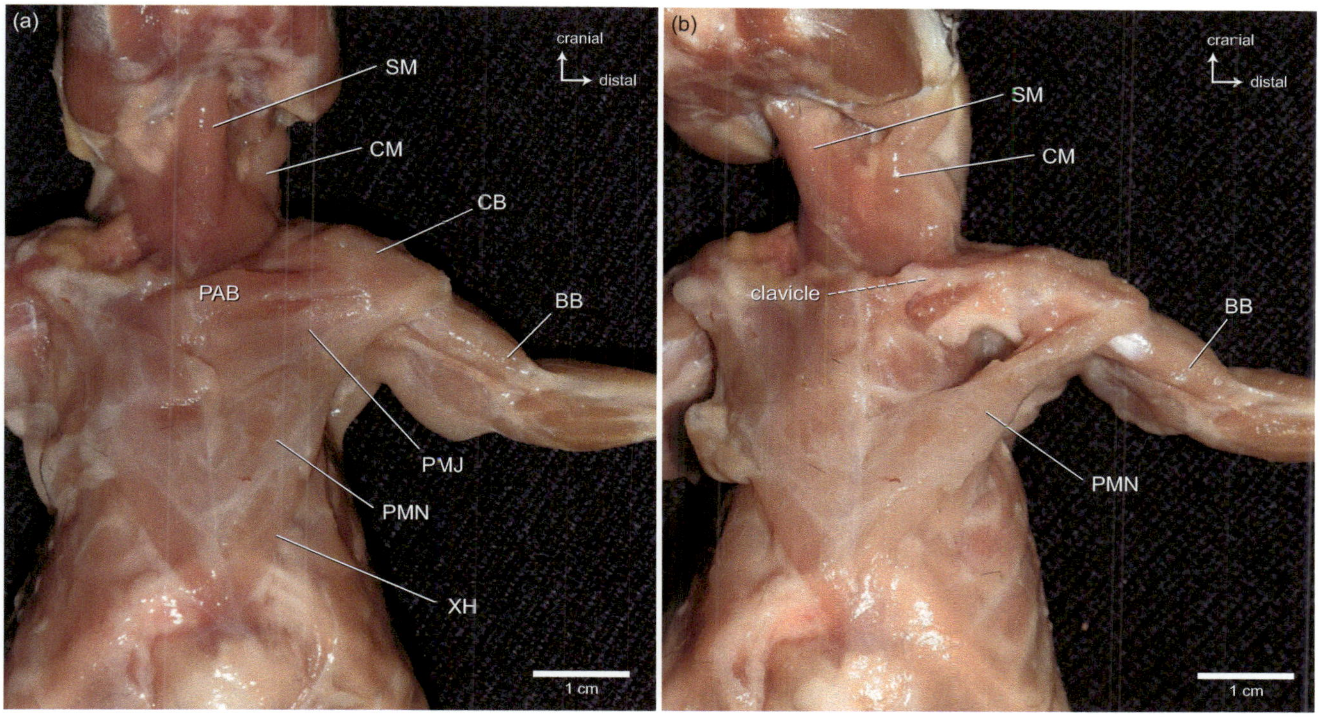
(a)
cranial
distal
SM
CM
CB
PAB
BB
PMJ
PMN
XH
1 cm
(b)
cranial
distal
SM
CM
clavicle
BB
PMN
1 cm

FIGURE II.19.3 **Deep neck and shoulder muscles. (a) In dorsolateral view.** Note the length of the ST, *M. spinotrapezius*. **(b) In craniolateral view and (c) dorsal view.** In the neck region, four *Mm. rhomboidei* are present (RCR, *M. rhomboideus cervicis*; RT, *M. rhomboideus thoracis*; RCP, *M. rhomboideus capitis*; and RPR, *M. rhomboideus profundus*). The RCR, *M. rhomboideus cervicis* does not reach the occiput.

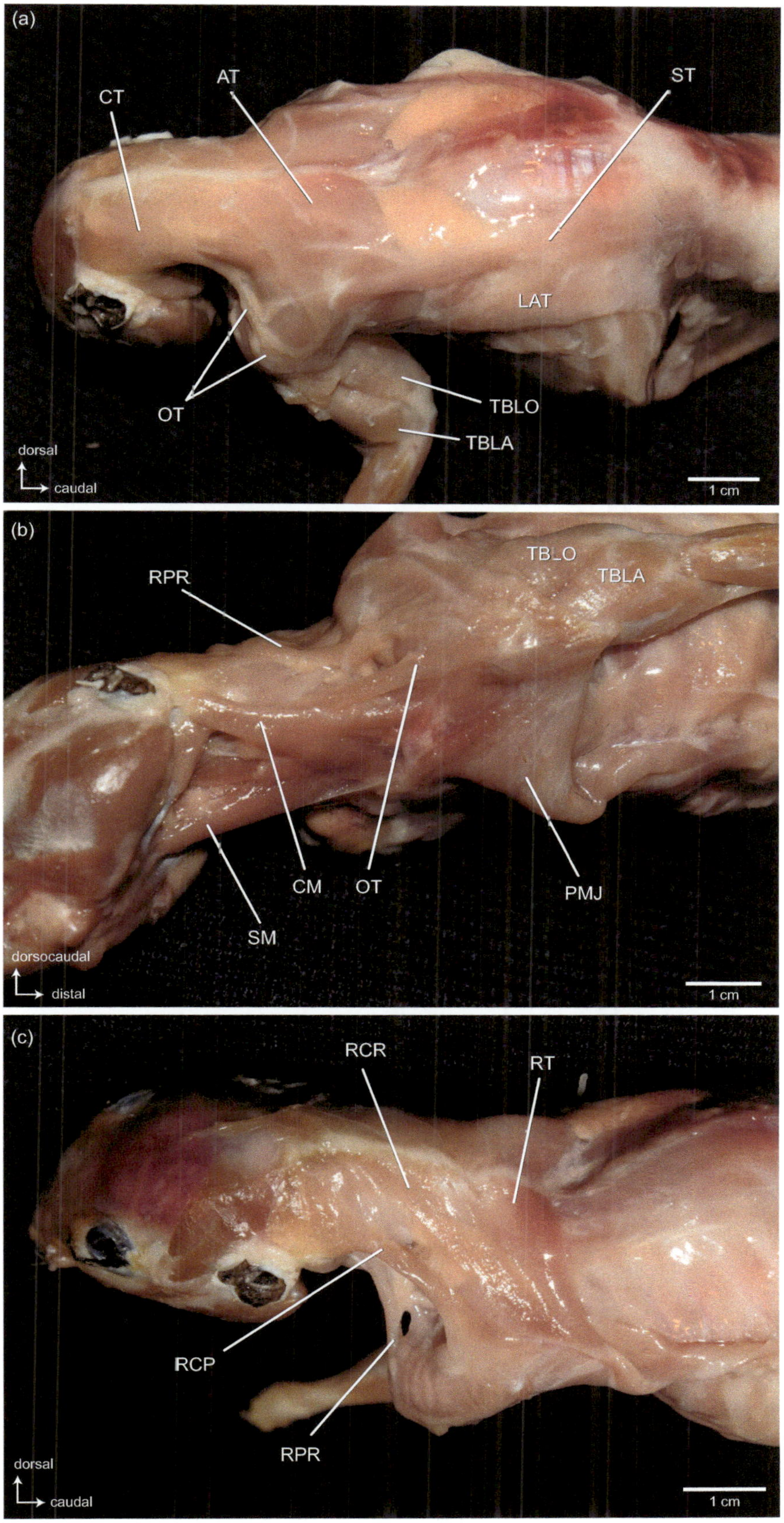
(a)
CT
AT
ST
LAT
OT
TBLO
TBLA
dorsal
caudal
1 cm
(b)
RPR
TBLO
TBLA
CM
OT
PMJ
SM
dorsocaudal
distal
1 cm
(c)
RCR
RT
RCP
RPR
dorsal
caudal
1 cm

FIGURE II.19.4 Intrinsic muscles in (a) proximolateral view and (b) proximomedial view. Intrinsic forelimb muscles in (c) lateral view and (d) medial view. The BB, *M. biceps brachii* has one distinct origin. The PL, *M. palmaris longus* inserts with four tendons on the distal phalanges of digits II through V.

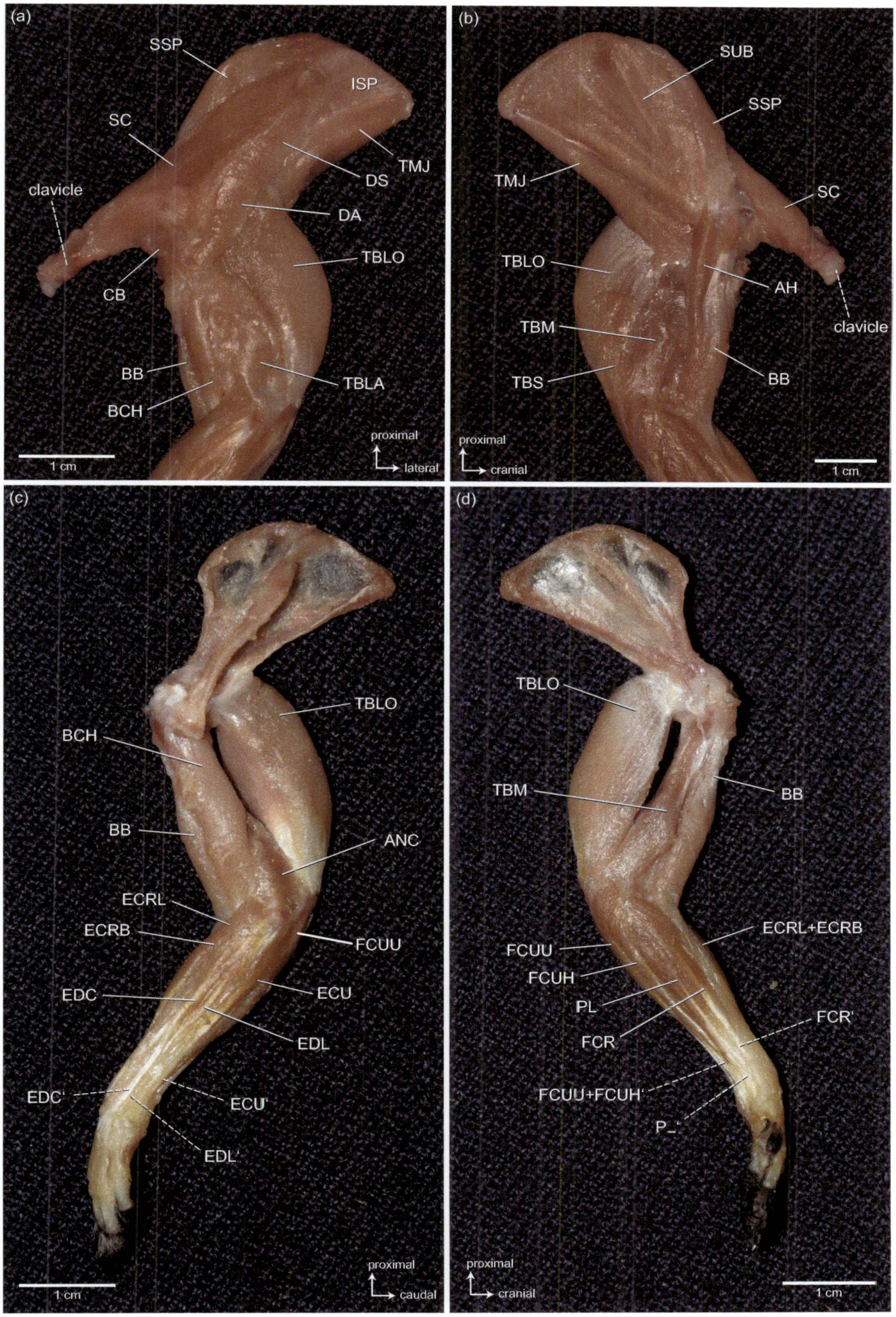
(a)
SSP
ISP
SC
TMJ
DS
clavicle
DA
TBLO
CB
BB
TBLA
BCH
1 cm
proximal
lateral
(b)
SUB
SSP
TMJ
SC
TBLO
AH
TBM
clavicle
TBS
BB
proximal
cranial
1 cm
(c)
TBLO
BCH
BB
ANC
ECRL
ECRB
FCUU
ECU
EDC
EDL
EDC'
ECU'
EDL'
1 cm
proximal
caudal
(d)
TBLO
TBM
BB
ECRL+ECRB
FCUU
FCUH
PL
FCR
FCR'
FCUU+FCUH'
P_'
proximal
cranial
1 cm

FIGURE II.19.5 **Distal forelimb muscles. (a, b) In dorsal view.** The SC, *M. subclavius* is well developed and covers the SSP, *M. supraspinatus*. The EDC, *M. extensor digitorum communis* inserts with four tendons on the distal phalanges of digits II through V. The EDL, *M. extensor digitorum lateralis* inserts with three tendons on the distal phalanges of digits III and V. The EI, *M. extensor digiti I and II* lies deep to the EDC, *M. extensor digitorum communis* and EDL, *M. extensor digitorum lateralis*. It inserts with two tendons on the distal phalanges of digits I and II. **(c, d) In ventral view.** The FDP, *M. flexor digitorum profundus* (four muscle bellies) inserts with four tendons on the distal phalanges of digits II through V.

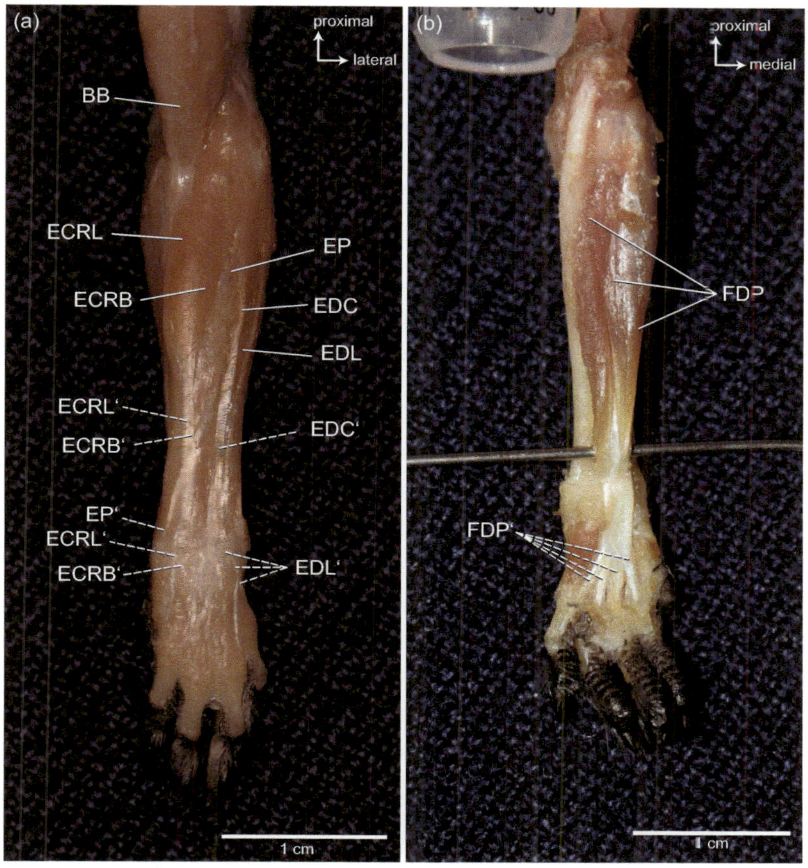
(a)
proximal
lateral
BB
ECRL
EP
ECRB
EDC
EDL
ECRL'
ECRB'
EDC'
EP'
ECRL'
ECRB'
EDL'
1 cm
(b)
proximal
medial
FDP
FDP'
1 cm

FIGURE II.19.6 **Deep rotator muscles. (a) Distal rotator muscles in dorsal view. (b) Proximal rotator muscles in dorsal view.** The tendon of the BB, *M. biceps brachii* inserts distally to the tendon of the BCH, *M. brachioradialis* into the coronoid process of the ulna. The tendon of the BCH, *M. brachioradialis* inserts medially into the bicipital tuberosity of the radius. **(c) Proximal rotator muscles in ventral view.**

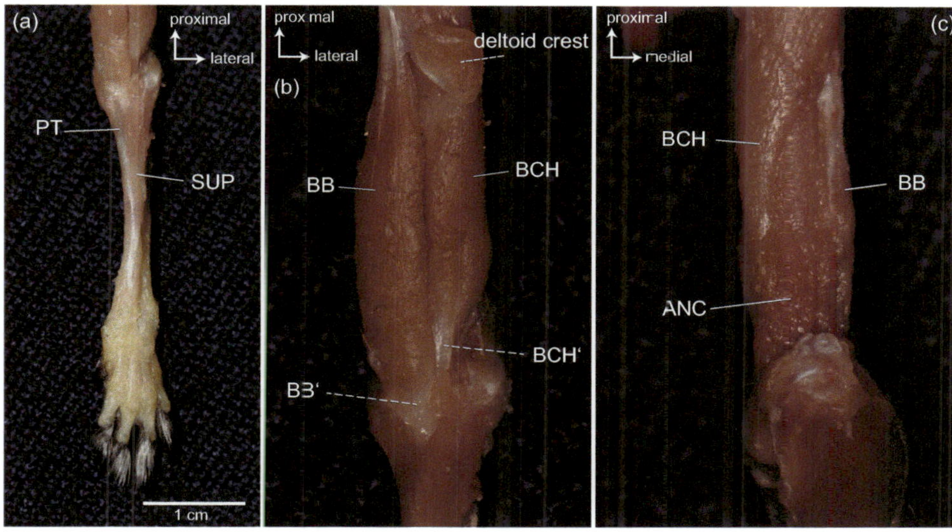
(a)
proximal
lateral
PT
SUP
1 cm
(b)
lateral
deltoid crest
BB
BCH
BCH'
(c)
BCH
BB
ANC

Sciurus vulgaris (Eurasian red squirrel)

Classification:	Rodentia, Sciuromorpha, Sciuridae
Mean body mass:	0.6 kg
Habitat:	Forest
Locomotor ecology:	Scansorial - saltatorial
Peculiarities:	Reversible hind feet

FIGURE II.20.1 Superficial muscles of forelimb and hind limb in lateral view. (a) Without labels. (b) With labels and muscle outlines. In the neck and shoulder region, the *Mm. trapezii* are divided into two parts (CT, *M. clavotrapezius* together with AT, *M. arcomiotrapezius*; ST, *M. spinotrapezius*). The EPI, *M. epitrochlearis* is well developed, and the BCR, *M. brachioradialis* is present. The OEA, *M. obliquus externus abdominis* originates from the thoracic ribs and inserts into the linea alba at the midline and on the pubis near the symphysis. In the distal hind limb, the four tendons of the EDLO, *M. extensor digitorum longus* insert on the distal phalanges of digits II through V. The EHL, *M. hallucis longus* lies deep to the TCR, *M. tibialis cranialis* and the EDLO, *M. extensor digitorum longus*. Note that the retinacula (ligaments) spanning the wrist and ankle joints have been largely removed.

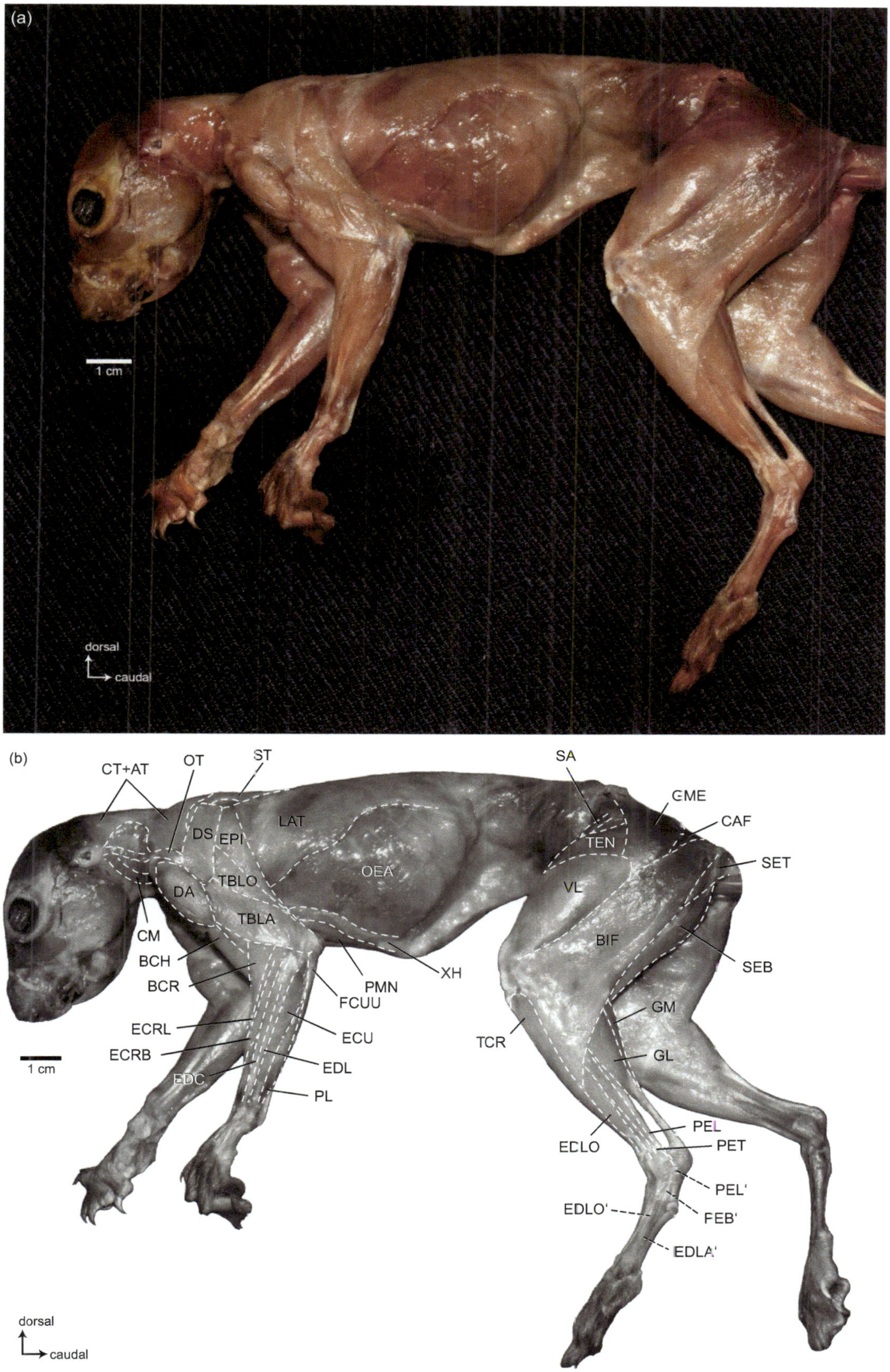
(a)
1 cm
dorsal
caudal
(b)
CT+AT
OT
ST
SA
GME
CAF
SET
LAT
DS
EPI
OEA
TEN
VL
DA
TBLO
TBLA
CM
BCH
BCR
BIF
SEB
PMN
XH
FCUU
ECRL
ECU
TCR
GM
GL
ECRB
EDL
EDC
PL
1 cm
PEL
PET
EDLO
PEL'
EDLO'
FEB'
EDLA'
dorsal
caudal

FIGURE II.20.2 **Pectoral muscles in ventral view. (a) Superficial view.** The XH, *M. xiphihumeralis* is clearly separated from the PMN, *M. pectoralis minor.* **(b) Deep view.** All four pectoral muscles (the PMJ, *M. pectoralis major*; the PAB, *M. pectoantebrachialis*; the PMN, *M. pectoralis minor*; and the XH, *M. xiphihumeralis*) have been removed. The SC, *M. subclavius* is weakly developed. The SV, *M. serratus ventralis* originates via distinct segments from the thoracic ribs and inserts into the craniomedial surface and vertebral border of the scapula.

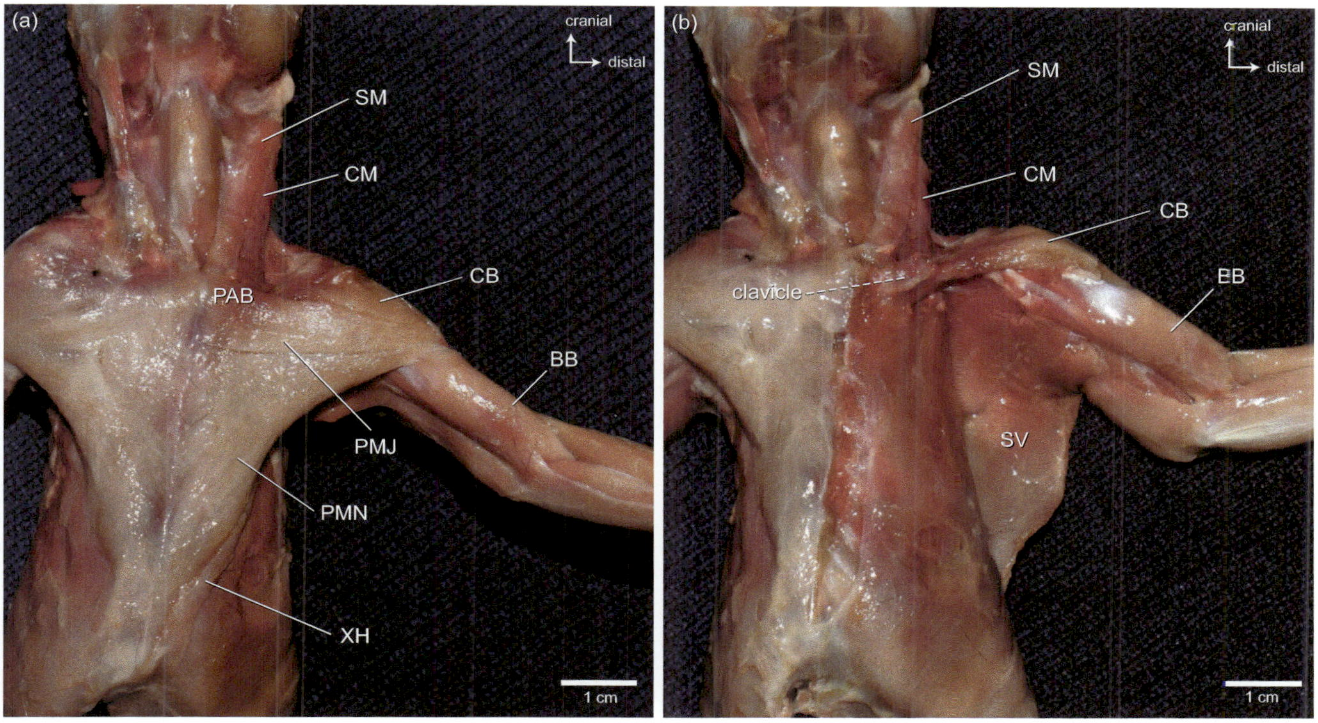
(a)
cranial
distal
SM
CM
CB
PAB
BB
PMJ
PMN
XH
1 cm
(b)
cranial
distal
SM
CM
CB
clavicle
BB
SV
1 cm

FIGURE II.20.3 **Deep neck and shoulder muscles. (a) In dorsolateral view.** The CT, *M. clavotrapezius* and the AT, *M. arcomiotrapezius* are fused together. **(b) In craniolateral view and (c) dorsal view.** In the neck region, two *Mm. rhomboidei* are present (RCR, *M. rhomboideus cervicis* and RPR, *M. rhomboideus profundus*). The RCR, *M. rhomboideus cervicis* does not reach the occiput. **(d) Cranial view.** The SC, *M. subclavius* has been removed.

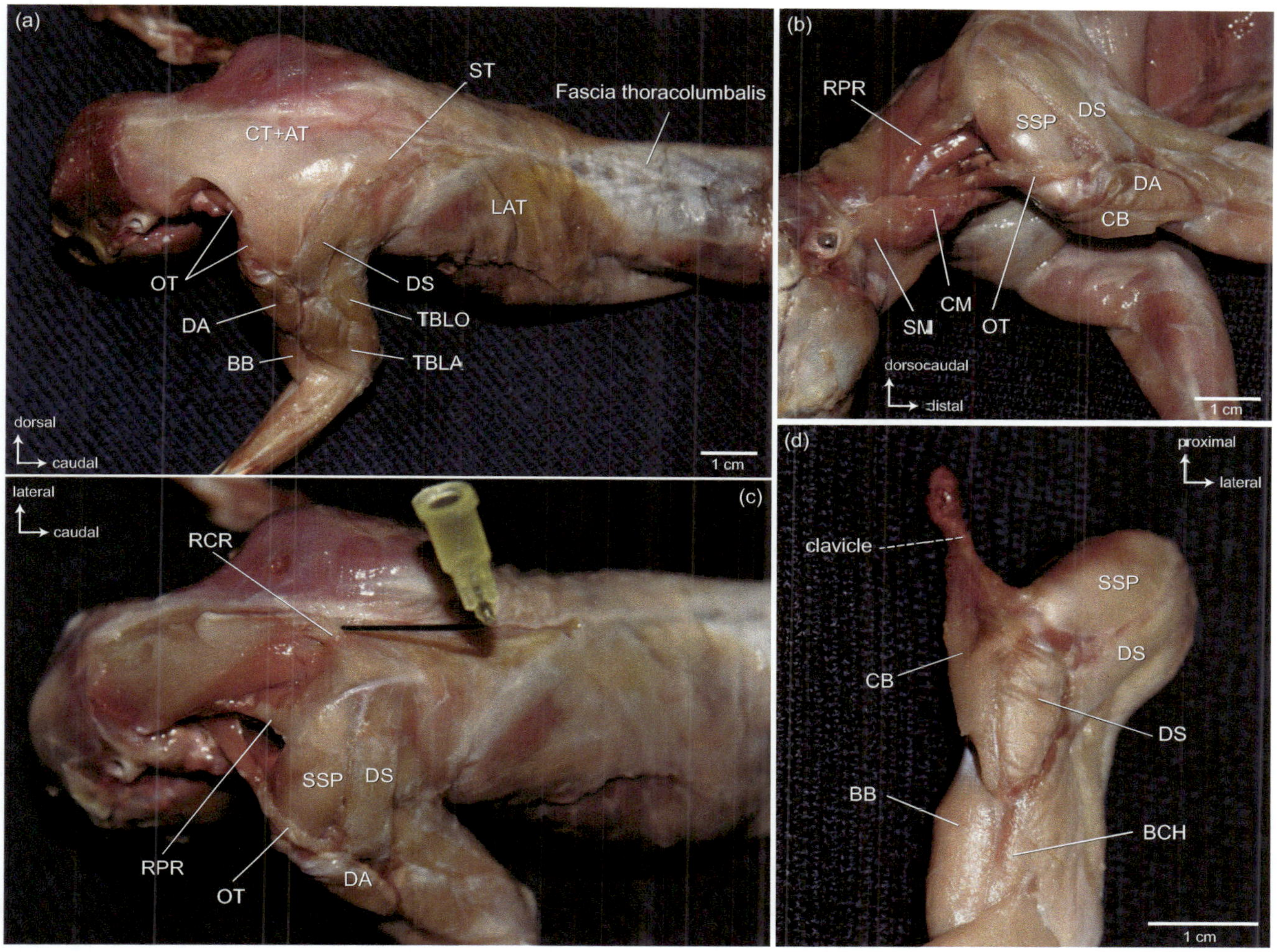
(a)
ST
Fascia thoracolumbalis
CT+AT
LAT
OT
DS
DA
TBLO
BB
TBLA
dorsal
caudal
1 cm
(b)
RPR
SSP
DS
DA
CB
CM
SM
OT
dorsocaudal
distal
1 cm
lateral
caudal
(c)
RCR
SSP
DS
RPR
OT
DA
(d)
proximal
lateral
clavicle
SSP
DS
CB
DS
BB
BCH
1 cm

FIGURE II.20.4 Intrinsic forelimb muscles in (a) proximolateral view and (b) proximomedial view. The BB, *M. biceps brachii* has two distinct origins. **Intrinsic muscles in (c) lateral view and (d) medial view.** The TMN, *M. teres minor* has been removed. **Distal intrinsic forelimb muscles in (e) lateral view and (f) medial view.** The EI, *M. extensor digiti I and II* inserts with two tendons on the distal phalanges of digits I and II. The EDL, *M. extensor digitorum lateralis* inserts with three tendons on the distal phalanges of digits III through V.

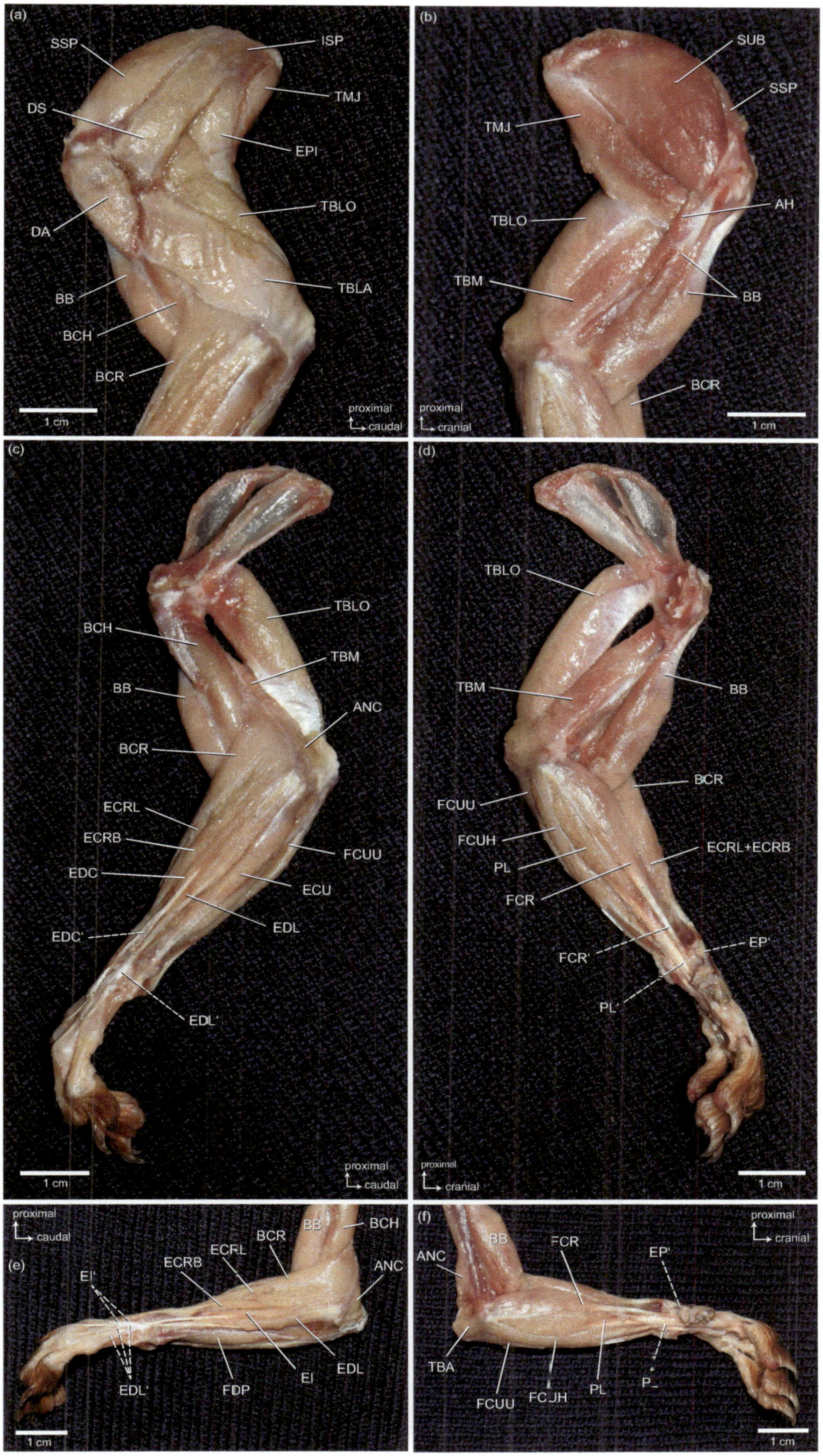
(a)
SSP
ISP
TMJ
DS
EPI
TBLO
DA
TBLA
BB
BCH
BCR
1 cm
proximal
caudal
(b)
SUB
SSP
TMJ
AH
TBLO
TBM
BB
BCR
proximal
cranial
1 cm
(c)
TBLO
BCH
TBM
BB
ANC
BCR
ECRL
ECRB
FCUU
EDC
ECU
EDL
EDC'
EDL'
1 cm
proximal
caudal
(d)
TBLO
TBM
BB
BCR
FCUU
FCUH
ECRL+ECRB
PL
FCR
FCR'
EP'
PL'
proximal
cranial
1 cm
proximal
caudal
(e)
BB
BCH
BCR
ECRL
ECRB
ANC
EI'
EDL'
FDP
EI
EDL
1 cm
(f)
proximal
cranial
ANC
BB
FCR
EP'
TBA
FCUU
FCUH
PL
PL'
1 cm

FIGURE II.20.5 **Distal forelimb muscles. (a, b) In dorsal view.** The EDC, *M. extensor digitorum communis* inserts with four tendons on the distal phalanges of digits II through V. The EDL, *M. extensor digitorum lateralis* inserts with three tendons on the distal phalanges of digits III through V. The BCR, *M. brachioradialis* inserts tendinously into the mesiodistal end of the radius. **(c, d) In ventral view.** The PL, *M. palmaris longus* inserts with five tendons on the distal phalanges of digits I through V. FDP, *M. flexor digitorum profundus* (four muscle bellies) inserts with four tendons on the distal phalanges of digits II through V.

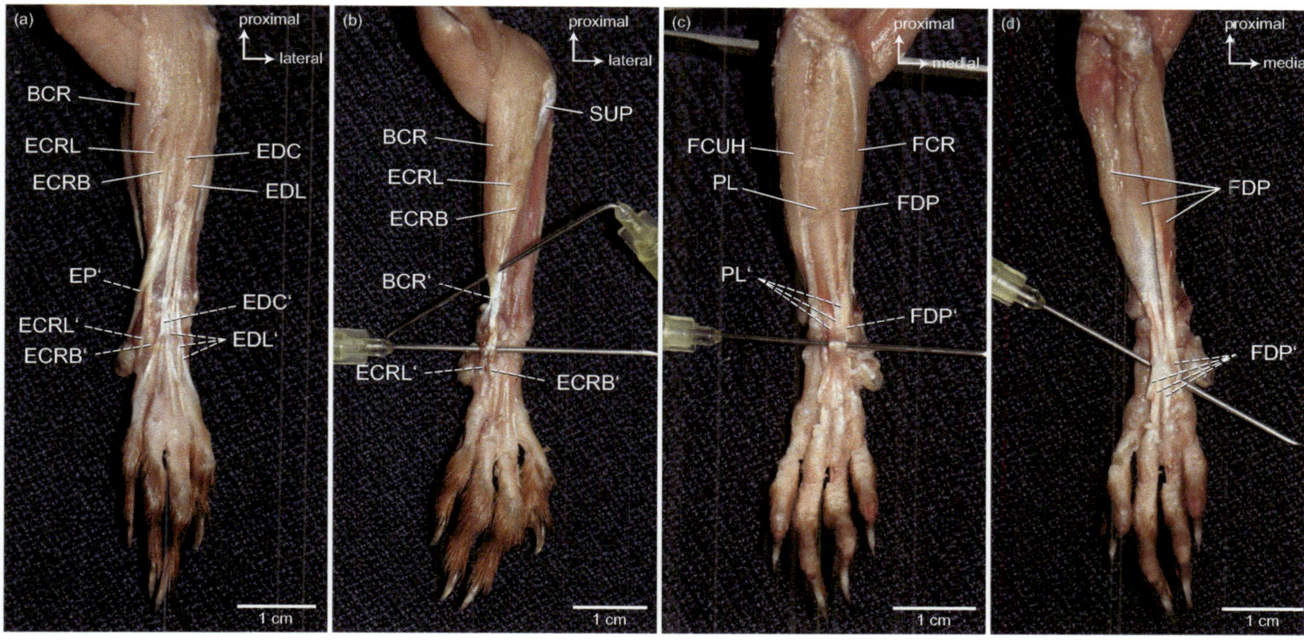
(a)
proximal
lateral
BCR
ECRL
EDC
ECRB
EDL
EP'
EDC'
ECRL'
EDL'
ECRB'
1 cm
(b)
proximal
lateral
SUP
BCR
ECRL
ECRB
BCR'
ECRL'
ECRB'
1 cm
(c)
proximal
medial
FCUH
FCR
PL
FDP
PL'
FDP'
1 cm
(d)
proximal
medial
FDP
FDP'
1 cm

FIGURE II.20.6 **Deep rotator muscles. (a) Distal rotator muscles in dorsal view. (b) Distal rotator muscles in ventral view.** Not visible in the picture, but the tendon of the BB, *M. biceps brachii* inserts into the bicipital tuberosity of the radius and the tendon of the BCH, *M. brachioradialis* inserts into the coronoid process of the ulna. The PQ, *M. pronator quadratus* is weakly developed.

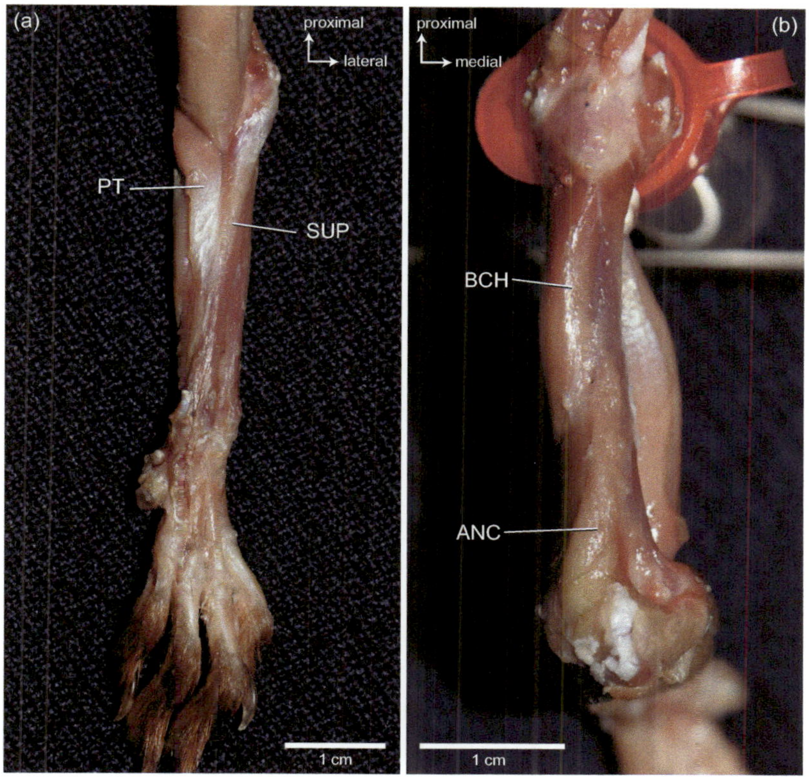
(a)
proximal
lateral
PT
SUP
1 cm
proximal
medial
(b)
BCH
ANC
1 cm

Pedetes capensis (Springhare)

Classification: Rodentia, Anomaluromorpha, Pedetidae

Mean body mass: 3.0–4.0 kg

Habitat: Desert, savanna

Locomotor ecology: Saltatorial

Peculiarities: Short forelimbs and long hind limbs

FIGURE II.21.1 Superficial muscles of forelimb and hind limb in lateral view. (a) Without labels. (b) With labels and muscle outlines. In the neck and shoulder region, the *Mm. trapezii* are divided into two parts (CT, *M. clavotrapezius* together with AT, *M. arcomiotrapezius*; ST, *M. spinotrapezius*). The EPI, *M. epitrochlearis* is absent. The BCR, *M. brachioradialis* is well developed. The clavicle is present.

(a)
dorsal
caudal
1 cm
(b)
GME
GSU
ST
CAF
CT+AT
LAT
ISP
TMJ
SA
TEN
CM
TBLO
SM
DS
VL
TBLA
OT
DA
PMN
BCH
BIF
SET
GL
SOL
TCR
dorsal
caudal
1 cm

FIGURE II.21.2 **Pectoral muscles in ventral view.** Three pectoral muscles are identified (the PMJ, *M. pectoralis major*; the PAB, *M. pectoantebrachialis*; and the PMN, *M. pectoralis minor*). The XH, *M. xiphihumeralis* is not clearly developed as a separate muscle, but part of the PMN, *M. pectoralis minor.* The SC, *M. subclavius* is well developed.

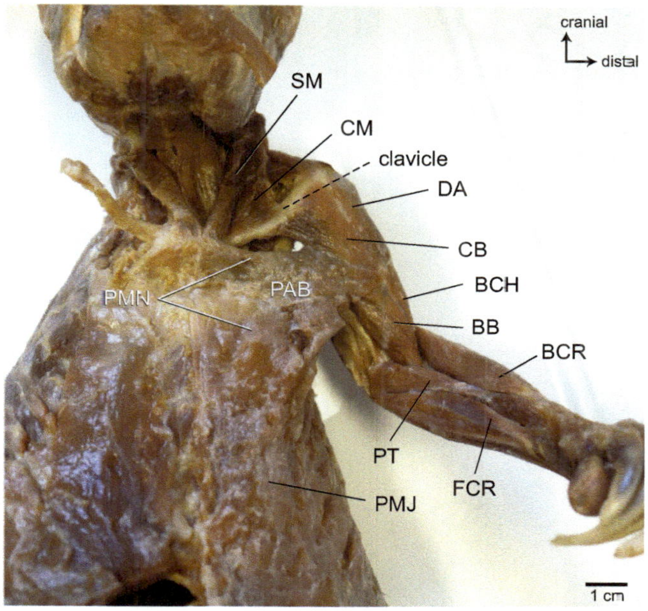

cranial
distal
SM
CM
clavicle
DA
CB
BCH
BB
BCR
PMN
PAB
PT
FCR
PMJ
1 cm

FIGURE II.21.3 **Deep neck and shoulder muscles. (a) In dorsolateral view.** In the neck region, three *Mm. rhomboidei* are present (RCR, *M. rhomboideus cervicis*; RT, *M. rhomboideus thoracis*; and RCP, *M. rhomboideus profundus*). The RCR, *M. rhomboideus cervicis* reaches the occiput. The RPR, *M. rhomboideus capitis* is absent. **(b) In cranial view.** The SC, *M. subclavius* has been removed.

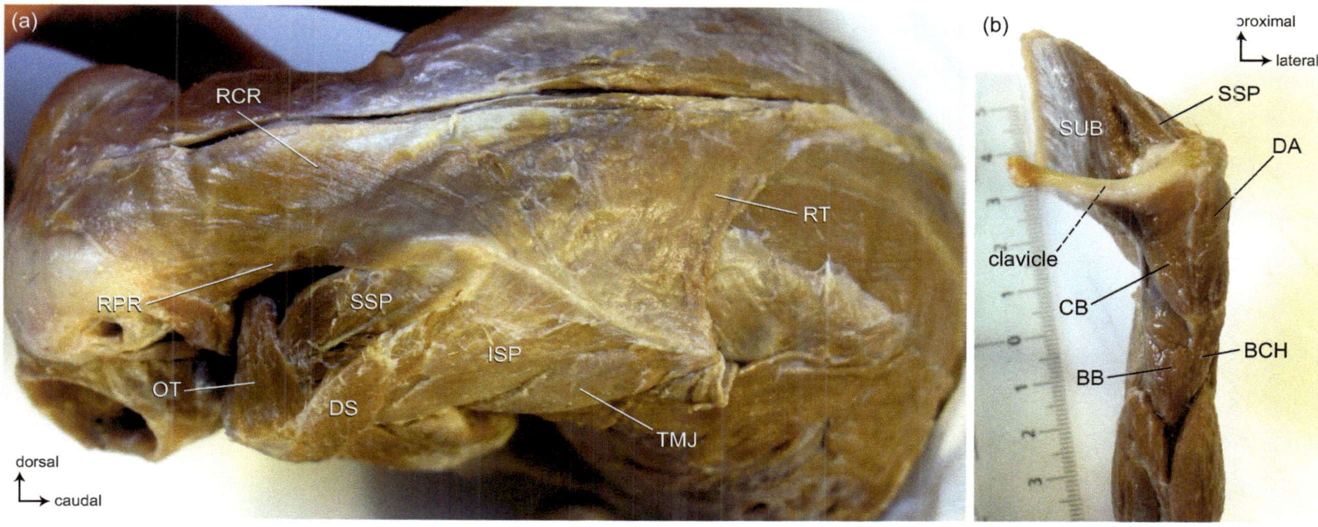
(a)
RCR
RT
RPR
SSP
ISP
OT
DS
TMJ
dorsal
caudal
(b)
proximal
lateral
SUB
SSP
DA
clavicle
CB
BCH
BB

FIGURE II.21.4 **Intrinsic forelimb muscles in (a) proximolateral view and (b) proximomedial view.** The BB, *M. biceps brachii* has two distinct origins. The PL, *M. palmaris longus* inserts into the palmar aponeurosis. **Intrinsic forelimb muscles in (c) lateral view and (d) medial view.** The EDL, *M. extensor digitorum lateralis* inserts with one tendon on the distal phalanx of digit V. The EI, *M. extensor digiti I and II* inserts with on tendon on the distal phalanx of digit I.

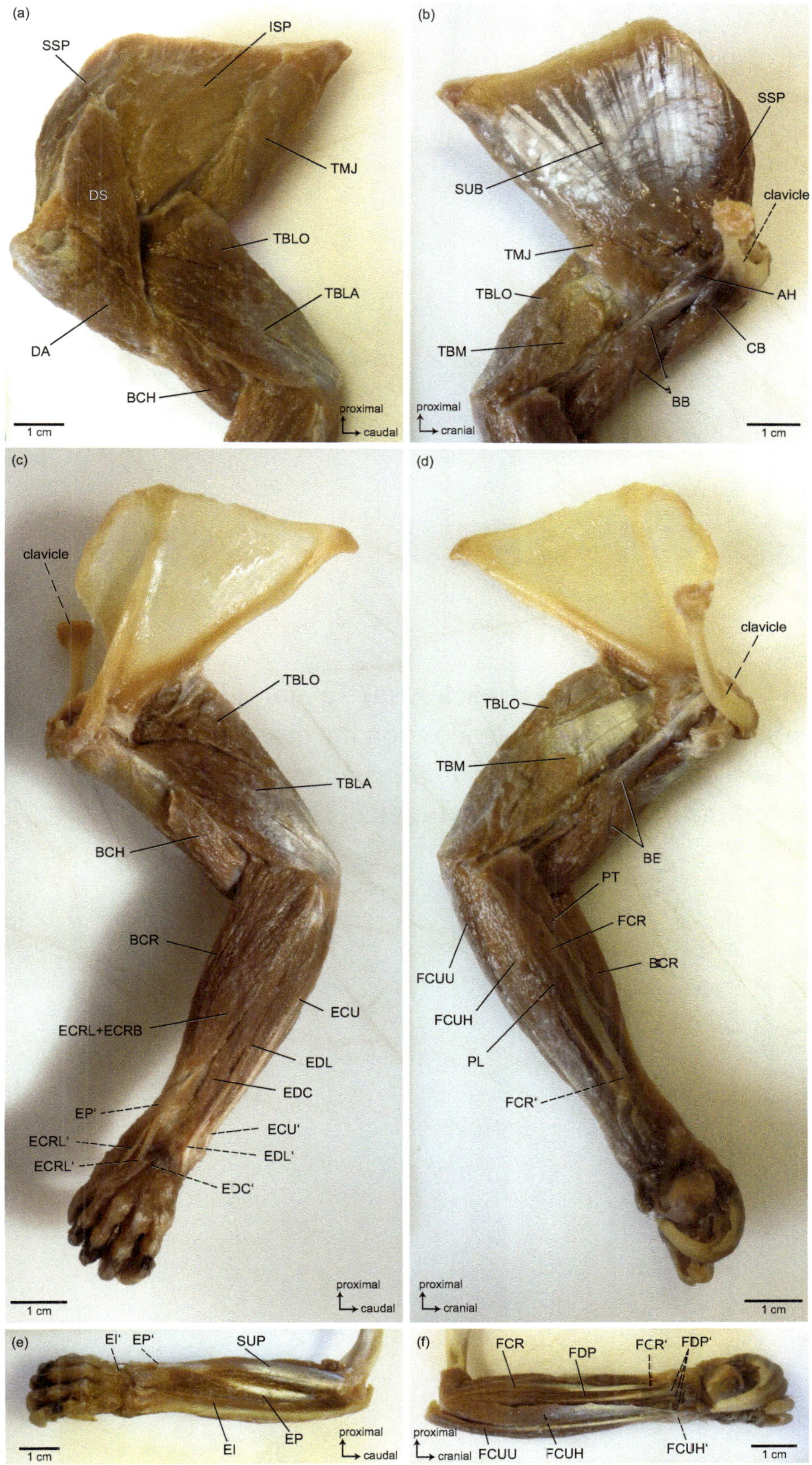
(a)
SSP
ISP
TMJ
DS
TBLO
TBLA
DA
BCH
proximal
caudal
1 cm
(b)
SSP
SUB
clavicle
TMJ
TBLO
AH
CB
TBM
BB
proximal
cranial
1 cm
(c)
clavicle
TBLO
TBLA
BCH
BCR
ECU
ECRL+ECRB
EDL
EDC
EP'
ECU'
ECRL'
EDL'
ECRL'
EDC'
proximal
caudal
1 cm
(d)
clavicle
TBLO
TBM
BE
PT
FCR
BCR
FCUU
FCUH
PL
FCR'
proximal
cranial
1 cm
(e)
EI'
EP'
SUP
EI
EP
proximal
caudal
1 cm
(f)
FCR
FDP
FCR'
FDP'
proximal
cranial
FCUU
FCUH
FCUH'
1 cm

FIGURE II.21.5 **Distal forelimb muscles. (a) In dorsal view.** The EDC, *M. extensor digitorum communis* inserts with four tendons on the distal phalanges of digits II through V. **(b) In ventral view.** The FDP, *M. flexor digitorum profundus* (four muscle bellies) inserts with five tendons on the distal phalanges of digits I through V.

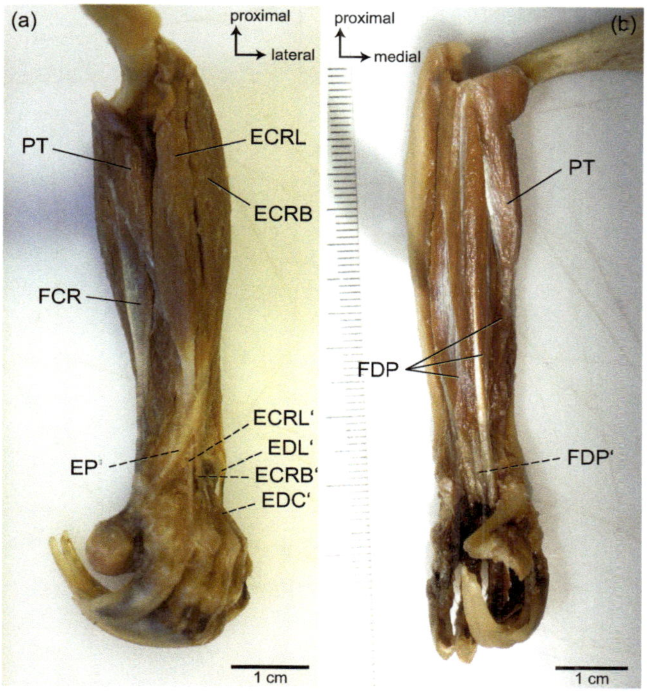
(a)
proximal
lateral
PT
ECRL
ECRB
FCR
ECRL'
EDL'
EP'
ECRB'
EDC'
1 cm
(b)
proximal
medial
PT
FDP
FDP'
1 cm

FIGURE II.21.6 **Deep rotator muscles. (a) Distal rotator muscles in dorsal view. (b) Distal rotator muscles in ventral view.** The tendon of the BCH, *M. brachioradialis* inserts distally to the tendon of the BB, *M. biceps brachii* into the radius. The tendon of the BB, *M. biceps brachii* inserts proximally into the coronoid process into the ulna. **(c) Proximal rotator muscles in ventral view.** Note that the ANC, *M. anconeus* extends medially.

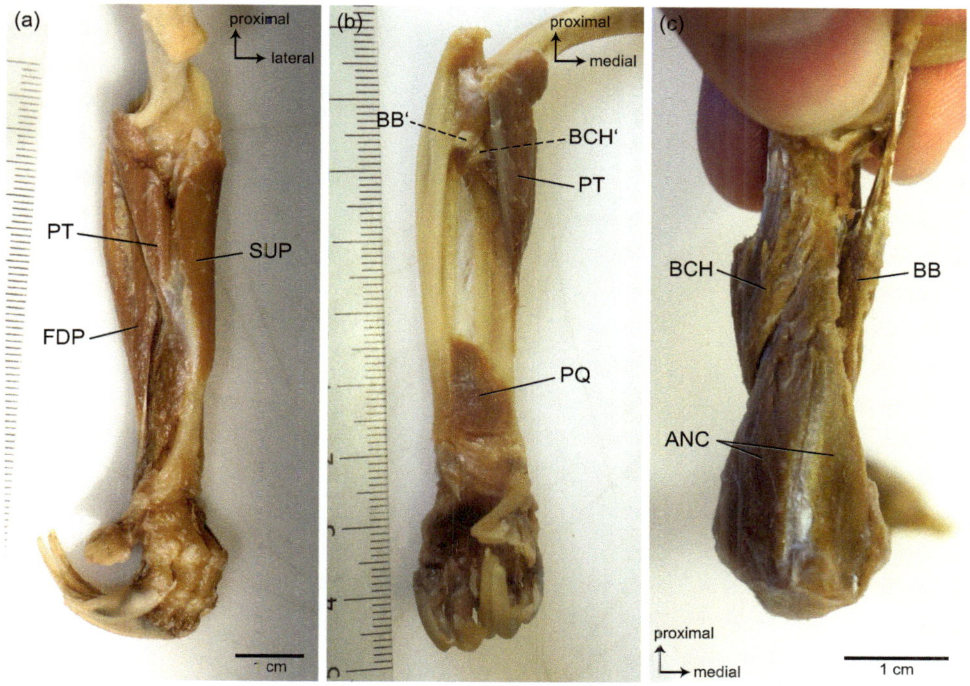
(a)
proximal
lateral
PT
SUP
FDP
cm
(b)
proximal
medial
BB'
BCH'
PT
PQ
(c)
BCH
BB
ANC
proximal
medial
1 cm

FIGURE II.21.7 **Hip muscles. (a) In ventral view. (b) In medial view.** The OE, *M. obturator externus* lies lateral to the OI, *M. obturator internus.*

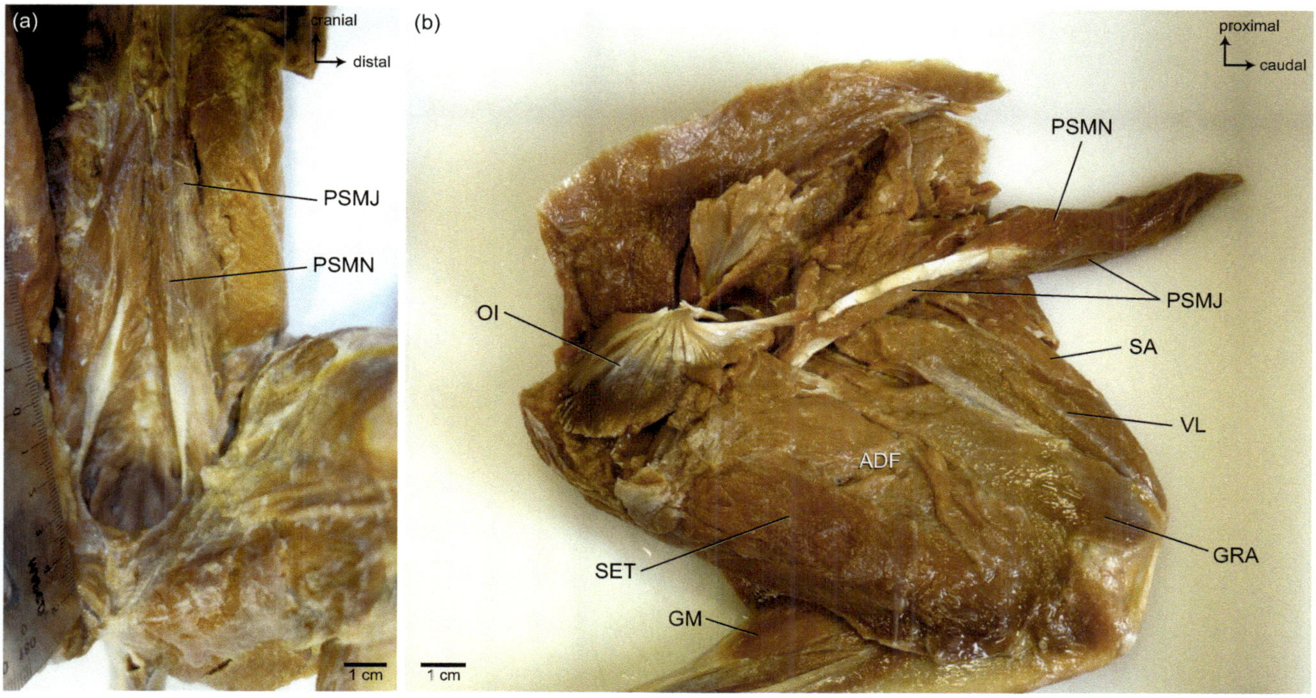
(a)
cranial
distal
PSMJ
PSMN
1 cm
(b)
proximal
caudal
PSMN
PSMJ
OI
SA
VL
ADF
SET
GRA
GM
1 cm

FIGURE II.21.8 Intrinsic hind limb muscles in (a) lateral view and (b) medial view. Distal intrinsic forelimb muscles in (c) lateral view and (d) medial view. The EDLA, *M. extensor digitorum lateralis* inserts on the distal phalanx of digit V. The EDLO, *M. extensor digitorum longus* inserts into the distal phalanges of digits II–V.

(a)
proximal
caudal
VL
GM
GL
TCR
EDLO
1 cm
(b)
proximal
cranial
VL
RF
VM
GM
FDLO
SOL
POP
1 cm
(c)
proximal
caudal
GM
GL
TCR
SOL
EDLO
FDLO
GL'+GM'+
SOL'+PL'
PEL
EDLO'
PEL'
EDLA'
(d)
proximal
cranial
GL
GM
PL
GL'+GM'+
SOL'+PL'
FHL
POP
FDLO
TCR'
FHL'
FDLO'

Mesocricetus auratus (Golden hamster)

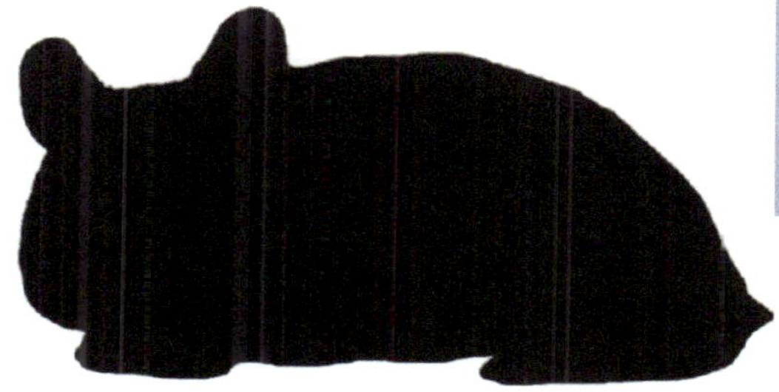

Classification:	Rodentia, Anomaluromorpha, Pedetidae
Mean body mass:	0.1 kg
Habitat:	Savanna, grassland
Locomotor ecology:	Ambulatorial, semifossorial
Peculiarities:	Scansorial ability

FIGURE II.22.1 **Superficial muscles of forelimb and hind limb in lateral view. (a) Without labels. (b) With labels and muscle outlines.** In the neck and shoulder region, the *Mm. trapezii* are divided into two parts (CT, *M. clavotrapezius* together with AT, *M. arcomiotrapezius*; and ST, *M. spinotrapezius*). The EPI, *M. epitrochlearis* and the BCR, *M. brachioradialis* are absent. In the distal hind limb, the four tendons of the EDLO, *M. extensor digitorum longus* insert on the distal phalanges of digits II through V. Note that the retinacula (ligaments) spanning the wrist and ankle joints have been partially removed.

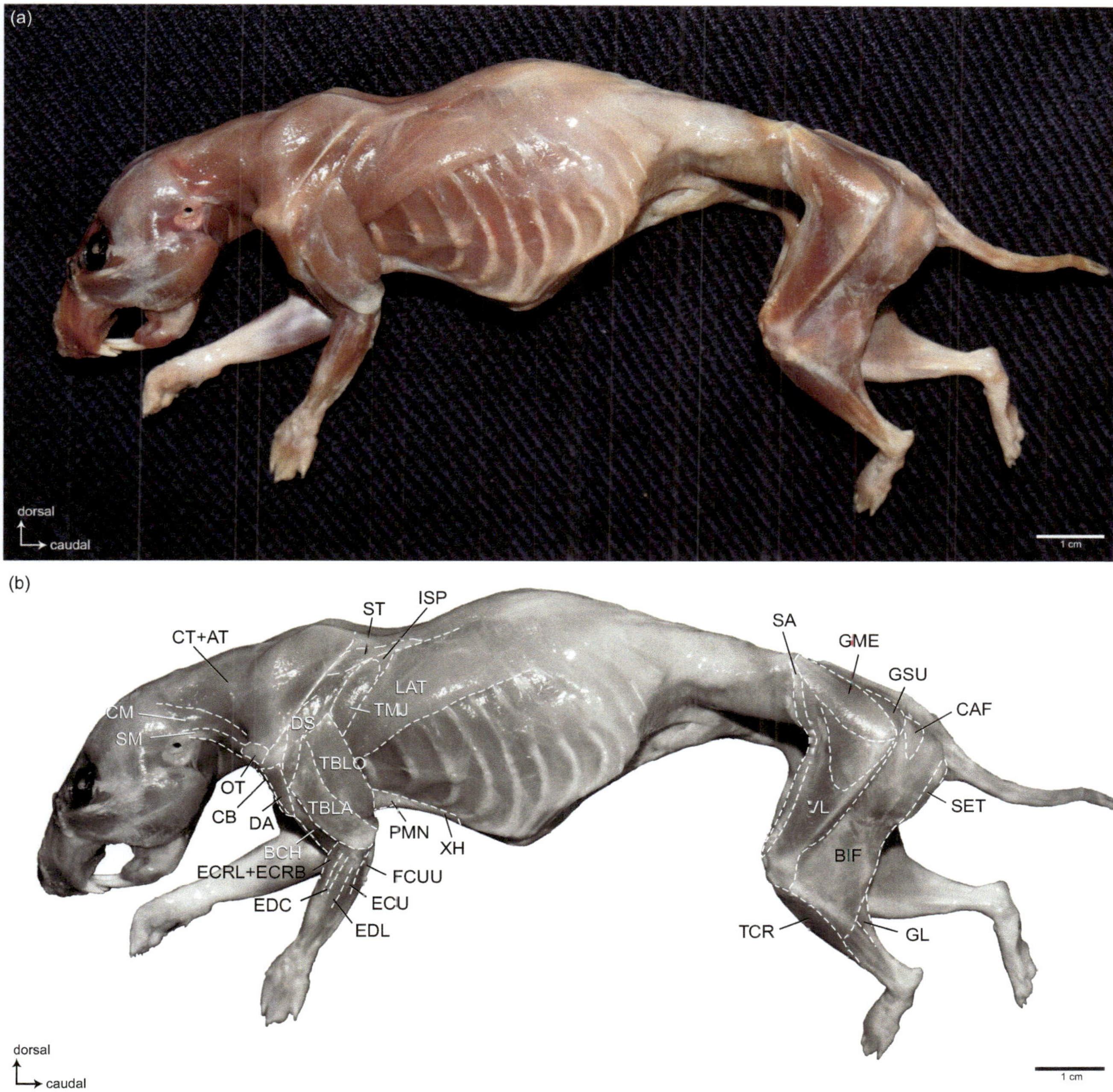
(a)
dorsal
caudal
1 cm
(b)
CT+AT
ST
ISP
SA
GME
GSU
CAF
CM
SM
LAT
TMJ
DS
TBLO
OT
CB
DA
TBLA
PMN
XH
BCH
ECRL+ECRB
FCUU
EDC
ECU
EDL
VL
BIF
SET
TCR
GL
dorsal
caudal
1 cm

FIGURE II.22.2 **Pectoral muscles in ventral view.** The clavicle is present. Four pectoral muscles are identified (the PMJ, *M. pectoralis major*; the PAB, *M. pectoantebrachialis*; the PMN; *M. pectoralis minor*; and the XH, *M. xiphihumeralis*).

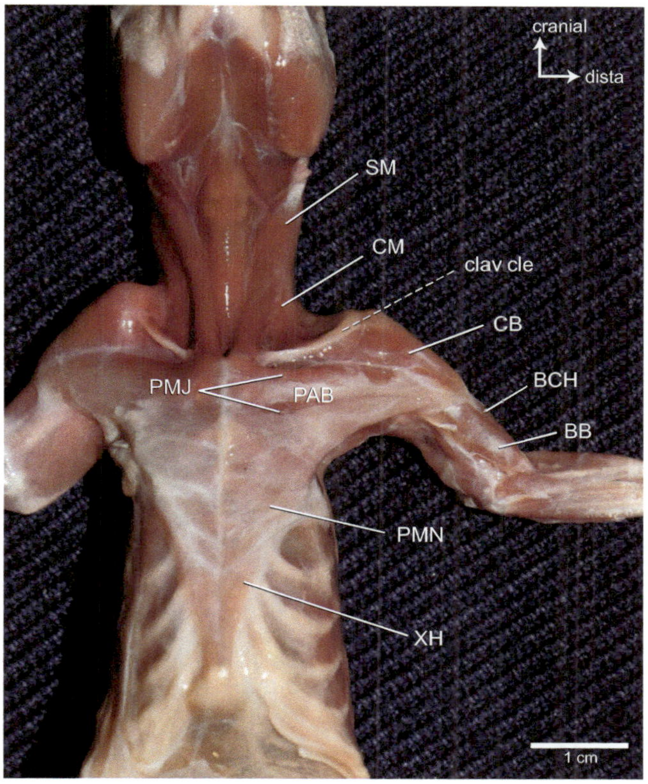
cranial
dista
SM
CM
clav cle
CB
BCH
PMJ
PAB
BB
PMN
XH
1 cm

FIGURE II.22.3 Deep neck and shoulder muscles in dorsal view. (a) In lateral view and (b) in dorsolateral view. In the neck region, two *Mm. rhomboidei* are present (RCR, *M. rhomboideus cervicis* and RCP, *M. rhomboideus capitis*). The RCR, *M. rhomboideus cervicis* does not reach the occiput.

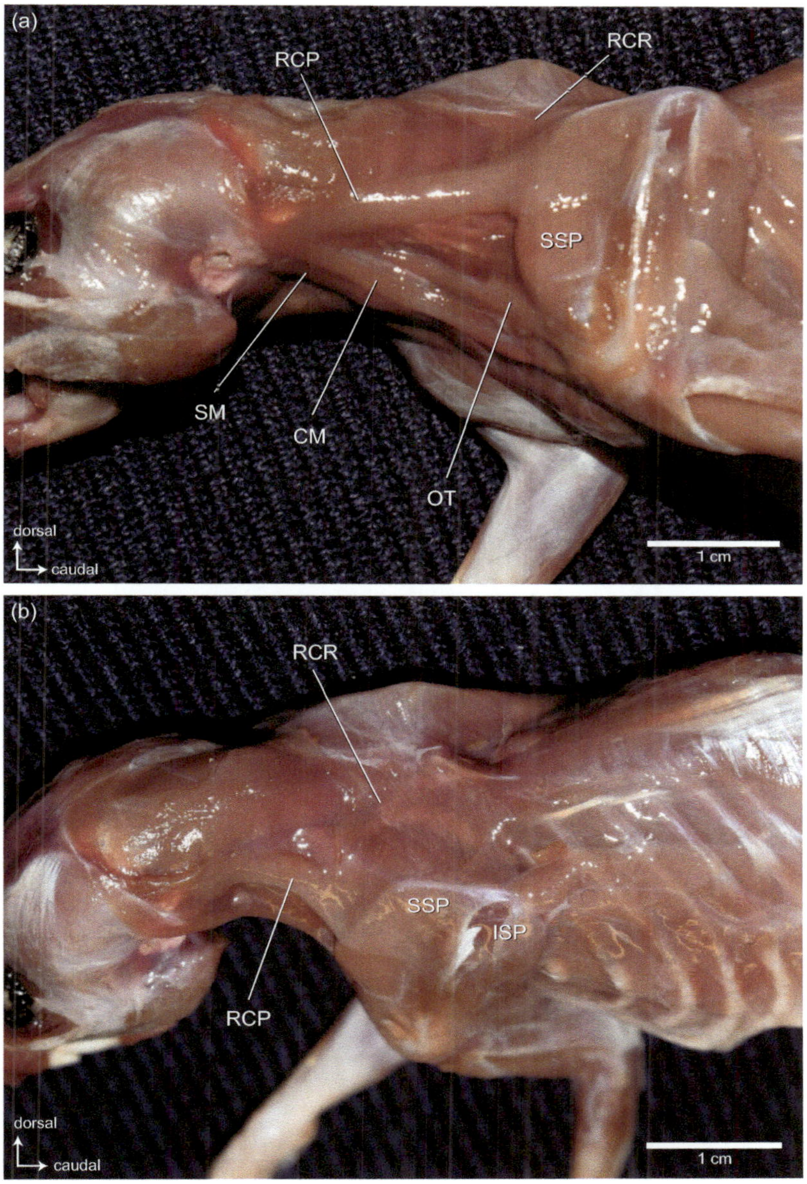
(a)
RCP
RCR
SSP
SM
CM
OT
dorsal
caudal
1 cm
(b)
RCR
SSP
ISP
RCP
dorsal
caudal
1 cm

FIGURE II.22.4 **Intrinsic forelimb muscles in (a) proximolateral view and (b) proximomedial view.** The BB, *M. biceps brachii* has one distinct origin. **Intrinsic forelimb muscles in (c) lateral view and (d) medial view.**

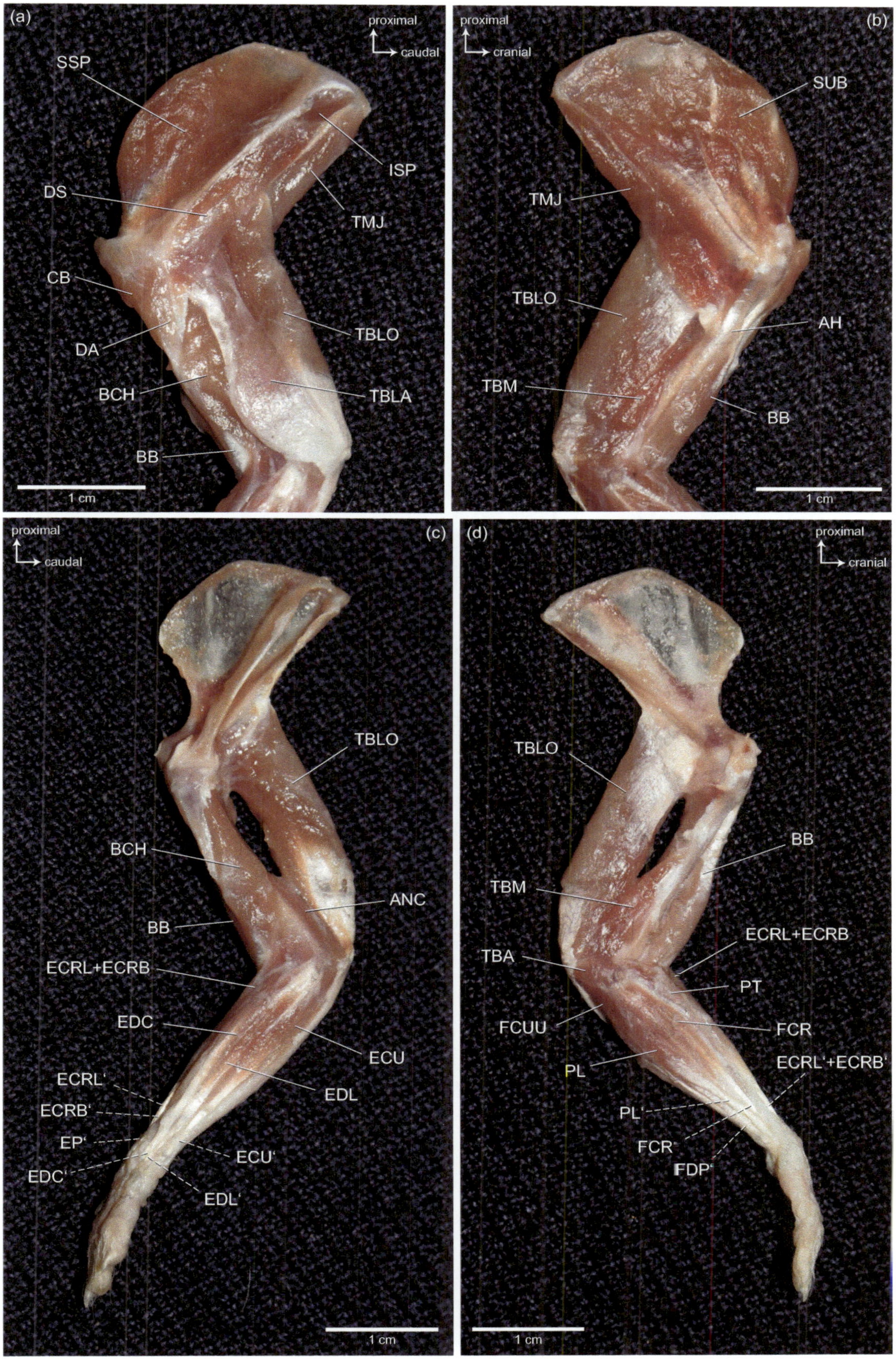
(a)
proximal
caudal
SSP
ISP
DS
TMJ
CB
TBLO
DA
BCH
TBLA
BB
1 cm
(b)
proximal
cranial
SUB
TMJ
TBLO
AH
TBM
BB
1 cm
(c)
proximal
caudal
TBLO
BCH
ANC
BB
ECRL+ECRB
EDC
ECU
EDL
ECRL'
ECRB'
EP'
ECU'
EDC'
EDL'
1 cm
(d)
proximal
cranial
TBLO
BB
TBM
ECRL+ECRB
TBA
PT
FCUU
FCR
PL
ECRL'+ECRB'
PL'
FCR'
FDP'
1 cm

FIGURE II.22.5 **Distal forelimb muscles. (a) In dorsal view.** The EDL, *M. extensor digitorum lateralis* inserts with three tendons on the distal phalanges of digits III through V. The EDC, *M. extensor digitorum communis* inserts with four tendons on the distal phalanges of digits II through V. The EI, M. extensor digiti I and II inserts with two tendons on the distal phalanges of digits I and II. **(b, c) In ventral view.** The PL, *M. palmaris longus* inserts with four tendons on the distal phalanges of digits II through V. The FDP, *M. flexor digitorum profundus* (four muscle bellies) inserts with five tendons on the distal phalanges of digits I through V.

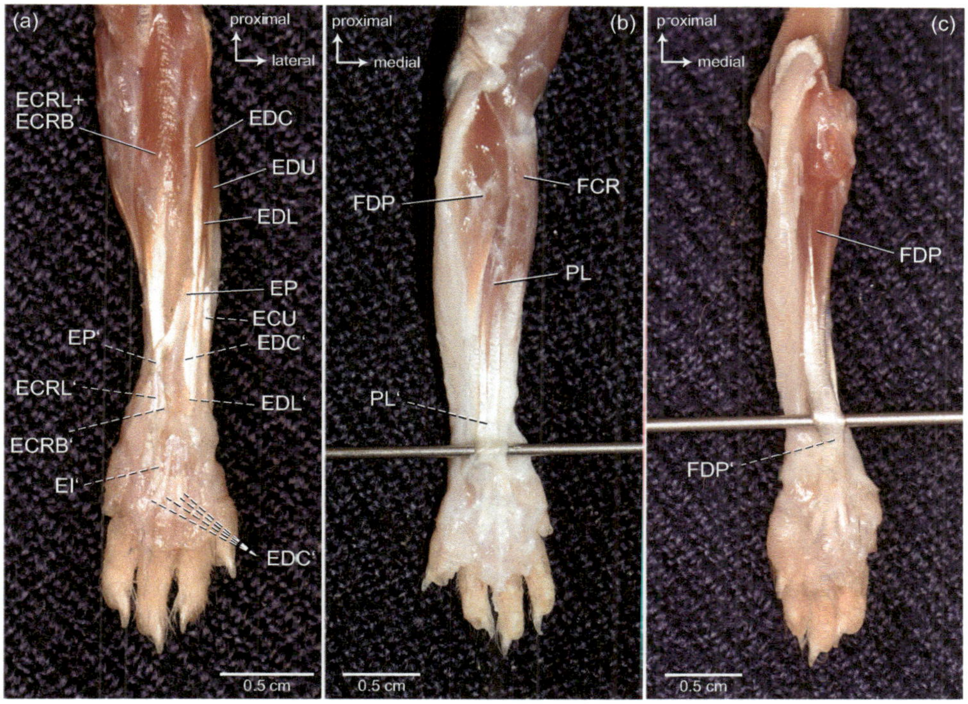
(a)
proximal
lateral
ECRL+
ECRB
EDC
EDU
EDL
EP
ECU
EP'
EDC'
ECRL'
EDL'
ECRB'
EI'
EDC'
0.5 cm
(b)
proximal
medial
FDP
FCR
PL
PL'
0.5 cm
(c)
proximal
medial
FDP
FDP'
0.5 cm

FIGURE II.22.6 Deep rotator muscles. (a) Distal rotator muscles in dorsal view. (b) Proximal rotator muscles in ventral view. The tendon of the BB, *M. biceps brachii* inserts into the radius. The tendon of the BCH, *M. brachioradialis* inserts into the coronoid process into the ulna.

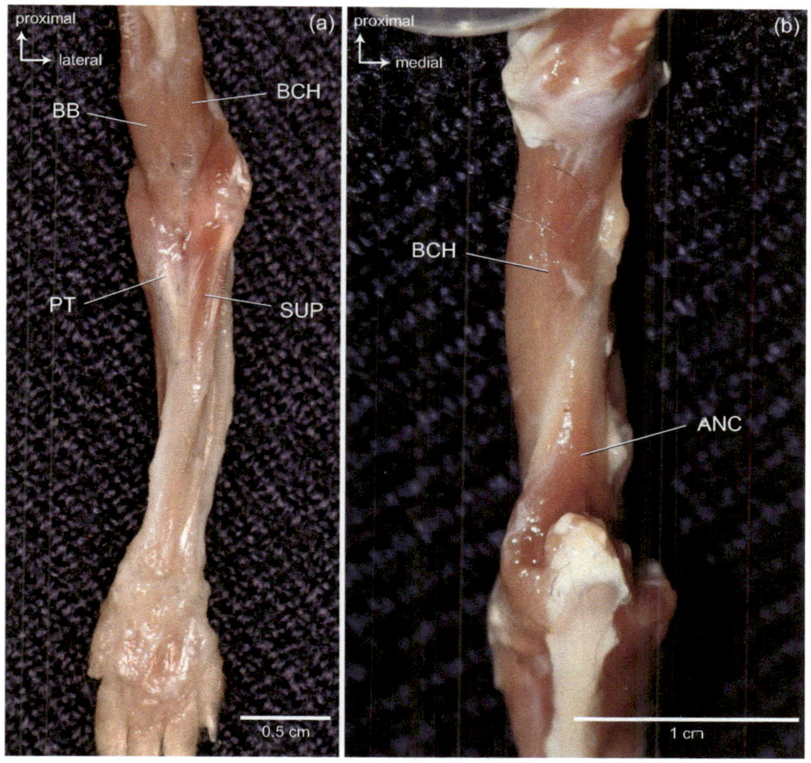
proximal
lateral
(a)
BCH
BB
PT
SUP
0.5 cm
proximal
medial
(b)
BCH
ANC
1 cm

Meriones unguiculatus (Mongolian gerbil)

Classification:	Rodentia, Myomorpha, Muridae
Mean body mass:	0.09 kg
Habitat:	Desert, savanna
Locomotor ecology:	Saltatorial
Peculiarities:	Short forelimbs and long hind limbs

FIGURE II.23.1 Superficial muscles of forelimb and hind limb in lateral view. (a) Without labels. (b) With labels and muscle outlines. In the neck and shoulder region, the *Mm. trapezii* are divided into two parts (CT, *M. clavotrapezius* together with AT, *M. arcomiotrapezius*; and ST, *M. spinotrapezius*). The EPI, *M. epitrochlearis* and the BCR, *M. brachioradialis* are absent. In the distal hind limb, the four tendons of the EDLO, *M. extensor digitorum longus* insert on the distal phalanges of digits II through V. Note that the retinacula (ligaments) spanning the wrist and ankle joints have been partially removed.

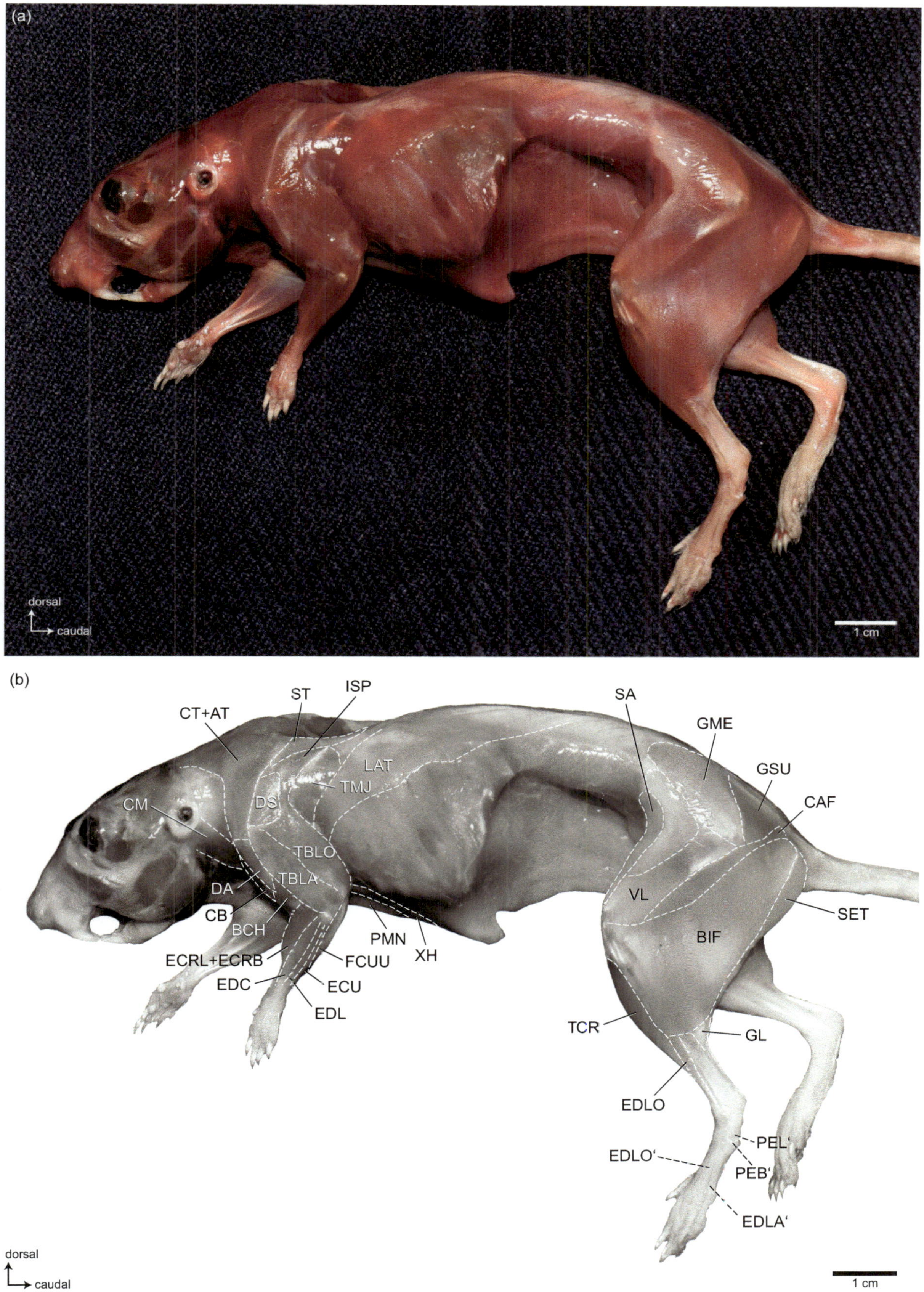
(a)
dorsal
caudal
1 cm
(b)
CT+AT
ST
ISP
SA
GME
GSU
CAF
LAT
TMJ
DS
CM
TBLO
TBLA
DA
CB
BCH
ECRL+ECRB
PMN
XH
FCUU
ECU
EDC
EDL
VL
BIF
SET
TCR
GL
EDLO
PEL'
EDLO'
PEB'
EDLA'
dorsal
caudal
1 cm

FIGURE II.23.2 **Pectoral muscles in ventral view.** The clavicle is present. Four pectoral muscles are identified (the PMJ, *M. pectoralis major*; the PAB, *M. pectoantebrachialis*; the PMN, *M. pectoralis minor*; and the XH, *M. xiphihumeralis*).

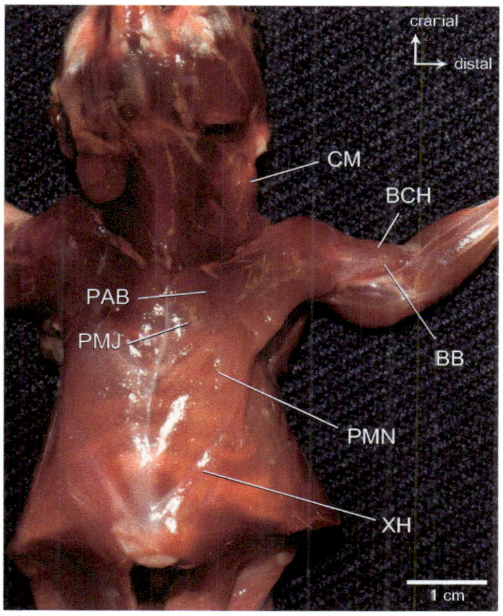
cranial
distal
CM
BCH
PAB
PMJ
BB
PMN
XH
1 cm

FIGURE II.23.3 Deep neck and shoulder muscles. (a, c) In craniolateral view. (b) In dorsal view. In the neck region, two *Mm. rhomboidei* are present (RCR, *M. rhomboideus cervicis* and RCP, *M. rhomboideus capitis*). The RCR, *M. rhomboideus cervicis* reaches the occiput. The SC, *M. subclavius* has been removed.

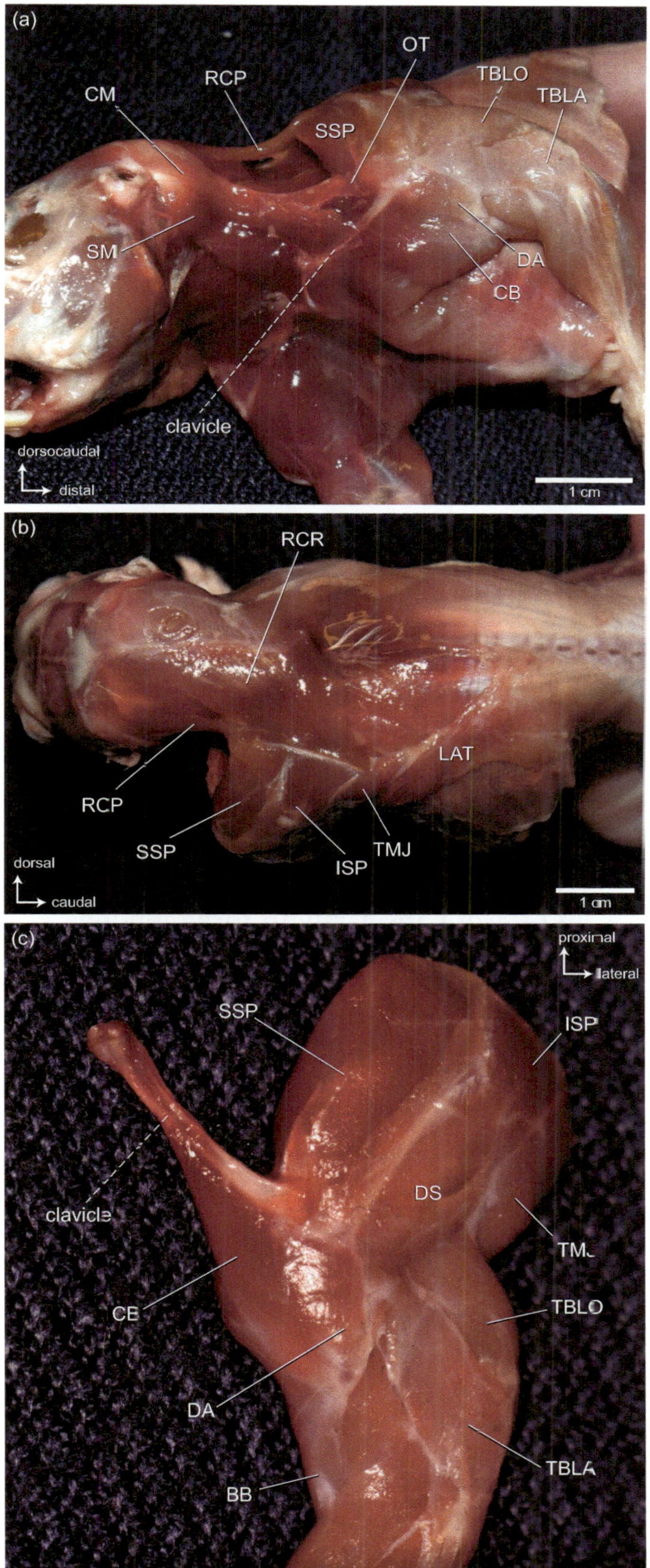
(a)
OT
RCP
TBLO
CM
TBLA
SSP
SM
DA
CB
clavicle
dorsocaudal
distal
1 cm
(b)
RCR
LAT
RCP
SSP
TMJ
ISP
dorsal
caudal
1 cm
(c)
proximal
lateral
SSP
ISP
clavicle
DS
TMJ
CB
TBLO
DA
TBLA
BB

FIGURE II.23.4 Intrinsic forelimb muscles in (a) proximolateral view and (b) proximomedial view. The *Mm. deltoidei* (DS, *M. spinodeltoideus* and DA, *M. acromiodeltoideus)* and the TMN, *M. teres minor* have been removed. The BB, *M. biceps brachii* has two distinct origins. **Intrinsic forelimb muscles in (c) lateral view and (d) medial view.** The EDL, *M. extensor digitorum lateralis* inserts with three tendons on the distal phalanges of digits III through V. The PL, *M. palmaris longus* inserts with four tendons on the distal phalanges of digits II through V.

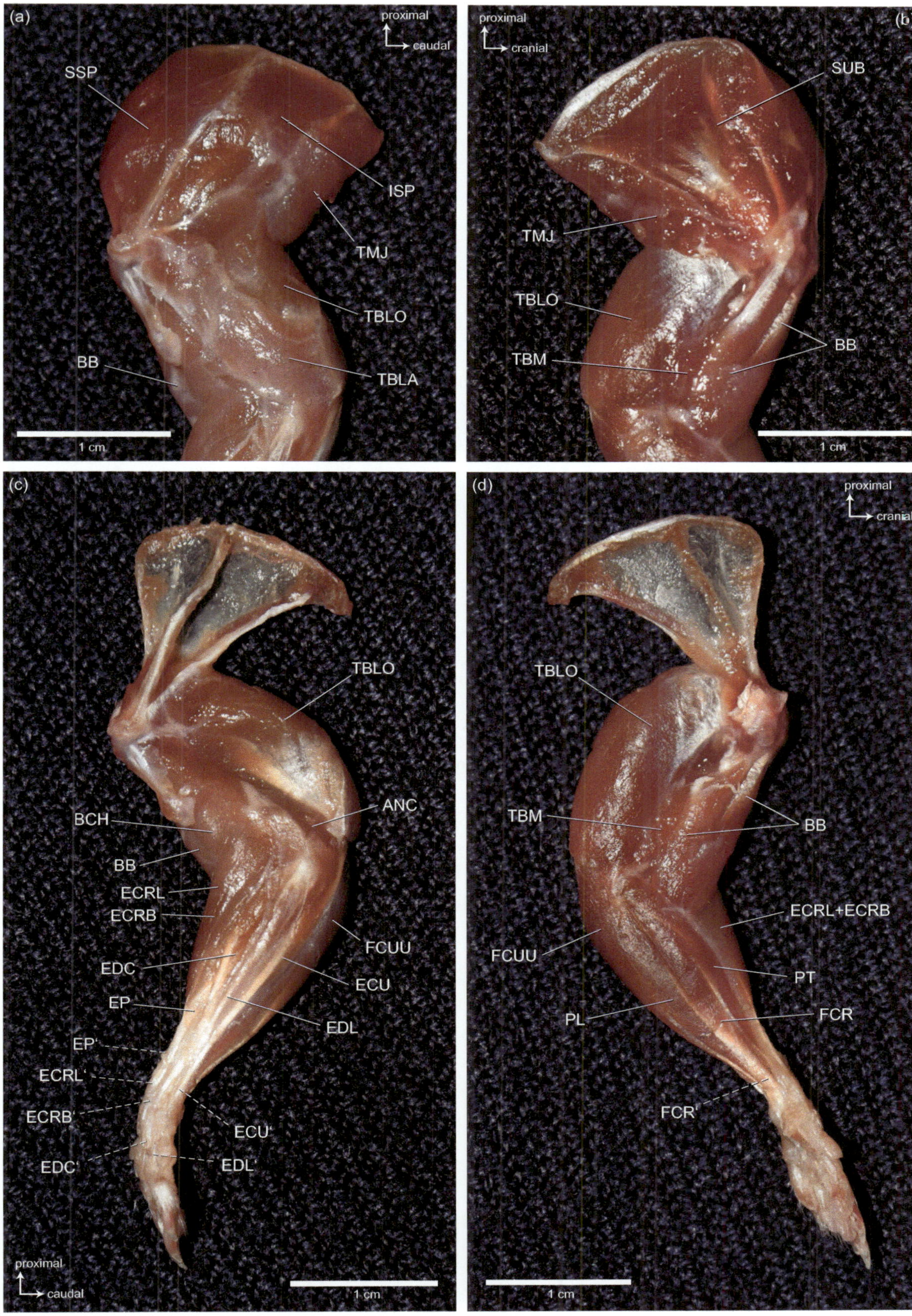
(a)
proximal
caudal
SSP
ISP
TMJ
TBLO
BB
TBLA
1 cm
(b)
proximal
cranial
SUB
TMJ
TBLO
BB
TBM
1 cm
(c)
TBLO
BCH
ANC
BB
ECRL
ECRB
FCUU
EDC
ECU
EP
EDL
EP'
ECRL'
ECRB'
ECU'
EDC'
EDL'
proximal
caudal
1 cm
(d)
proximal
cranial
TBLO
TBM
BB
ECRL+ECRB
FCUU
PT
PL
FCR
FCR'
1 cm

FIGURE II.23.5 **Distal forelimb muscles. (a) In dorsal view.** The EDC, *M. extensor digitorum communis* inserts with four tendons on the distal phalanges of digits II through V. The EI, *M. extensor digiti I and II* inserts with two tendons on the distal phalanges of digits I and II. **(b, c) In ventral view.** The FDP, *M. flexor digitorum profundus* (four muscle bellies) inserts with four tendons on the distal phalanges of digits II through V.

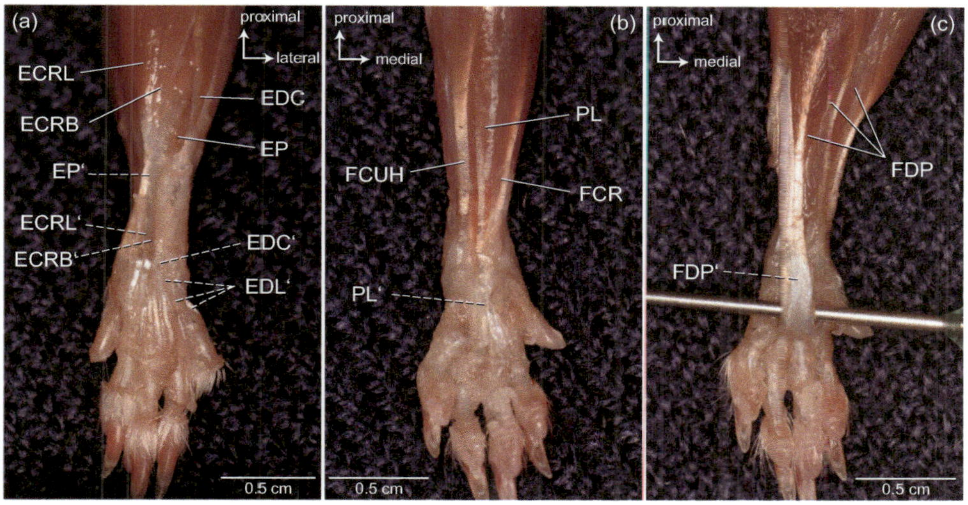
(a)
proximal
lateral
ECRL
ECRB
EDC
EP
EP'
ECRL'
ECRB'
EDC'
EDL'
0.5 cm
(b)
proximal
medial
PL
FCUH
FCR
PL'
0.5 cm
(c)
proximal
medial
FDP
FDP'
0.5 cm

FIGURE II.23.6 Deep rotator muscles. (a) Distal rotator muscles in dorsal view. (b) Proximal rotator muscles in ventral view. The tendon of the BB, *M. biceps brachii* inserts into the radius. The tendon of the BCH, *M. brachioradialis* inserts into the coronoid process into the ulna.

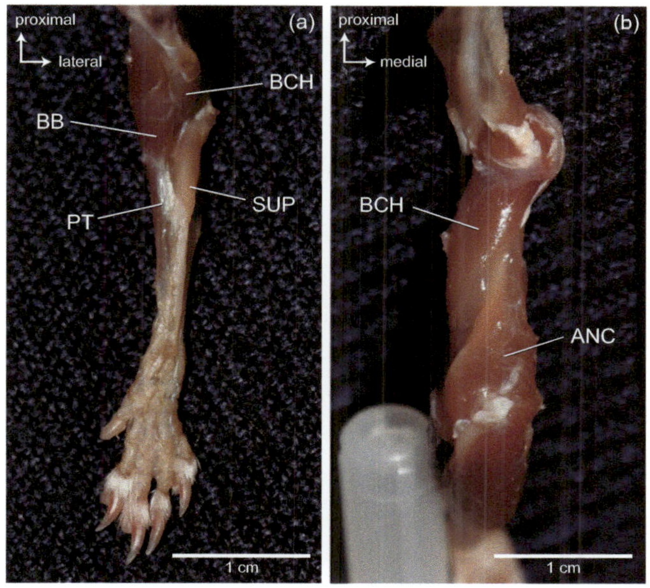
proximal
lateral
(a)
BCH
BB
SUP
PT
1 cm
proximal
medial
(b)
BCH
ANC
1 cm

Rattus norvegicus (Norway rat)

Classification:	Rodentia, Myomorpha, Muridae
Mean body mass:	0.3 kg
Habitat:	Variety of habitats (Native to forests and brushy areas)
Locomotor ecology:	Ambulatorial – scansorial
Peculiarities:	Use their tail as an aid in climbing and jumping

FIGURE II.24.1 Superficial muscles of forelimb and hind limb in lateral view. (a) Without labels. (b) With labels and muscle outlines. In the neck and shoulder region, the *Mm. trapezii* are divided into two parts (CT, *M. clavotrapezius* together with AT, *M. arcomiotrapezius*; and ST, *M. spinotrapezius*). The EPI, *M. epitrochlearis* and the BCR, *M. brachioradialis* are absent. In the distal hind limb, the four tendons of the EDLO, *M. extensor digitorum longus* insert on the distal phalanges of digits II through V. Note that the retinacula (ligaments) spanning the wrist and ankle joints have been partially removed.

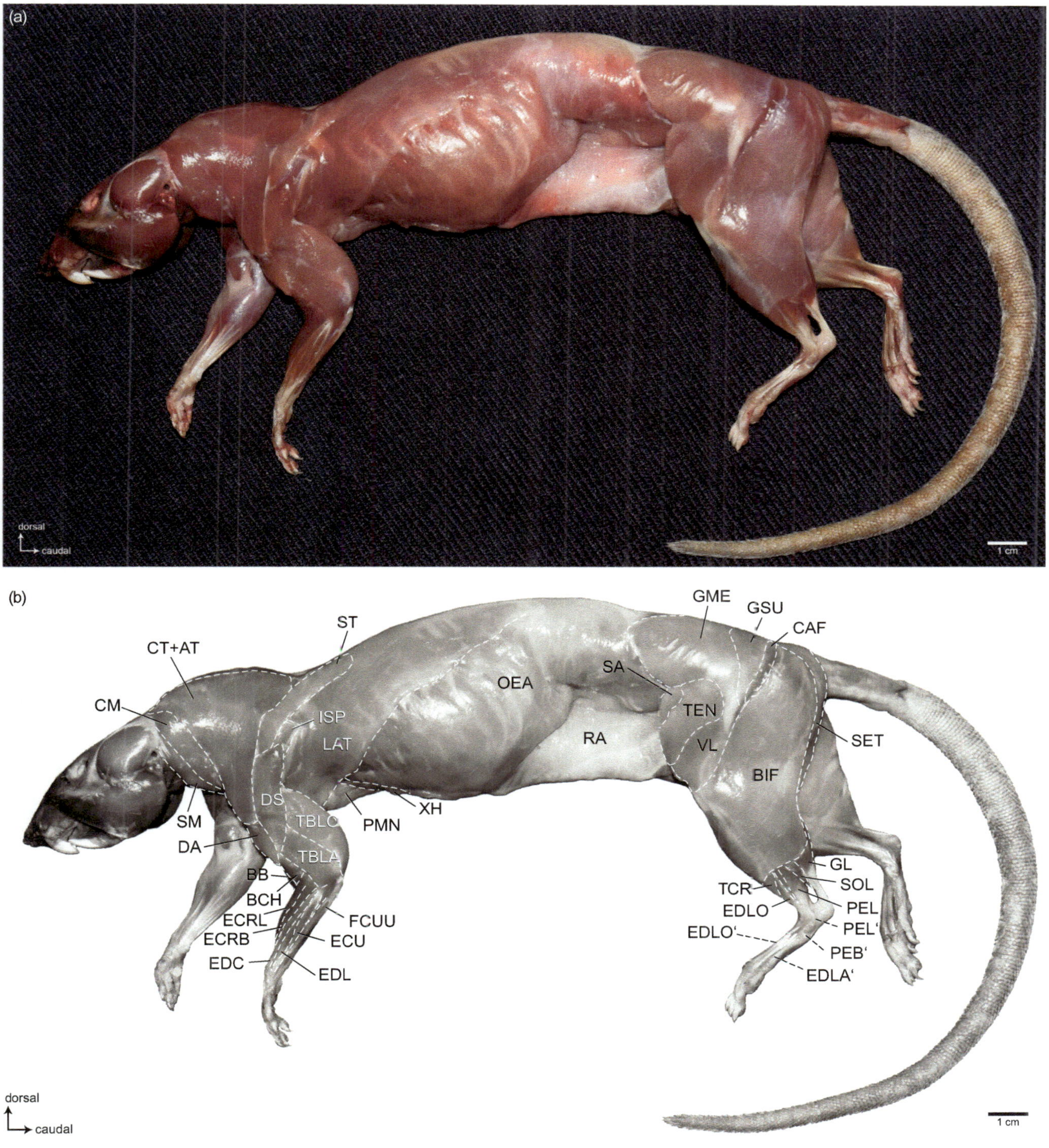
(a)
dorsal
caudal
1 cm
(b)
ST
CT+AT
CM
GME
GSU
CAF
SA
OEA
ISP
LAT
TEN
RA
VL
SET
BIF
DS
SM
TBLC
PMN
XH
DA
TBLA
BB
BCH
ECRL
ECRB
FCUU
ECU
EDC
EDL
GL
TCR
SOL
EDLO
PEL
EDLO'
PEL'
PEB'
EDLA'
dorsal
caudal
1 cm

FIGURE II.24.2 **Pectoral muscles in ventral view.** The clavicle is present. Three pectoral muscles are identified (the PMJ, *M. pectoralis major* together with the PMN, *M. pectoralis minor*; the PAB, *M. pectoantebrachialis*; and the XH, *M. xiphihumeralis*).

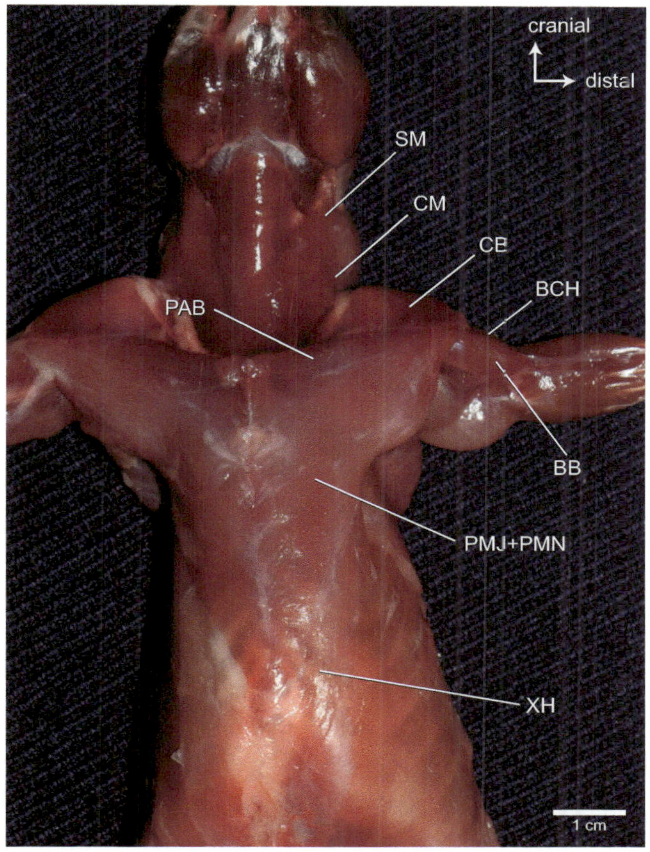
cranial
distal
SM
CM
CB
BCH
PAB
BB
PMJ+PMN
XH
1 cm

FIGURE II.24.3 Deep neck and shoulder muscles. (a, c) In craniolateral view. (b) In lateral view. In the neck region, two *Mm. rhomboidei* are present (RCR, *M. rhomboideus cervicis*; and RCP, *M. rhomboideus capitis*). The RCR, *M. rhomboideus cervicis* does not reach the occiput.

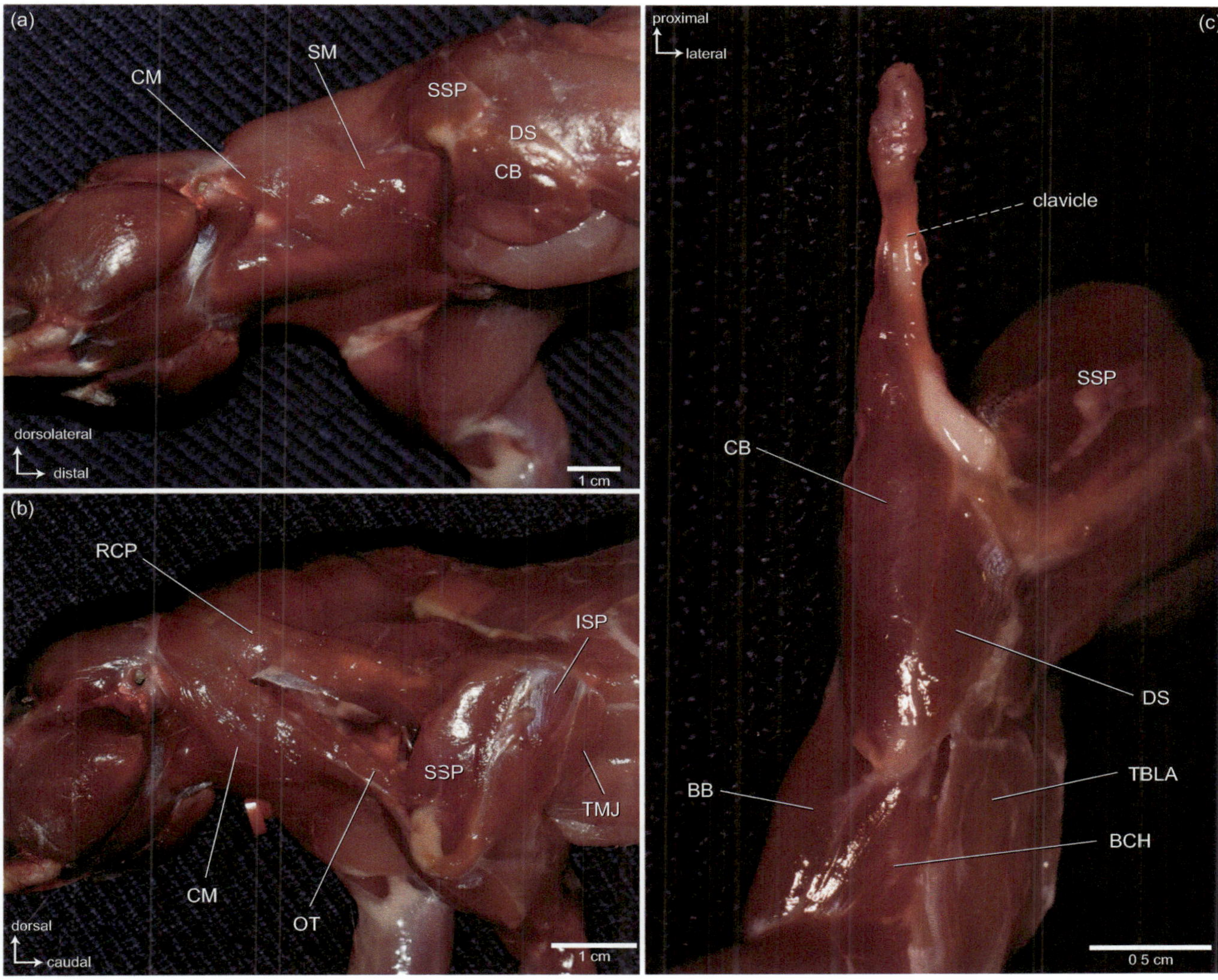
(a)
CM
SM
SSP
DS
CB
dorsolateral
distal
1 cm
(b)
RCP
ISP
SSP
TMJ
CM
OT
dorsal
caudal
1 cm
proximal
lateral
(c)
clavicle
SSP
CB
DS
TBLA
BB
BCH
0 5 cm

FIGURE II.24.4 **Intrinsic forelimb muscles in (a) proximolateral view and (b) proximomedial view**. The BB, *M. biceps brachii* has one distinct origin. **Intrinsic forelimb muscles in (c, e) lateral view and (d, f) medial view**.

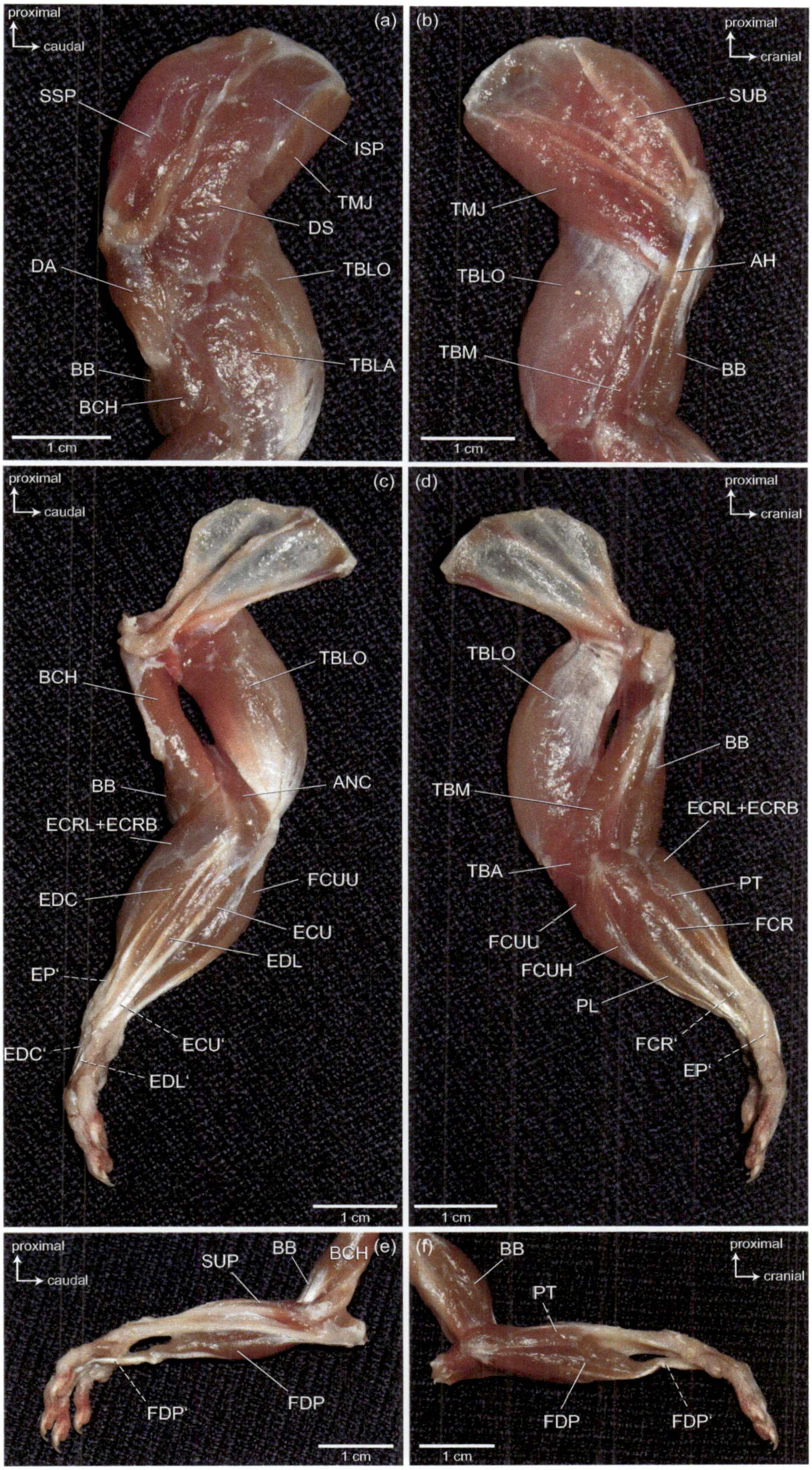

proximal
caudal
(a)
SSP
ISP
TMJ
DS
DA
TBLO
BB
TBLA
BCH
1 cm
(b)
proximal
cranial
SUB
TMJ
AH
TBLO
TBM
BB
1 cm
(c)
proximal
caudal
TBLO
BCH
BB
ANC
ECRL+ECRB
FCUU
EDC
ECU
EDL
EP'
ECU'
EDC'
EDL'
1 cm
(d)
proximal
cranial
TBLO
BB
TBM
ECRL+ECRB
TBA
PT
FCR
FCUU
FCUH
PL
FCR'
EP'
1 cm
(e)
proximal
caudal
SUP
BB
BCH
FDP'
FDP
1 cm
(f)
proximal
cranial
BB
PT
FDP
FDP'
1 cm

FIGURE II.24.5 **Distal forelimb muscles. (a) In dorsal view.** The EDL, *M. extensor digitorum lateralis* inserts with three tendons on the distal phalanges of digits III through V. The EDC, *M. extensor digitorum communis* inserts with four tendons on the distal phalanges of digits II through V. The EI, *M. extensor digiti I and II* inserts with two tendons on the distal phalanges of digits I and II. **(b, c) In ventral view.** The PL, *M. palmaris longus* inserts with four tendons on the distal phalanges of digits II through V. The FDP, *M. flexor digitorum profundus* (four muscle bellies) inserts with four tendons on the distal phalanges of digits II through V.

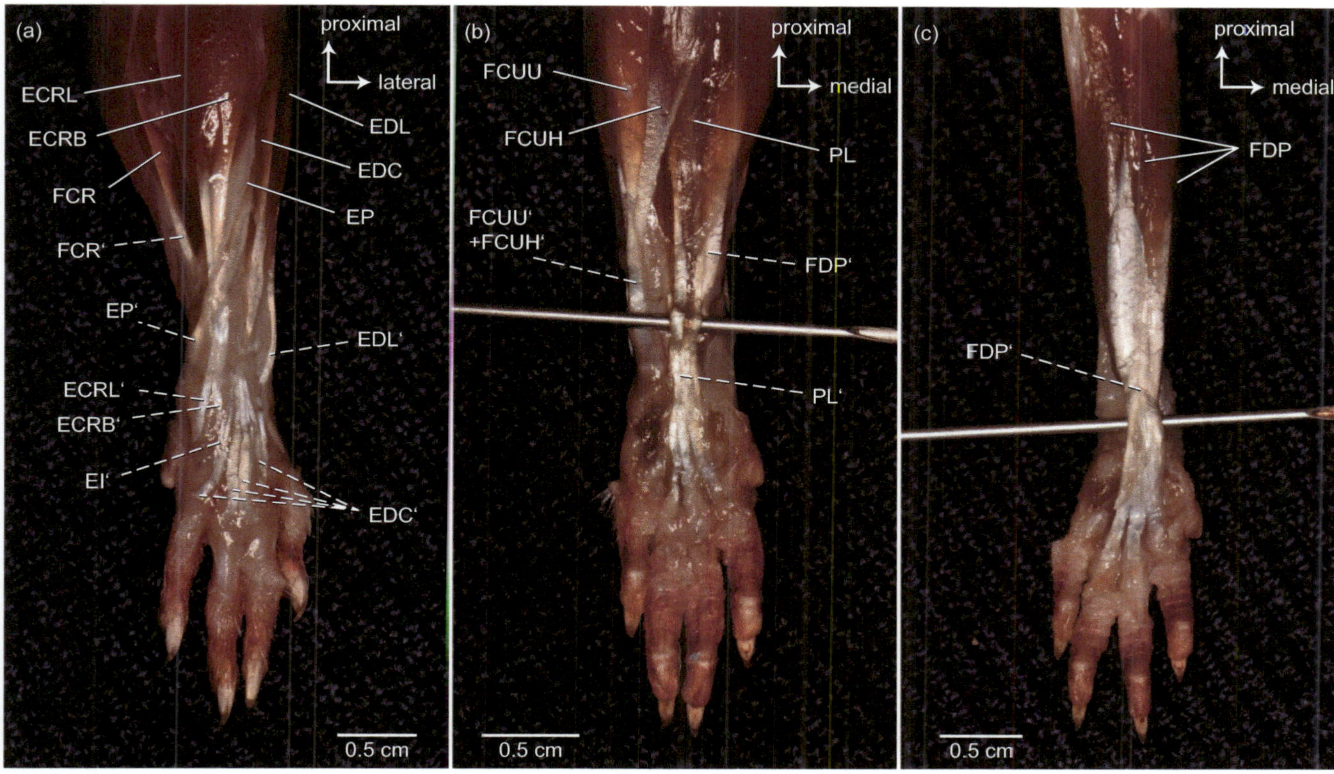
(a)
proximal
lateral
ECRL
ECRB
FCR
FCR'
EDL
EDC
EP
EP'
EDL'
ECRL'
ECRB'
EI'
EDC'
0.5 cm
(b)
proximal
medial
FCUU
FCUH
PL
FCUU'
+FCUH'
FDP'
PL'
0.5 cm
(c)
proximal
medial
FDP
FDP'
0.5 cm

FIGURE II.24.6 Deep rotator muscles. Deep rotator muscles. (a) Distal rotator muscles in dorsal view. (b) Proximal rotator muscles in ventral view. The tendon of the BB, *M. biceps brachii* inserts into the radius. The tendon of the BCH, *M. brachioradialis* inserts into the coronoid process into the ulna.

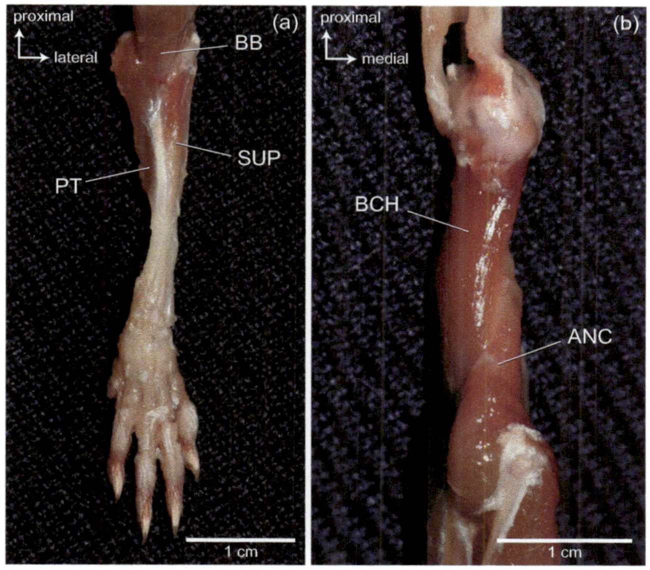
proximal
lateral
(a)
BB
SUP
PT
1 cm
proximal
medial
(b)
BCH
ANC
1 cm

Elephantulus brachyrhynchus (Short-snouted elephant shrew)

Classification:	Macroscelidea
Mean body mass:	0.05 kg
Habitat:	Savanna, grassland
Locomotor ecology:	Cursorial – saltatorial
Peculiarities:	Short forelimbs and long hind limbs

FIGURE II.25.1 Superficial muscles of forelimb and hind limb in lateral view. (a) Without labels. (b) With labels and muscle outlines. In the neck and shoulder region, the *Mm. trapezii* are clearly divided into three parts (CT, *M. clavotrapezius*; AT, *M. arcomiotrapezius*; and ST, *M. spinotrapezius*). Note the length of the ST, *M. spinotrapezius*. The EPI, *M. epitrochlearis* and the BCR, *M. brachioradialis* are absent. The clavicle is present. The SV, *M. serratus ventralis* originates via distinct segments from the thoracic ribs and inserts into the craniomedial surface and vertebral border of the scapula. The OEA, *M. obliquus externus abdominis* originates from the thoracic ribs and inserts into the linea alba at the midline and on the pubis near the symphysis. In the distal hind limb, the four tendons of the EDLO, *M. extensor digitorum longus* insert on the distal phalanges of digits II through V. Note that the retinacula (ligaments) spanning the wrist and ankle joints have been partially removed.

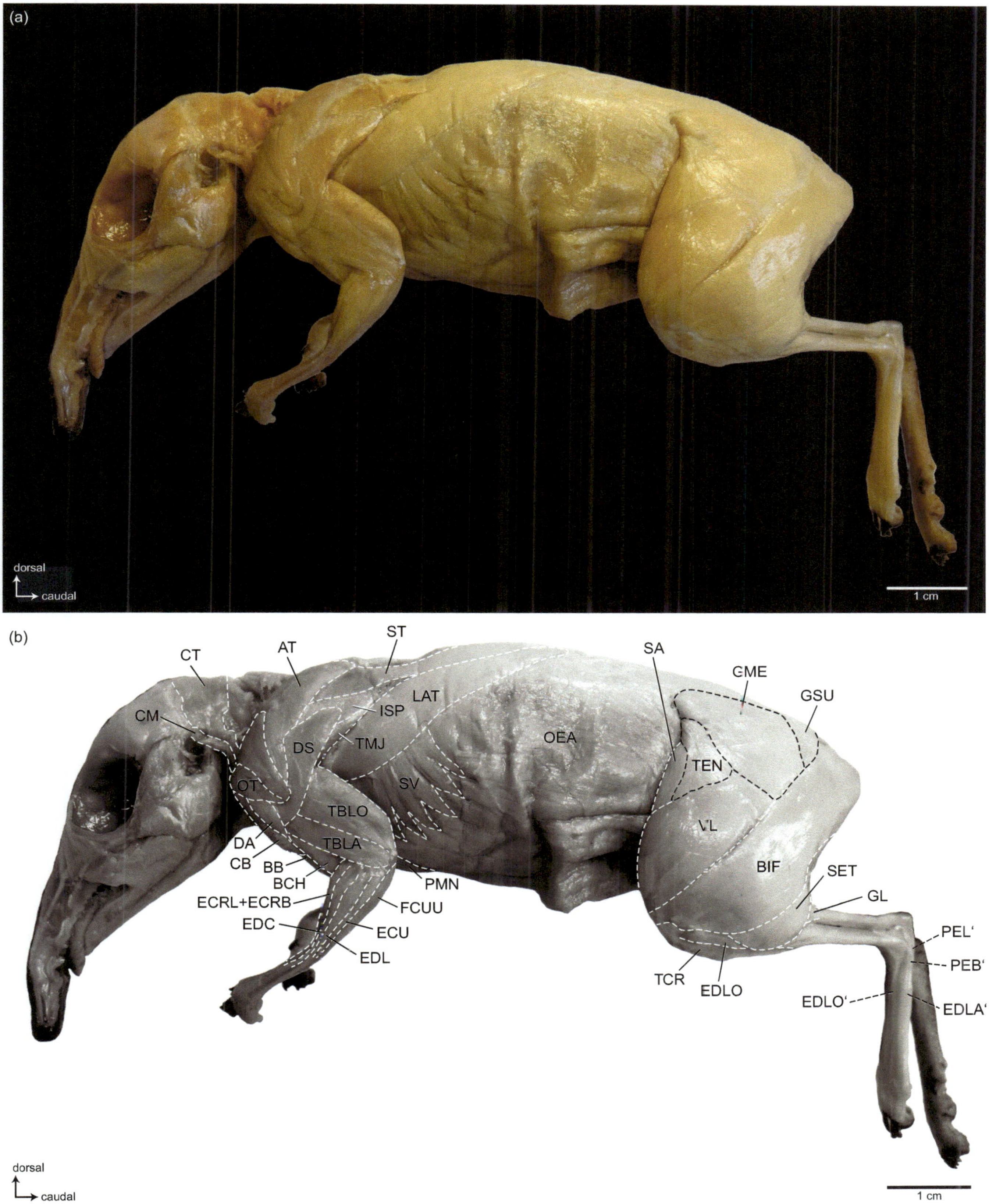
(a)
dorsal
caudal
1 cm
(b)
CT
AT
ST
SA
CME
GSU
CM
ISP
LAT
OEA
DS
TMJ
TEN
OT
SV
TBLO
VL
DA
TBLA
CB
BB
BCH
PMN
BIF
SET
GL
ECRL+ECRB
FCUU
EDC
ECU
EDL
PEL'
PEB'
TCR
EDLO
EDLO'
EDLA'
dorsal
caudal
1 cm

FIGURE II.25.2 **Pectoral muscles.** The XH, *M. xiphihumeralis* is not clearly developed as a separate muscle, but part of the PMN, *M. pectoralis minor.* The clavicle is present.

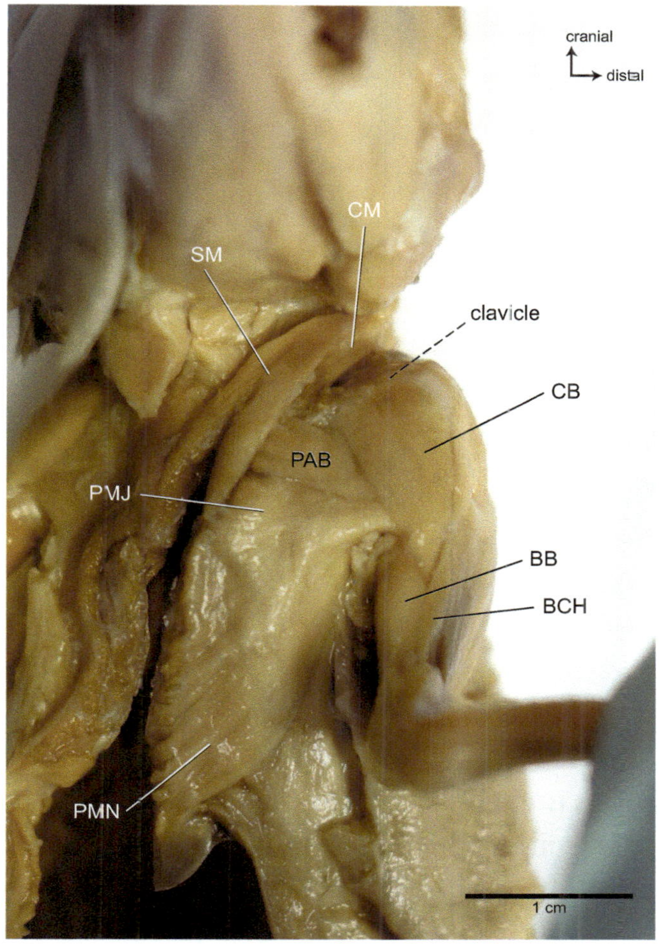
cranial
distal
CM
SM
clavicle
CB
PAB
PMJ
BB
BCH
PMN
1 cm

FIGURE II.25.3 Deep neck and shoulder muscles in dorsolateral view. In the neck region, two *Mm. rhomboidei* are present (RCR, *M. rhomboideus cervicis* and RT, *M. rhomboideus thoracis*). The RCP, *M. rhomboideus profundus* and the RPR, *M. rhomboideus capitis* appear to be absent. The RCR, *M. rhomboideus cervicis* does not reach the occiput.

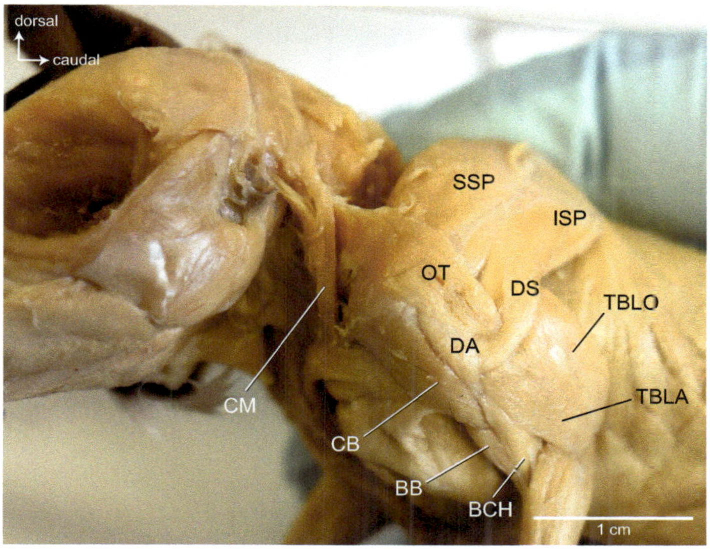
dorsal
caudal
SSP
ISP
OT
DS
TBLO
DA
TBLA
CM
CB
BB
BCH
1 cm

FIGURE II.25.4 Intrinsic forelimb muscles in (a) lateral view and (b) medial view. The SC, *M. subclavius* has been removed. The BB, *M. biceps brachii* has two distinct origins.

(a)
dorsal
caudal
TBLO
ANC
clavicle
BB
BCH
ECRL+ECRB
EDC
FCUU
ECU
ECRL'+ECRB'
EDL
1 cm
(b)
dorsal
cranial
TBLO
TBM
BB
TBA
ECRL+ECRB
PT
FCUU
+FCUH
1 cm

FIGURE II.25.5 Distal forelimb muscles. In dorsal view. The EDL, *M. extensor digitorum lateralis* inserts with two tendons on the distal phalanges of digits IV and V. The EDC, *M. extensor digitorum communis* inserts with four tendons on the distal phalanges of digits II through V. The EI, *M. extensor digiti I and II* inserts with one tendon on the distal phalanx of digits I.

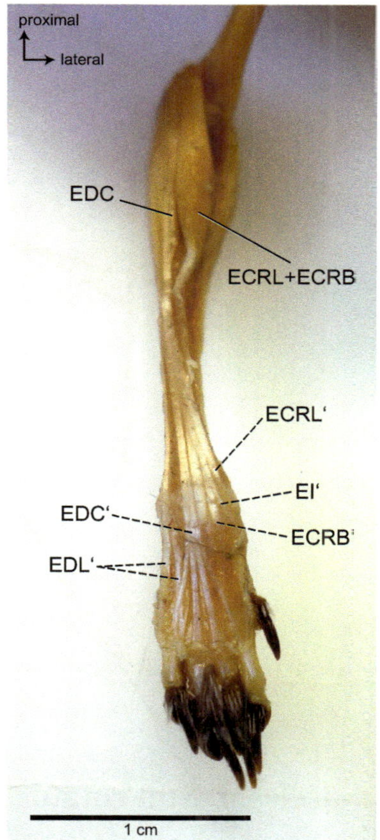
proximal
lateral
EDC
ECRL+ECRB
ECRL'
EI'
EDC'
ECRB'
EDL'
1 cm

FIGURE II.25.6 Deep rotator muscles. Distal rotator muscles in dorsal view.

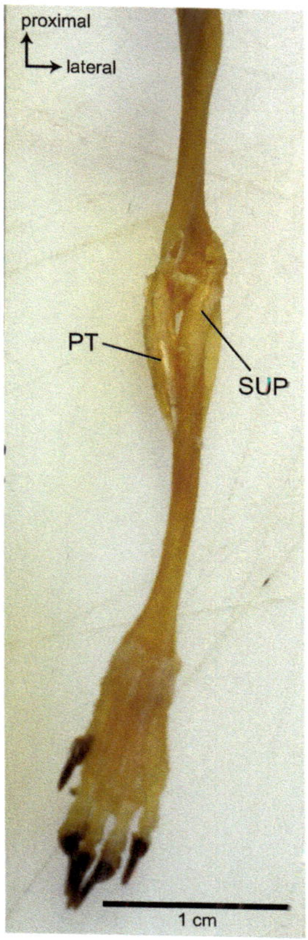
proximal
lateral
PT
SUP
1 cm

Bradypus tridactylus (Pale-throated three-toed sloth)

Classification:	Pilosa, Xenarthra, Bradypodidae
Mean body mass:	2.3–5.5 kg
Habitat:	Rainforest
Locomotor ecology:	Scansorial (Suspension)
Peculiarities:	Long neck (nine cervical vertebrae); hook-like hand with three digits; long, curved claws

FIGURE II.26.1 Superficial muscles of forelimb and hind limb in lateral view. (a) Without labels. (b) With labels and muscle outlines. In the neck and shoulder region, the three parts of the *Mm. trapezii* (CT, *M. clavotrapezius*; AT, *M. arcomiotrapezius*; and ST, *M. spinotrapezius*) are present, but not clearly separated in this specimen. In general, the muscular anatomy in sloths is variable, and many left-right asymmetries in the musculature have been identified. Note that the CT, *M. clavotrapezius* does not reach the occiput. A rudimentary clavicle is present. The EPI, *M. epitrochlearis* is absent. The BCR, *M. brachioradialis* is well developed.

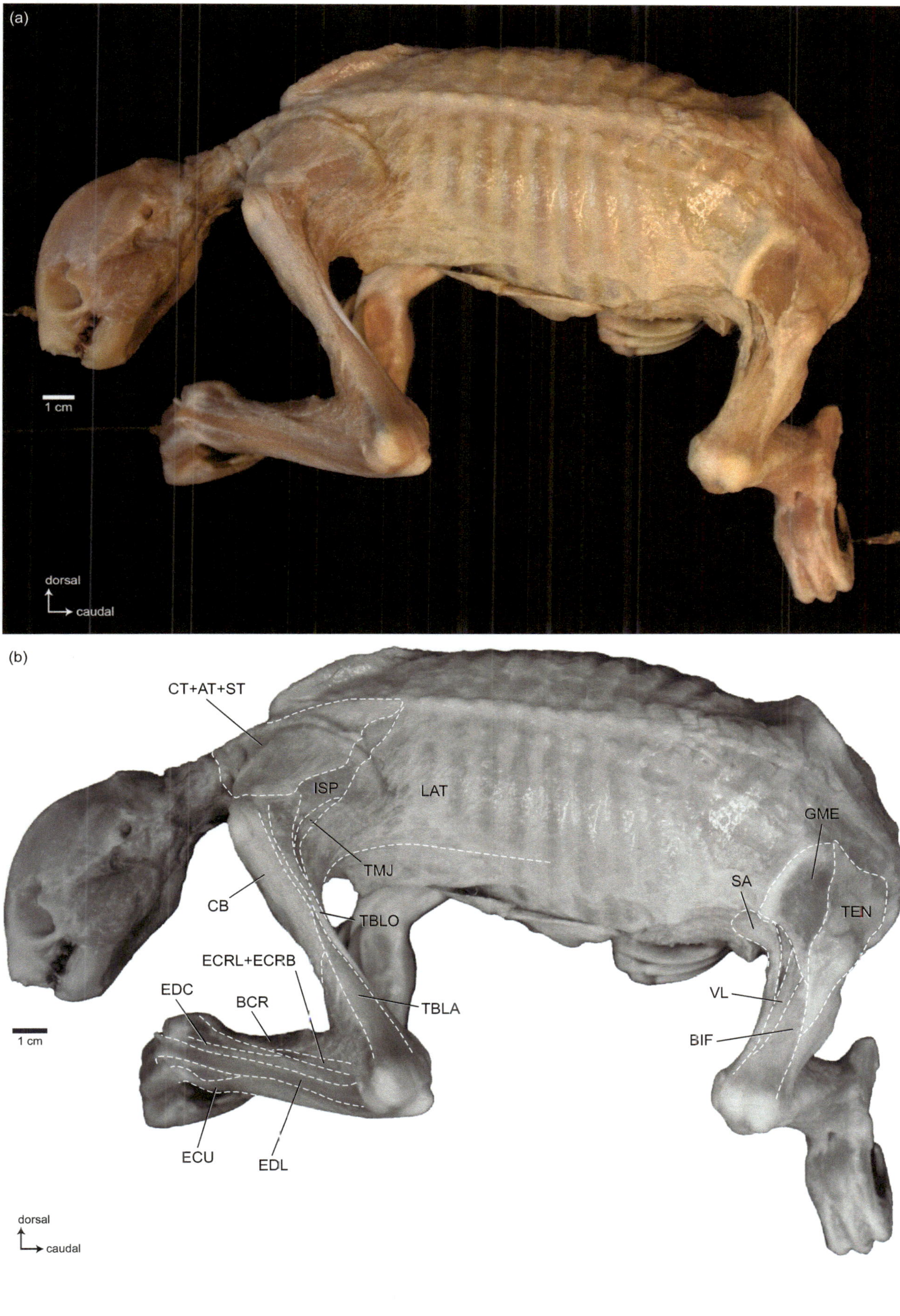
(a)
1 cm
dorsal
caudal
(b)
CT+AT+ST
ISP
LAT
TMJ
CB
TBLO
ECRL+ECRB
EDC
BCR
TBLA
ECU
EDL
GME
SA
TEN
VL
BIF
1 cm
dorsal
caudal

FIGURE II.26.2 **Deep neck and shoulder muscles in dorsal view. (a)** The three parts of the *Mm. trapezii* (CT, *M. clavotrapezius*; AT, *M. arcomiotrapezius*; and ST, *M. spinotrapezius*) are partially separated in this specimen. Note that the CT, *M. clavotrapezius* does not reach the occiput. **(b)** The RT, *M. rhomboideus thoracis* is not clearly developed as a separate muscle, but part of the RCR, *M. rhomboideus cervicis*. The RCR, *M. rhomboideus cervicis* does not reach the occiput. The RCP, *M. rhomboideus capitis* and the RPR, *M. rhomboideus profundus* are absent.

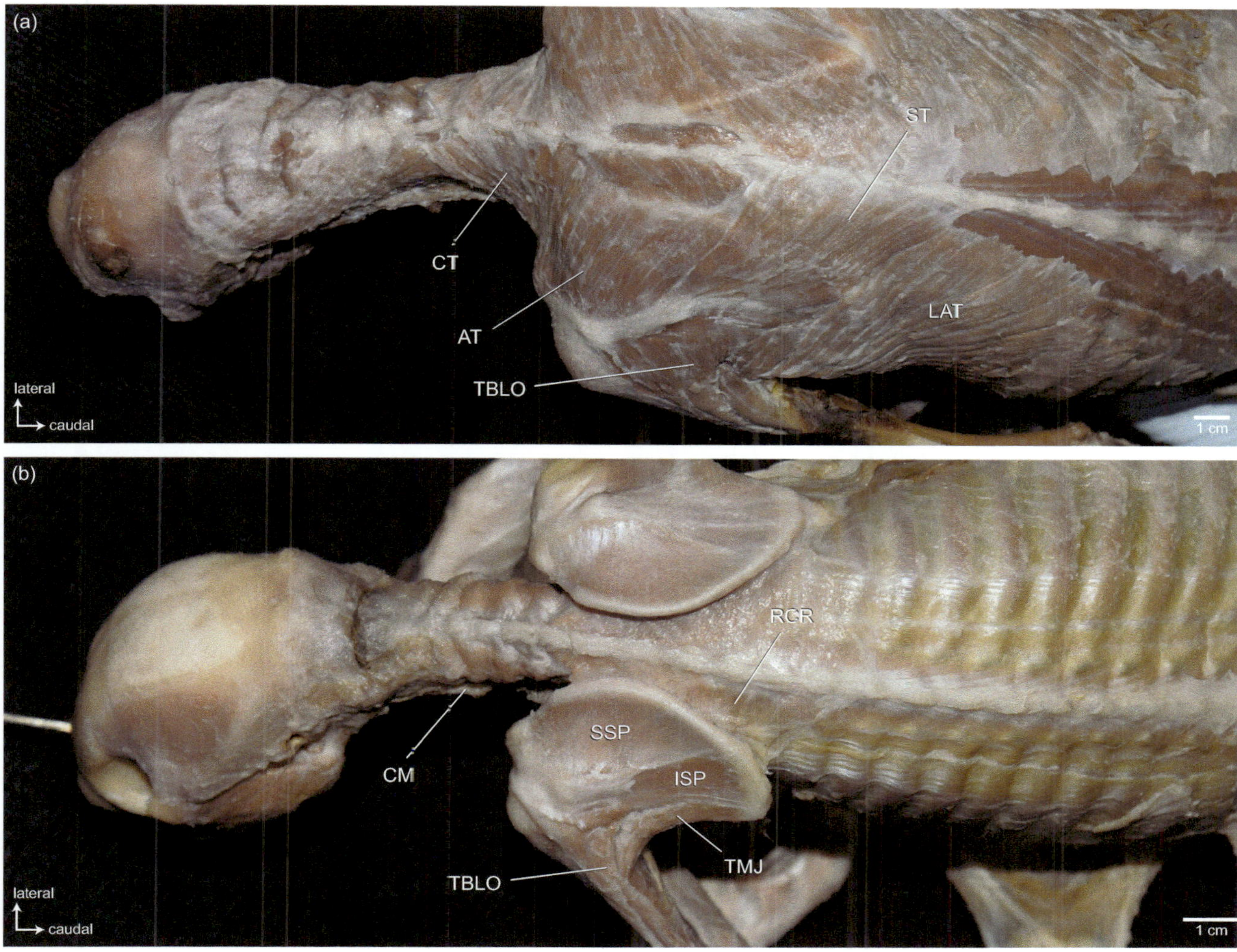
(a)
ST
CT
AT
LAT
TBLO
lateral
caudal
1 cm
(b)
RCR
SSP
CM
ISP
TMJ
TBLO
lateral
caudal
1 cm

FIGURE II.26.3 Intrinsic forelimb muscles in (a) lateral view and (b) medial view.

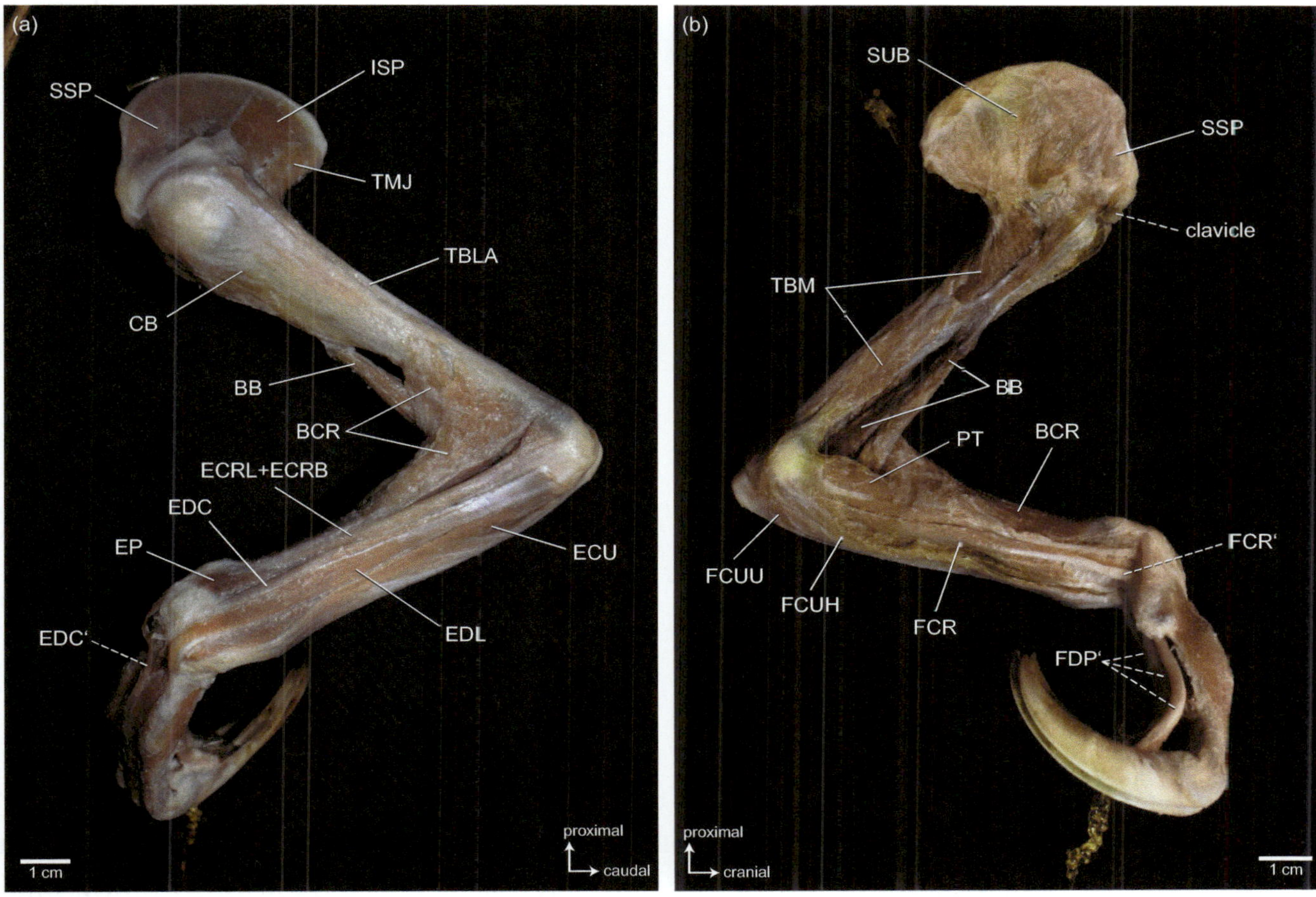

(a)
SSP
ISP
TMJ
TBLA
CB
BB
BCR
ECRL+ECRB
EDC
EP
ECU
EDC'
EDL
1 cm
proximal
caudal
(b)
SUB
SSP
clavicle
TBM
BB
PT
BCR
FCR'
FCUU
FCUH
FCR
FDP'
proximal
cranial
1 cm

FIGURE II.26.4 **Distal forelimb muscles. (a, b) In dorsal view.** The EDC, *M. extensor digitorum communis* inserts with three tendons on the distal phalanges of digits II through IV. The EDL, *M. extensor digitorum lateralis* inserts with one tendon on the proximal phalanx of digit IV. Note that the fleshy part extends very distally. The ECRL, *M. extensor carpi radialis longus* and the ECRB, *M. extensor carpi radialis brevis* are a single muscle and insert with one tendon on the radial carpal bones. Note the fleshy part of between the two tendons of the EI, *M. extensor digiti I and II.* **(c) In ventral view.** The FDP, *M. flexor digitorum profundus* inserts with three tendons on the distal phalanges of digits II through IV.

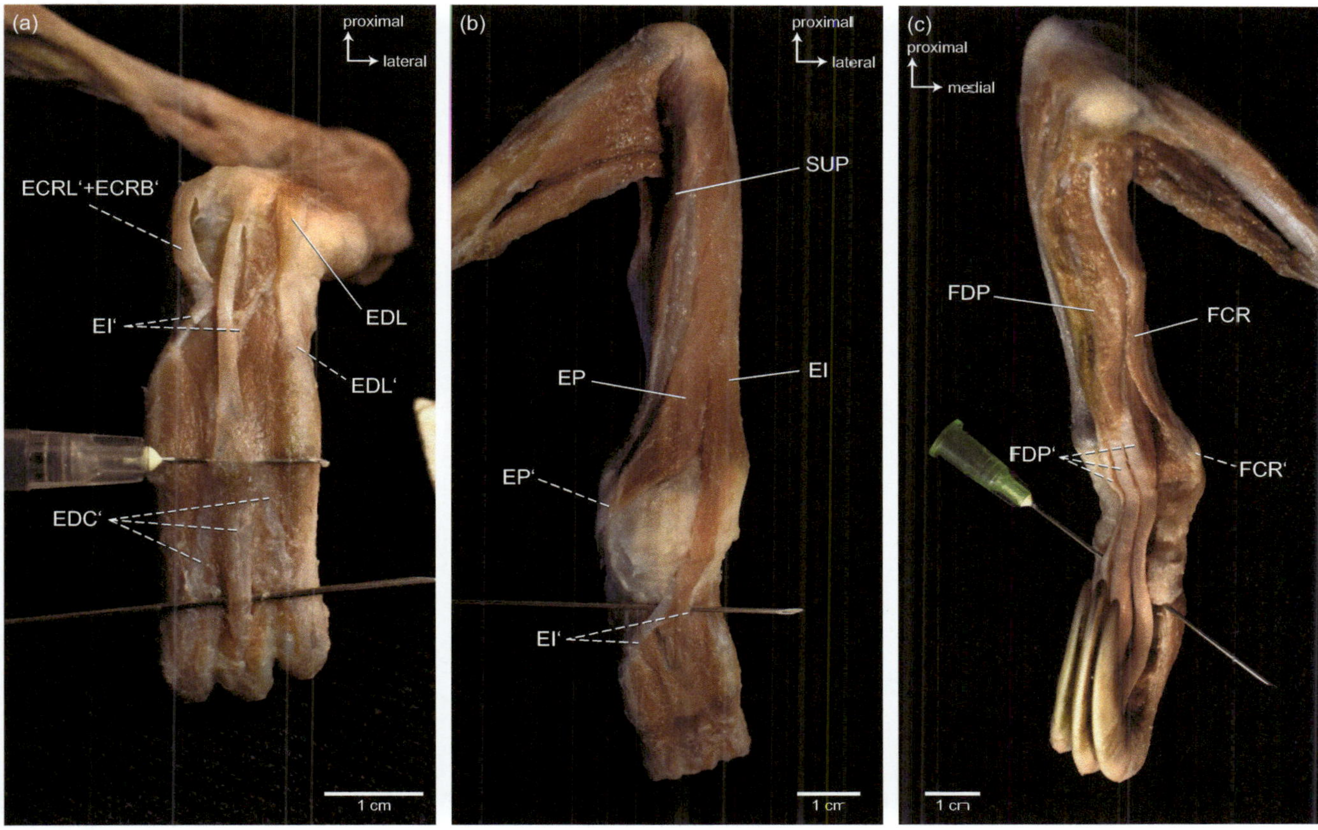
(a)
proximal
lateral
ECRL'+ECRB'
EDL
EI'
EDL'
EDC'
1 cm
(b)
proximal
lateral
SUP
EP
EI
EP'
EI'
1 cm
(c)
proximal
medial
FDP
FCR
FDP'
FCR'
1 cm

FIGURE II.26.5 **Deep rotator muscles. (a) Distal rotator muscles in dorsal view.** Note the short SUP, *M. supinator.* **(b) Distal rotator muscles in ventral view.** The BB, *M. biceps brachii* is distally separated into two parts. They insert into the radius and ulna. The tendon of the BCH, *M. brachioradialis* inserts into radius. **(c) Proximal rotator muscles in ventral view.**

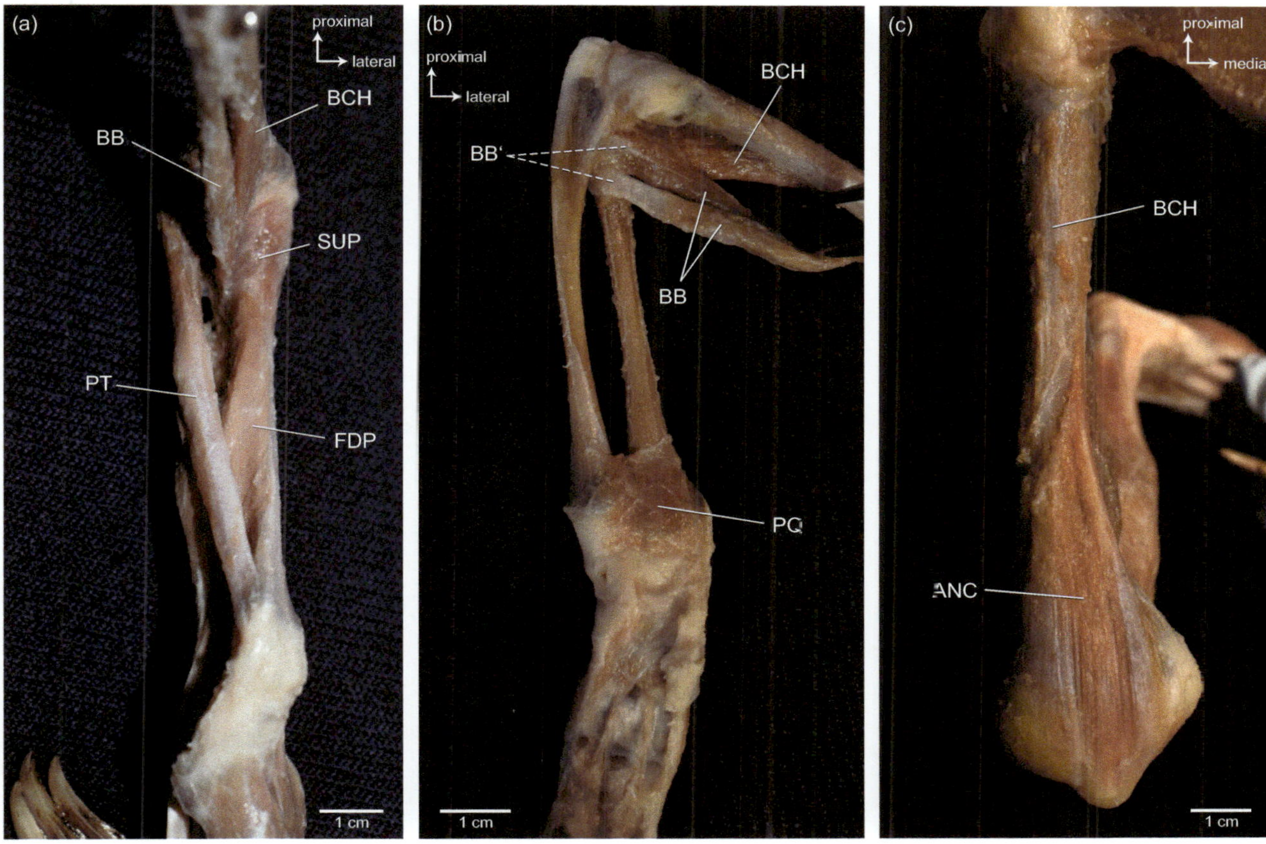
(a)
proximal
lateral
BCH
BB
SUP
PT
FDP
1 cm
(b)
proximal
lateral
BCH
BB'
BB
PQ
1 cm
(c)
proximal
medial
BCH
ANC
1 cm

Dasyurus viverrinus (Eastern quoll)

Classification: Australidelphia, Dasyuromorphia, Dasyuridae

Mean body mass: 0.6–1.6 kg

Habitat: Forest

Locomotor ecology: Ambulatorial – scansorial

Peculiarities: Digging ability

Pectoral muscles. Two pectoral muscles are identified (the PMJ, *M. pectoralis major* and the PMN, *M. pectoralis minor*).

Deep neck and shoulder muscles. In the neck region, three *Mm. rhomboidei* are present (RCR, *M. rhomboideus cervicis*; RT, *M. rhomboideus thoracis*; and RCP, *M. rhomboideus profundus*). The RCR, *M. rhomboideus cervicis* reaches the occiput. The RPR, *M. rhomboideus capitis* is absent. The SC, *M. subclavius* is well developed.
Intrinsic forelimb. The BB, *M. biceps brachii* has two distinct origins. The EDL, *M. extensor digitorum lateralis* inserts with three tendons on the distal phalanges of digits IV and V. The EDC, *M. extensor digitorum communis* inserts with four tendons on the distal phalanges of digits II through V. The EI, *M. extensor digiti I and II* inserts with two tendons on the distal phalanges of digits I and II. The M. *extensor carpi radialis* is identified as one muscle belly, but inserts with one tendon into the base of metacarpal II (ECRL, *M. extensor carpi radialis longus*) and with one tendon into the base of metacarpal III (ECRB, *M. extensor carpi radialis brevis*). The PL, *M. palmaris longus* inserts into the palmar aponeurosis. The FDP, *M. flexor digitorum profundus* (four muscle bellies) inserts with five tendons on the distal phalanges of digits I through V.

FIGURE II.27.1 Superficial muscles of forelimb and hind limb in lateral view. (a) Without labels. (b) With labels and muscle outlines. In the neck and shoulder region, the *Mm. trapezii* are divided into two parts (CT, *M. clavotrapezius* together with AT, *M. arcomiotrapezius*; and ST, *M. spinotrapezius*). The EPI, *M. epitrochlearis* is absent. The BCR, *M. brachioradialis* is present. The clavicle is present. The TEN, *M. tensor fasciae latae* is present, but has been removed. In the distal hind limb, the four tendons of the EDLO, *M. extensor digitorum longus* insert on the distal phalanges of digits II through V.

(a)
dorsal
caudal
1 cm
(b)
ST
CT+AT
ISP
TMJ
SA
GME
GSU
CAF
CM
DS
LAT
OT
SM
DA
TBLO
RF
CB
TBLA
PMN
VL
SET
SET
BB
BCH
FCUU
BIF
BCR
ECRL+ECRB
ECU
GL
EDC
PEL
EDL
TCR
EDLO
dorsal
caudal
1 cm

Philander opossum (Gray four-eyed opossum)

Classification:	Ameridelphia, Didelphimorphia, Didelphidae
Mean body mass:	0.4 kg
Habitat:	Tropical forest, lowland area, swamp
Locomotor ecology:	Ambulatorial – scansorial
Peculiarities:	Prehensile tail

FIGURE II.28.1 Superficial muscles of forelimb and hind limb in lateral view. (a) Without labels. (b) With labels and muscle outlines. In the neck and shoulder region, the *Mm. trapezii* are divided into two parts (CT, *M. clavotrapezius* together with AT, *M. arcomiotrapezius*; and ST, *M. spinotrapezius*). The EPI, *M. epitrochlearis* is absent. The BCR, *M. brachioradialis* is present. The OEA, *M. obliquus externus abdominis* originates from the thoracic ribs and inserts into the linea alba at the midline and on the pubis near the symphysis. The RA, *M. rectus abdominis* originates from the costal cartilages and inserts into the prebubic tendon. The presence of the *M. pyramidalis* is linked to the pouch that is a distinctive characteristic common to marsupials. The TEN, *M. tensor fasciae latae* is present, but has been removed.

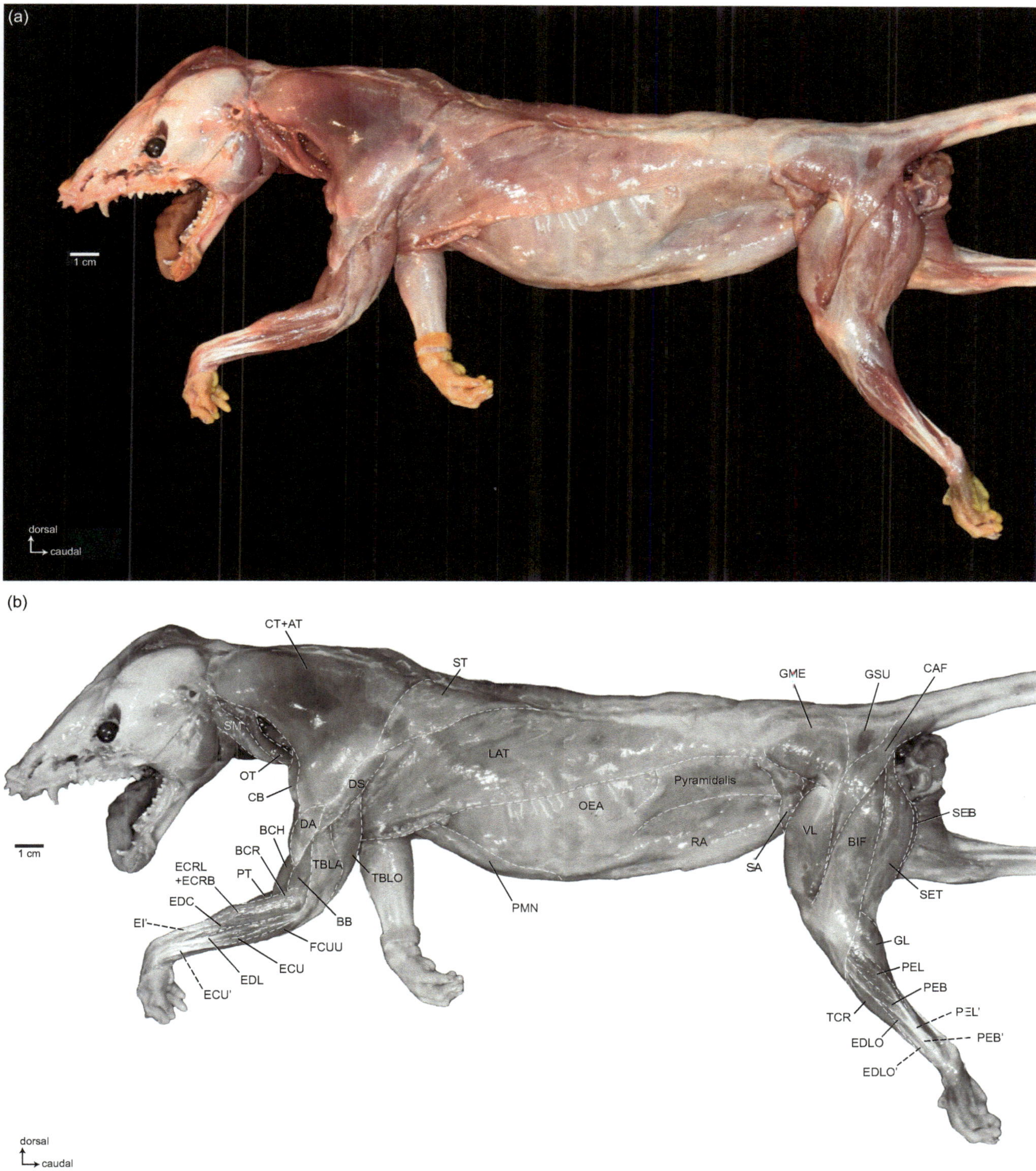
(a)
1 cm
dorsal
caudal
(b)
CT+AT
ST
GME
GSU
CAF
SM
OT
CB
LAT
Pyramidalis
DS
OEA
SEB
DA
BCH
VL
BIF
RA
BCR
1 cm
TBLA
ECRL
+ECRB
PT
TBLO
SA
SET
PMN
EDC
BB
EI'
FCUU
GL
ECU
PEL
EDL
ECU'
PEB
TCR
PEL'
EDLO
PEB'
EDLO'
dorsal
caudal

FIGURE II.28.2 **Pectoral muscles in ventral view. (a) Superficial view.** The XH, *M. xiphihumeralis* is not clearly developed as a separate muscle, but part of the PMN, *M. pectoralis minor.* **(b) Deep view.** The CM, *M. cleidomastoideus* and the OT, *M. omotransverarius* have been removed. The PAB, *M. pectoantebrachialis* is not clearly developed as a separate muscle, but part of the PMJ, *M. pectoralis major.*

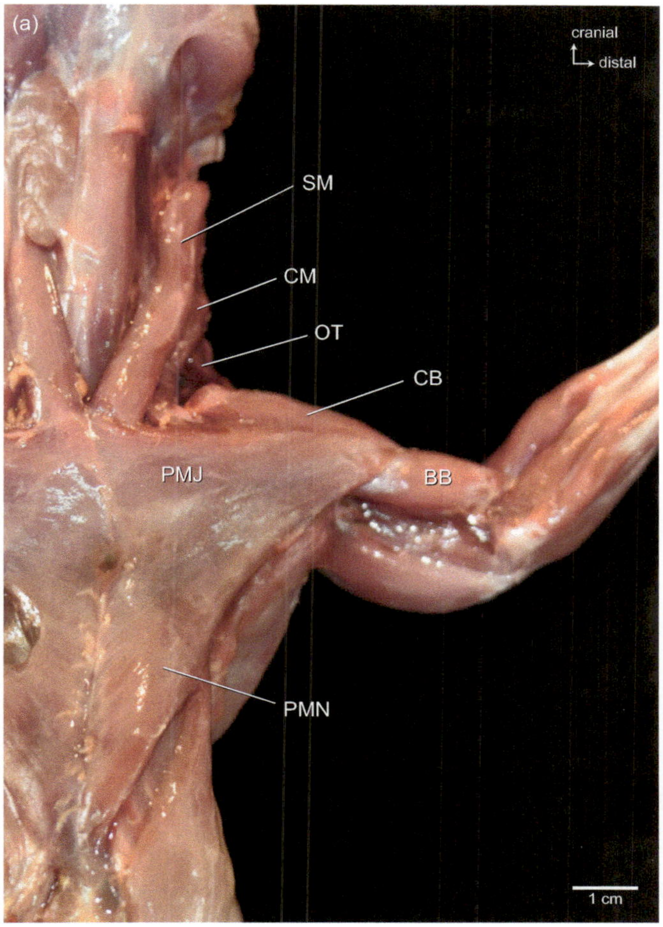
(a)
cranial
distal
SM
CM
OT
CB
PMJ
BB
PMN
1 cm

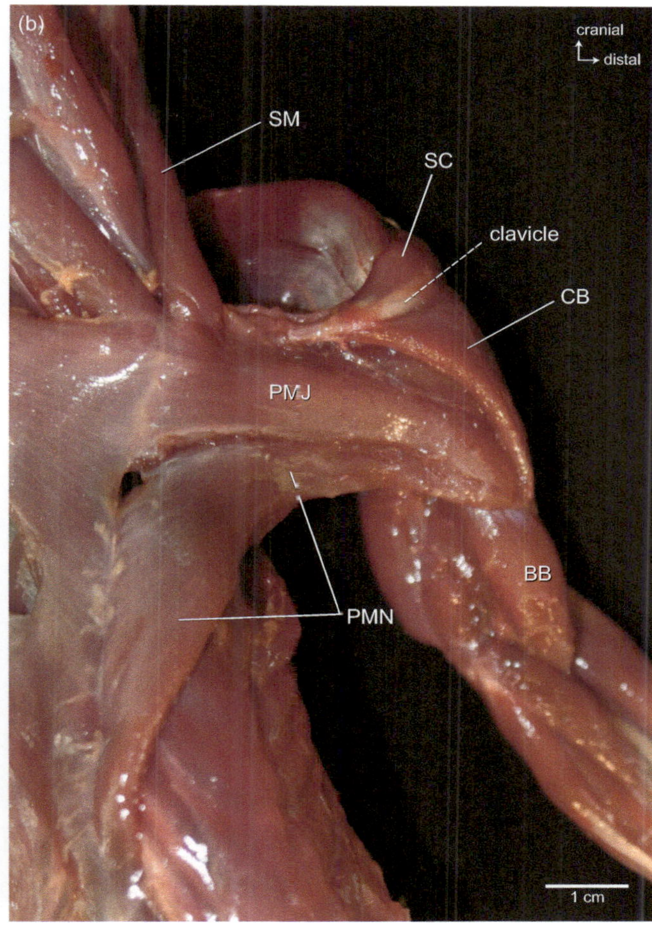
(b)
cranial
distal
SM
SC
clavicle
CB
PMJ
BB
PMN
1 cm

FIGURE II.28.3 Deep neck and shoulder muscles. (a) In dorsolateral view and (b) dorsal view. In the neck region, two *Mm. rhomboidei* are present (RCR, *M. rhomboideus cervicis* and RPR, *M. rhomboideus profundus*). The RCR, *M. rhomboideus cervicis* does not reach the occiput. **(c) In cranial view.**

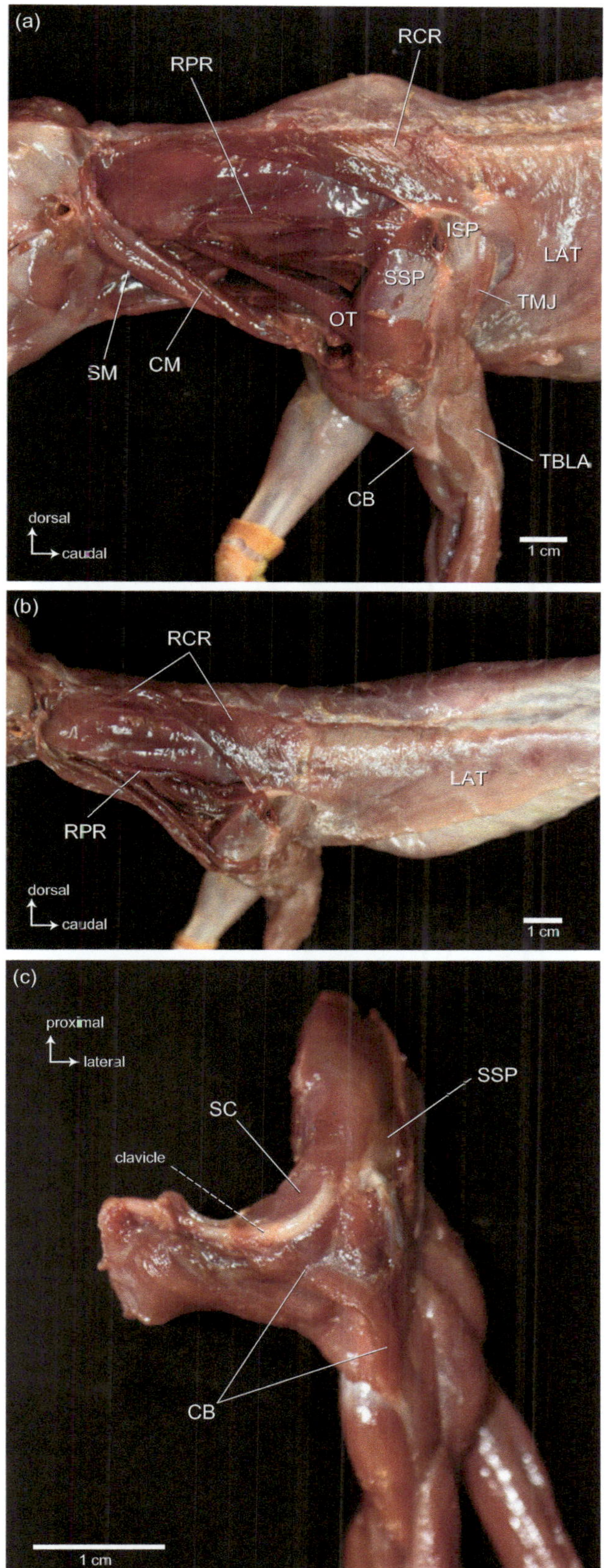
(a)
RCR
RPR
ISP
LAT
SSP
TMJ
OT
SM
CM
TBLA
CB
dorsal
caudal
1 cm
(b)
RCR
LAT
RPR
dorsal
caudal
1 cm
(c)
proximal
lateral
SSP
SC
clavicle
CB
1 cm

FIGURE II.28.4 Intrinsic forelimb muscles in (a) lateral view and (b) medial view. The TMJ, *M. teres major* is present, but has been removed. **Distal intrinsic forelimb muscles in (c) lateral view and (d) medial view**.

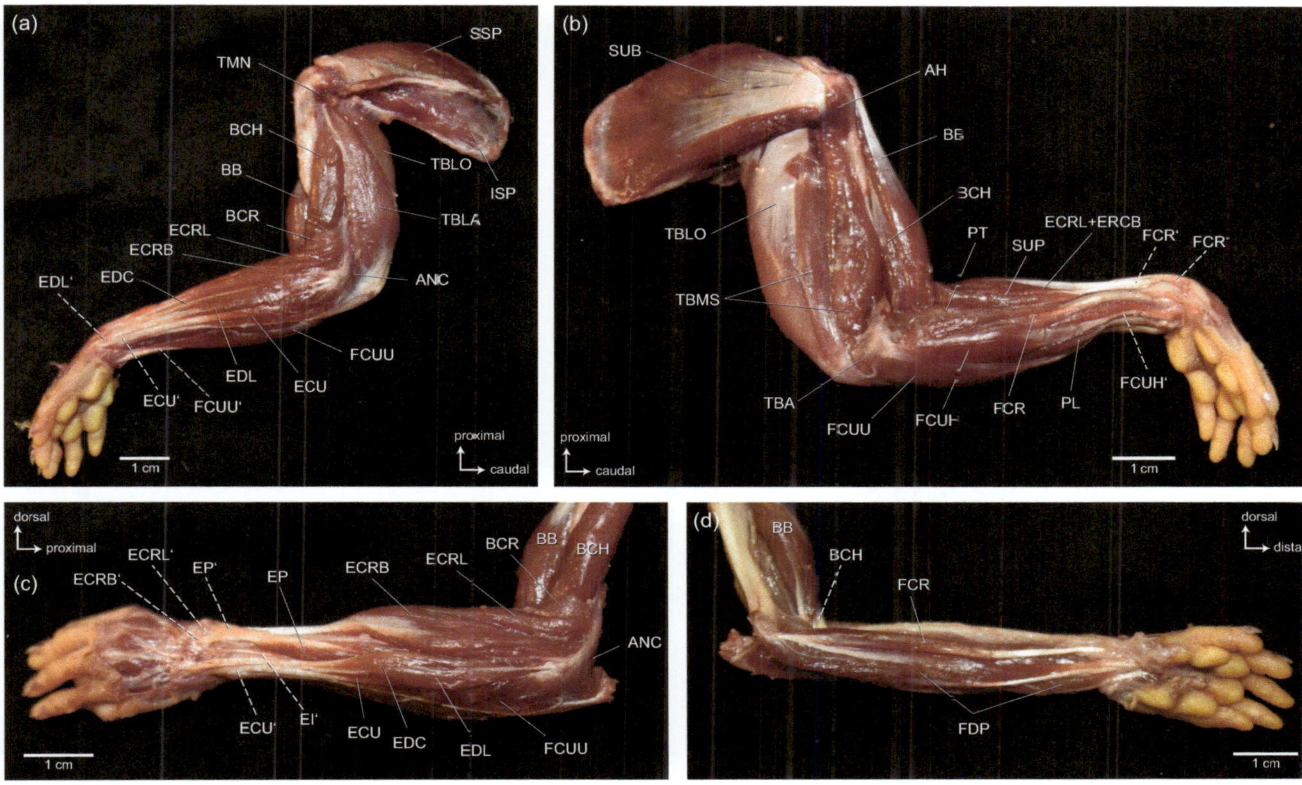
(a)
SSP
TMN
BCH
BB
TBLO
ISP
BCR
TBLA
ECRL
ECRB
EDL'
EDC
ANC
FCUU
EDL
ECU
ECU'
FCUU'
1 cm
proximal
caudal
(b)
SUB
AH
BB
BCH
TBLO
PT
SUP
ECRL+ERCB
FCR'
FCR'
TBMS
FCUH'
TBA
FCUU
FCUH
FCR
PL
proximal
caudal
1 cm
dorsal
proximal
(c)
ECRL'
ECRB'
EP'
EP
ECRB
ECRL
BCR
BB
BCH
ANC
ECU'
EI'
ECU
EDC
EDL
FCUU
1 cm
(d)
BB
BCH
FCR
FDP
dorsal
distal
1 cm

FIGURE II.28.5 **Distal forelimb muscles. (a) In dorsal view.** The EDC, *M. extensor digitorum communis* inserts with three tendons on the distal phalanges of digits I through III. The EDL, *M. extensor digitorum lateralis* inserts with three tendons on the distal phalanges of digits III through V. **(b) In ventral view.**

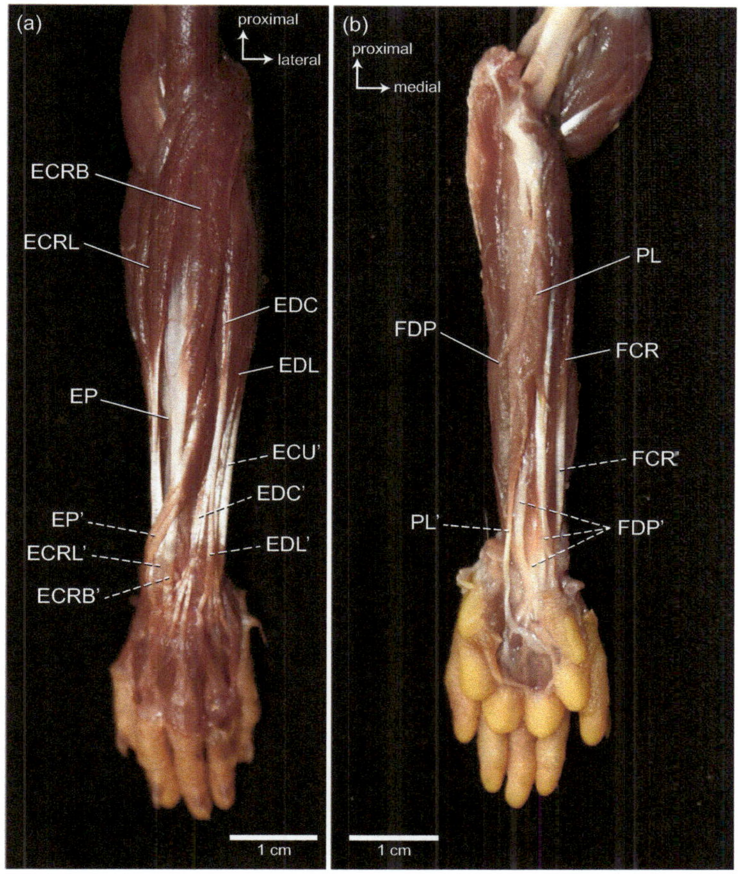
(a)
proximal
lateral
ECRB
ECRL
EDC
EDL
EP
ECU'
EDC'
EP'
EDL'
ECRL'
ECRB'
1 cm
(b)
proximal
medial
PL
FDP
FCR
FCR'
PL'
FDP'
1 cm

FIGURE II.28.6 **Deep rotator muscles. (a) Distal rotator muscles in dorsal view. (b) Distal rotator muscles in ventral view.**

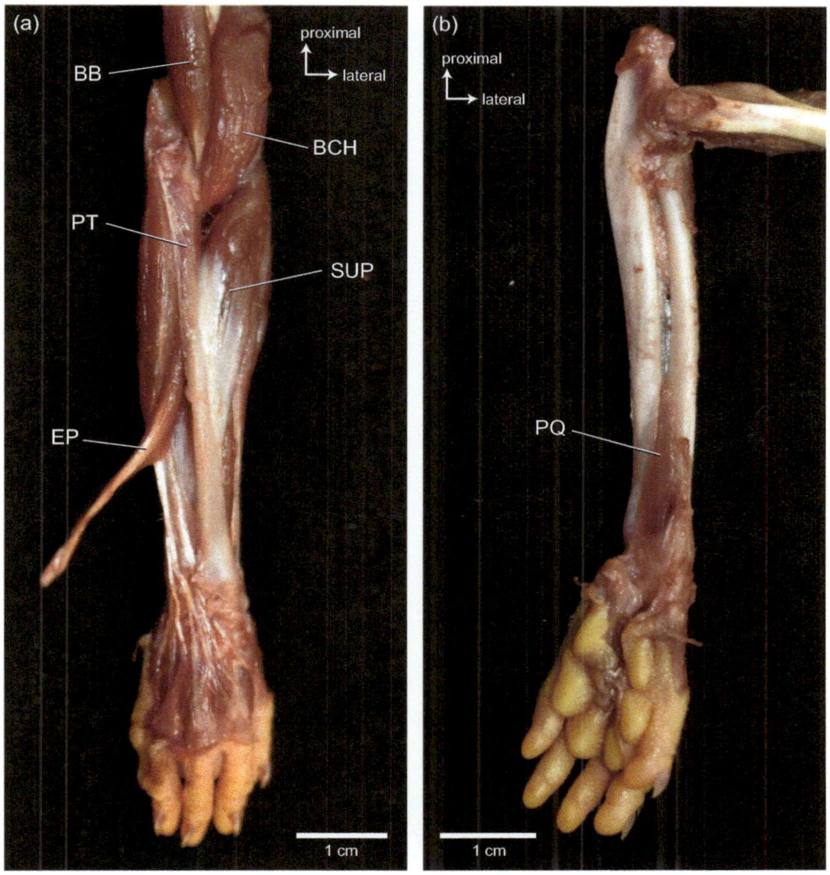
(a)
proximal
lateral
BB
BCH
PT
SUP
EP
1 cm
(b)
proximal
lateral
PQ
1 cm

Part III

Muscle synonyms

A standardized anatomical terminology is the basis for precise and interdisciplinary communication. The *Nomina Anatomica Veterinaria* (NAV, 2017) represents a general agreement on the nomenclature of veterinary anatomy, but there are different terms for muscles in the various anatomical sciences. To facilitate comparisons with previous works on limb myology in mammals, the following two tables provide an overview of the most common muscle synonyms found in the literature (Tables III.1.1 and III.1.2).

Shoulder and forelimb

TABLE III.1.1

Synonyms. Common Muscle Synonyms of the Shoulder and Forelimb in Mammals

Muscle	Comparative Anatomy	Human Anatomy	Old Textbooks	Miscellaneous
		Superficial dorsal neck and shoulder		
Mm. trapezii				
M. clavotrapezius	*M. trapezius pars cervicalis*	*M. trapezius pars descendens*	*M. trapezius superior*	*M. cleidocervicalis*
M. acromiotrapezius		*M. trapezius pars transversa*		*M. trapezius cervicis*
M. spinotrapezius	*M. trapezius pars thoracis*	*M. trapezius pars ascendens*	*M. trapezius inferior*	*M. trapezius thoracis*
M. latissimus dorsi	-	-	-	-
		Ventral neck and shoulder		
M. sternomastoideus	*M. sternocephalicus*	*M. sternocleidomastoideus*	*M. sternocephalicus pars mastoidea*	*M. cleidocephalicus pars mastoidea*
M. cleidomastoideus	*M. brachiocephalicus pars cleidocephalica*		-	*M. cleidocephalicus pars cervicalis*
M. clavobrachialis	*M. brachiocephalicus pars cleidobrachialis*	*M. deltoideus clavicularis*	-	*M. deltoideus clavicularis*
Mm. pectorales				
M. pectoantebrachialis	*M. pectoralis descendens*	*M. pectoralis major*	*M. pectoantibrachialis*	*M. pectoralis superficialis*
M. pectoralis major	*M. pectoralis transversus*		-	
M. pectoralis minor	*M. pectoralis profundus*	*M. pectoralis minor*	*M. xiphihumeralis*	-
M. xiphihumeralis	-		*M. pectoralis minor (entopectoralis)*	-
M. serratus ventralis	-	*M. serratus anterior*	*M. serratus anterior* *M. levator scapulae*	*M. serratus anterior*
		Deep dorsal neck and shoulder		
Mm. rhomboidei				
M. rhomboideus cervicis	-	*M. rhomboideus minor*	*M. rhomboideus*	*M. rhomboideus minor*
M. rhomboideus thoracis	-	*M. rhomboideus major*		*M. rhomboideus major*
M. rhomboideus capitis	-		*M. occipitoscapularis (M. levator scapulae dorsalis)*	*M. rhomboideus occipitalis/capitis*

(*Continued*)

TABLE III.1.1 (*Continued*)
Synonyms. Common Muscle Synonyms of the Shoulder and Forelimb in Mammals

Muscle	Comparative Anatomy	Human Anatomy	Old Textbooks	Miscellaneous
M. rhomboideus profundus	-	-	-	-
M. omotransversarius	-	*M. levator scapulae*	*M. levator scapulae ventralis (M. levator claviculae)*	-
M. subclavius	-	-	-	-
		Lateral shoulder and arm		
M. supraspinatus	-	-	-	-
M. infraspinatus	-	-	-	-
Mm. deltoidei				
M. spinodeltoideus	-	*M. deltoideus spinalis*	-	*M. deltoideus (pars) scapularis*
M. acromiodeltoideus	-	*M. deltoideus acromialis*	-	*M. deltoideus (pars) acromialis*
M. teres minor	-	-	-	-
M. epitrochlearis	*M. tensor fasciae antebrachii*	*M. omo-anconeus*	-	*M. dorsoepitrochlearis*
Mm. triceps brachii				
M. triceps brachii caput longum	-	-	*M. anconeus longus*	*M. meditriceps (caput longum)*
M. triceps brachii caput laterale	-	-	*M. anconeus laterale*	*M. ectotriceps (caput laterale)*
M. triceps brachii caput mediale, short portion	*M. triceps brachii caput mediale*	*M. triceps brachii caput mediale; caput profundum*	*M. anconeus posterior*	
M. triceps brachii caput medial, intermediate and long portion				
M. brachialis	-	-	*M. supinator longus*	-
M. anconeus	*M. anconeus*	-	-	*M. anconeus externus*
		Medial shoulder and arm		
M. subscapularis	-	-	-	-
M. teres major	-	-	-	-
M. articularis humeri	-	*M. coracobrachialis*	*M. coracobrachialis*	*M. articularis humeri* *M. coracobrachialis*
M. triceps brachii caput accessorium	-	*M. articularis cubiti*	*M. epitrochleoanconeus*	*M. anconeus internus*
M. biceps brachii	-	*M. biceps brachii caput longum* *M. biceps brachii caput breve*	-	-
		Laterodorsal distal arm		
M. brachioradialis	-	-	*M. brachialis anticus*	*M. supinator longus*
M. extensor carpi radialis longus	-	-	*M. extensor carpi radialis longior*	*M. extensor carpi radialis*
M. extensor carpi radialis brevis	-	-	*M. extensor carpi radialis brevior*	

(*Continued*)

TABLE III.1.1 (*Continued*)

Synonyms. Common Muscle Synonyms of the Shoulder and Forelimb in Mammals

Muscle	Comparative Anatomy	Human Anatomy	Old Textbooks	Miscellaneous
M. abductor pollicis	*M. abductor pollicis longus*	*M. extensor pollicis longus* *M. extensor pollicis brevis*	*M. extensor brevis pollicis*	*M. abductor digiti I longus*
M. extensor digitorum communis	-	*M. extensor digitorum*	-	*M. extensor digitorum* *M. extensor digiti minimi*
M. extensor digitorum lateralis	-	-	-	-
M. extensor indicis	*M. extensor pollicis longus et indicis proprius*	-	*M. extensor indicis (proprius)*	*M. extensor digit I and II*
M. extensor carpi ulnaris	-	-	-	-
		Medioventral distal arm		
M. flexor carpi radialis	-	-	-	-
M. flexor carpi ulnaris, caput ulnare	-	-	*M. flexor carpi ulnaris*	*M. flexor carpi ulnaris*
M. flexor carpi ulnaris, caput humerale	-	-		
M. palmaris longus	*M. flexor digitalis superficialis*	-	-	-
M. flexor digitorum profundus (4 heads)	*M. flexor digitalis profundus*	*M. perforans*	*M. flexor communis digitorum*	*M. flexor profundus digitorum*
		Deep distal arm		
M. supinator	-		*M. supinator brevis*	*M. supinator brevis*
M. pronator teres	-	*M. pronator teres caput humerale* *M. pronator teres caput ulnare*	-	-
M. pronator quadratus	-	-	-	-

Hip and hind limb

TABLE III.1.2

Synonyms. Common Muscle Synonyms of the Hip and Hind Limb in Mammals

Muscle	Comparative Anatomy	Human Anatomy	Old Textbooks	Miscellaneous
		Superficial hip		
M. sartorius	-	-	-	-
M. tensor fasciae latae	-	-	-	-
M. caudofemoralis	*M. abductor cruris cranialis*	-	*M. abductor cruris cranialis*	*M. gluteofemoralis/ M. femorococcygeus*
M. biceps femoris	*M. glutaeobiceps if fused together with M. gluteus superficialis*	-	*M. biceps femoris anticus* *M. biceps femoris posticus*	*M. biceps femoris, long head* *M. biceps femoris, short head*

(*Continued*)

TABLE III.1.2 (*Continued*)

Synonyms. Common Muscle Synonyms of the Hip and Hind Limb in Mammals

Muscle	Comparative Anatomy	Human Anatomy	Old Textbooks	Miscellaneous
M. semitendinosus	-	-	-	-
M. semimembranosus	-	-	*M. semimembranosus anticus* *M. semimembranosus posticus*	-
Mm. glutei				
M. gluteus superficialis	*M. glutaeus superficialis*	*M. gluteus maximus*	-	-
M. gluteus medius	*M. glutaeus medius*	-	-	-
M. gluteus profundus	*M. glutaeus profundus*	*M. gluteus minimus*	-	-
		Deep ventral hip		
M. psoas minor	-	-	-	-
M. iliopsoas				
M. psoas major	-	-	-	-
M. iliacus	-	-	-	-
		Deep medial hip		
M. gracilis	-	-	-	-
M. adductor femoris (magnus et brevis)	*M. adductor magus* *M. adductor brevis*	*M. adductor magus* *M. adductor brevis*	*M. adductor femoris (superficial and deep portion)*	-
M. obturator externus	*M. obturatorius externus*	*M. obturatorius externus*	-	-
M. obturator internus	-*M. obturatorius internus*	*M. obturatorius internus*	-	-
		Deep lateral hip		
M. piriformis	-	-	*M. pyriformis*	-
Mm. gemelli (M. gemellus superioris, M. gemellus inferioris)	*Mm. gemelli*	*Mm. gemelli (M. gemellus superior, M. gemellus inferior)*	*Mm gemelli (M. gemellus cranialis, M. gemellus caudalis)*	-
M. quadratus femoris	-	-	-	-
M. tenuissimus	*M. abductor cruris caudalis*	-	*M. tensor fasciae cruris*	-
M. adductor longus	-	-	-	-
M. pectineus	-	-	-	-
Mm. quadriceps femoris				
M. rectus femoris	-	-	-	-
M. vastus lateralis	-	-	-	-
M. vastus intermedius	-	-	-	-
M. vastus medialis	-	-	-	-
		Craniolateral distal leg		
M. tibialis cranialis	-	*M. tibialis anterior*	*M. tibialis anticus*	-
M. extensor digitorum longus	*M. extensor digitalis longus*	-	-	-
M. extensor digitorum lateralis	*M. extensor digitalis lateralis*	-	-	-

(*Continued*)

TABLE III.1.2 (*Continued*)
Synonyms. Common Muscle Synonyms of the Hip and Hind Limb in Mammals

Muscle	Comparative Anatomy	Human Anatomy	Old Textbooks	Miscellaneous
M. extensor hallucis longus	-	-	*M. extensor digiti I longus*	*M. extensor hallucis proprius (longus)*
Mm. peronei				
M. peroneus longus	*M. fibularis longus*	*M. fibularis longus*	*M. peroneus longus*	-
M. peroneus tertius	*M. fibularis tertius*	*M. fibularis tertius*	*M. peroneus tertius*	*M. peroneus digiti quarti and M. peroneus digit quinti*
M. peroneus brevis	*M. fibularis brevis*	*M. fibularis brevis*	*M. peroneus brevis*	-
		Caudal distal leg		
Mm. gastrocnemii				
M. gastrocnemius caput laterale	-	-	-	*M. triceps surae (together with M. soleus)*
M. gastrocnemius caput mediale	-	-	-	
M. plantaris	*M. flexor digitorum pedis superficialis*	-	-	*M. flexor digitorum superficialis*
M. soleus	-	-	-	-
M. popliteus	-	-	-	-
M. flexor digitorum longus	*M. flexor digitorum pedis longus*	-	-	-
M. flexor hallucis longus	-	-	-	-
M. tibialis caudalis	-	*M. tibialis posterior*	*M. tibialis posticus*	-

NON-EXHAUSTIVE LIST OF REFERENCES ON MAMMALIAN LIMB MUSCLE ANATOMY (IN ALPHABETICAL ORDER)

Anatomy (Comparative)

Ariano MA, Armstrong PB, Edgerton VP. 1973. Hind limb muscle fiber populations of five mammals. *J. Histochem. Cytochem.* **21**: 51–55.

Arlamowska-Palider A, Zablocki J. 1972. Musculus omotransversarius in the light of comparative anatomy. *Acta Theriol. (Warsz.)* **17**: 381–398.

Barone R. 1966. *Anatomie comparée des mammifères domestiques. Tome I. Ostéologie.* Vigot Frères Editeurs: Paris.

Barone R. 1968. *Anatomie comparée des mammifères domestiques. Tome 2: Arthrologie et myologie.* Vigot Frères Editeurs: Paris.

Campbell B. 1939. The comparative anatomy of the dorsal interosseous muscles. *Anat. Rec.* **73**: 115–125.

Chauveau A. 1891. *The Comparative Anatomy of the Domesticated Animals.* 2nd ed. D. Appleton: New York.

Dastugue J. 1963. Anatomie des muscles pronateurs et supinateurs de quelques mammifères. *Mammalia* **27**: 256–284.

De Iuliis GD, Pulerà D. 2007. *The Dissection of Vertebrates. A Laboratory Manual.* Academic Press: Amsterdam.

Diogo R, Abdala V. 2010. *Muscles of Vertebrates – Comparative Anatomy, Evolution, Homologies and Development.* Taylor and Francis: Oxford.

Diogo R, Molnar J. 2014. Comparative anatomy, evolution, and homologies of tetrapod hindlimb muscles, comparison with forelimb muscles, and destruction of the forelimb–hindlimb serial homology hypothesis. *Anat. Rec.* **297**: 1047–1075

Diogo R, Bello-Hellegouarch G, Kohlsdorf T, Esteve-Altava B, Molnar JL. 2016. Comparative myology and evolution of marsupials and other vertebrates, with notes on complexity, bauplan, and "scala naturae". *Anat Rec (Hoboken)* **299**: 1224–1255.

Ellenberger W, Baum H. 1974. *Handbuch der vergleichenden Anatomie der Haustiere.* Springer-Verlag: Berlin.

Ellenberger W, Baum H, Dittrich H. 1901. *Handbuch der Anatomie der Tiere für Künstler.* Dieterich'sche Verlagsbuchhandlung Theodor Weicher: Leipzig.

English AW, Letbetter WD. 1982. A histochemical analysis of identified compartments of cat lateral gastrocnemius muscle. *Anat. Rec.* **204**: 123–130.

Getty R. 1975a. Carnivore myology. In *Sisson & Grossman's the Anatomy of Domestic Animals* (ed. Getty R), pp. 1507–1537. W.B. Saunders Company: Philadelphia.

Getty R. 1975b. Ruminant myology. In *Sisson and Grossman's the Anatomy of the Domestic Animals* (ed. Getty R), pp. 791–860. W. B. Saunders Company: Philadelphia.

Haines RW. 1950. The flexor muscles of the forearm and hand in lizards and mammals. *J. Anat.* **84**: 13–29.

Howell AB. 1936. The phylogenetic arrangement of the muscular system. *Anat. Rec.* **66**: 295–316.

Jouffroy FK. 1971. Musculature des members. In *Traite de zoologie. Anatomie, Systematique, Biologie, Tome XVI, Fascicule III* (ed. Grasse PP), pp. 1–475. Masson et Cie. Editeurs: Paris.

Kerr NS. 1955. The homologies and nomenclature of the thigh muscles of the opossum, cat, rabbit, and rhesus monkey. *Anat. Rec.* **121**: 481–493.

König HE, Liebich HG. 2012. *Anatomie der Haussäugetiere.* Schattauer: Stuttgart.

Leach WJ. 1946. *Functional Anatomy of the Mammal. A Guide to the Dissection of the Cat and an Introduction to the Structural and Functional Relationship Between the Cat and Man.* McGraw-Hill Book Company: New York.

Macalister A. 1869. On the arrangement of the pronator muscles in the limbs of vertebrate animals. *J. Anat. Physiol.* **3**: 335–340.

Marchi D, Hartstone-Rose A. 2018. Functional morphology and behavioral correlates to postcranial musculature. *Anat. Rec.* **301**: 419–423.

Mendez J, Keys A. 1960. Density and composition of mammalian muscle. *Metabolism* **9**: 184–188.

Molnar JL, Diogo R, Hutchinson JR, Pierce SE. 2018. Reconstructing pectoral appendicular muscle anatomy in fossil fish and tetrapods over the fins-to-limbs transition. *Biol. Rev.* **93**: 1077–1107.

Nickel R, Schummer A, Seiferle E. 2003. *Lehrbuch der Anatomie der Haustiere I: Bewegungsapparat.* Stuttgart: Parey.

Owen R. 1868. *On the Antomy of Vertebrates.* Vol. III Mammals. Longmans, Green, and Co.: London.

Panyutina AA, Korzun LP, Kuznetsov AN. 2015. *Flight of Mammals: From Terrestrial Limbs to Wings.* Springer: London.

Paul AC. 2001. Muscle length affects the architecture and pattern of innervation differently in leg muscles of mouse, guinea pig, and rabbit as compared to those of human and monkey muscles. *Anat. Rec.* **262**: 301–309.

Popesko, P. 1972. *Atlas of Topographic Anatomy of Domestic Animals.* Priroda: Bratislava.

Putz R, Pabst R eds. 2001. *Sobotta Atlas of Human Anatomy.* Urban & Fischer: Munich.

Romer AS. 1924. Pectoral limb musculature and shoulder-girdle structure in fish and tetrapods. *Anat Rec (Hoboken)* **27**: 119–143.

Slijper EJ. 1946. Comparative biologic-anatomical investigations on the vertebral column and spinal musculature of mammals. *Kon. Ned. Akad. Wet.* **42**: 1–128.

Standring S ed. 2015. *Gray's Anatomy. The Anatomical Basis of Clinical Practice.* Elsevier: Edinburgh.

Windle BCA. 1889. The flexors of the digits of the hand. I. The muscular masses in the fore-arm. *J. Anat. Physiol.* **24**: 72–84.

Wood JT. 1870. On a group of varieties of the muscles of the human neck, shoulder, and chest, with their transitional forms and homologies in the Mammalia. *Phil. Trans. R. Soc. Lond.* **160**: 83–116.

Ziermann JM, Diaz Jr RE, Diogo R. 2019. *Heads, Jaws, and Muscles: Anatomical, Functional, and Developmental Diversity in Chordate Evolution.* Springer: Cham.

Afrotheria

Domning DP. 1977. Observations on the myology of *Dugong dugong* (Müller). *Smithsonian Contrib. Zool.* **226**: 1–57.

Domning DP. 1978. The myology of the Amazonian manatee, *Trichechus inunguis* (Natterer) (Mammalia: Sirenia). *Acta. Amaz.* **8**: 1–81.

Nagel RM, Forstenpointner G, Soley JT, Weissengruber GE. 2018. Muscles and fascial elements of the antebrachium and manus of the African elephant (*Loxodonta africana*, Blumenbach 1797): Starring comparative and functional considerations. *Anat., Histol., Embryol.* **47**: 195–205.

Puttick GM, Jarvis JUM. 1977. The functional anatomy of the neck and forelimbs of the Cape golden mole, *Chrysochloris asiatica* (Lipotyphla: Chrysochloridae). *Zool. Afr.* **12**: 445–458.

Thewissen JG, Badoux DM. 1986. The descriptive and functional myology of the fore-limb of the Aardvark (*Orycteropus afer*, Pallas 1766). *Anat. Anz.* **162**: 109–123.

Artiodactyla

Budras K-D, Habel RE. 2010. *Bovine Anatomy*. Schlütersche: Hannover.

Campbell B. 1936. The comparative myology of the forelimb of the hippopotamus, pig and tapir. *Am. J. Anat.* **59**: 201–247.

Fisher RE, Scott KM, Adrian B. 2010. Hind limb myology of the common hippopotamus, *Hippopotamus amphibius* (Artiodactyla: Hippopotamidae). *Zool. J. Linn. Soc.* **158**: 661–682.

Fisher RE, Scott KM, Naples VL. 2007. Forelimb myology of the pygmy hippopotamus (*Choeropsis liberiensis*). *Anat. Rec.* **290**: 673–693.

Kneepkens A, Badoux D, MacDonald A. 1989. Descriptive and comparative myology of the forelimb of the babirusa (*Babyrousa babyrussa* L. 1758). Anat. Histol. Embryol. **18**: 349–365.

Macalister A. 1873c. The anatomy of *Chaeropsis liberiensis. Proc. R. Irish Acad. Sci.* **1**: 494–500.

Macdonald AA, Kneepkens AFLM, Van Kolfschoten T, Houtekamer JL, Sondaar PY, Badoux DM. 1985. Comparative anatomy of the limb musculature of some Suina. *Fortschr. Zool.* **30**: 95–97.

Smuts MM, Bezuidenhout AJ. 1987. *Anatomy of the Dromedary*. Clarendon Press: Oxford.

Wareing K, Tickle PG, Stokkan K-A, Codd JR, Sellers WI. 2011. The musculoskeletal anatomy of the reindeer (*Rangifer tarandus*): fore- and hindlimb. *Polar Biol.* **34**: 1571–1578.

Windle BCA, Parsons FG. 1901. On the muscles of the Ungulata. Part I. Muscles of the head, neck, and fore-limb. *Proc. Zool. Soc. Lond.* **2**: 656–704.

Carnivora

Alix ME. 1876. Mémoire sur la myologie du putois (*Putorius communis*, Cuv.). J. Zool. Paris **5**: 153–188.

Allen H. 1882. The muscles of the limbs of the raccoon (*Procyon lotor*). *Proc. Acad. Nat. Sci. Philadelphia* **34**: 115–144.

Anton M, Salesa M, Pastor JF, Peigné S, Morales J. 2006. Implications of the functional anatomy of the hand and forearm of *Ailurus fulgens* (Carnivora, Ailuridae) for the evolution of the 'false-thumb' in pandas. *J. Anat.* **209**: 757–764.

Barone R. 1967. La myologie du lion (*Panthera leo*). *Mammalia* **31**: 459–514.

Barone R, Deutsch H. 1953. Myologie des membres du furet (*Putorius furo*). *Bull. Soc. Sci. Vet. Lyon* **55**: 445–454.

Beddard FE. 1895. On the visceral and muscular anatomy of *Cryptoprocta ferox*. *Proc. Zool. Soc. Lond.* **1895**: 430–437.

Beddard FE. 1900. On the anatomy of *Bassaricyon alleni*. *Proc Zool. Soc. Lond.* **69**: 661–675.

Beddard FE. 1905. Some notes on the anatomy of the ferret badger *Helictis personata*. *Proc Zool. Soc. Lond.* **75**: 21–29.

Beswick-Perrin J. 1871. On the myology of the limbs in the kinkajou (*Cercoleptes caudivolvulus*). *Proc Zool. Soc. Lond.* **1871**: 547–559.

Böhmer C, Fabre A-C, Herbin M, Peigné S, Herrel A. 2018. Anatomical basis of differences in locomotor behavior in martens: A comparison of the forelimb musculature between two sympatric species of *Martes*. *Anat. Rec.* **301**: 449–472.

Böhmer C, Fabre AC, Taverne M, Herbin M, Peigné S, Herrel A. 2019. Functional relationship between myology and ecology in carnivores: do forelimb muscles reflect adaptations to prehension? *Biol. J. Linn. Soc.* **127**: 661.

Brown IE, Satoda T, Richmond FJ, Loeb GE. 1998. Feline caudofemoralis muscle. Muscle fibre properties, architecture, and motor innervation. *Exp. Brain Res.* **121**: 76–91.

Carlon B, Hubbard C. 2012. Hip and thigh anatomy of the clouded leopard (*Neofelis nebulosa*) with comparison to the domestic cat (*Felis catus*). *Anat. Rec.* **295**: 577–589.

Carlsson A. 1900. Über die systematische Stellung der *Nandina binotata. Zoologische Jahrbücher. Abteilung für Systematik, Geographie und Biologie der Tiere* **13**: 509–528.

Carlsson A. 1902. Über die systematische Stellung vn *Eupleres goudoti. Zoologische Jahrbücher Jahrb.* **16**: 217–242.

Carlsson A. 1911. Über *Cryptoprocta ferox. Zoologische Jahrbücher. Abteilung für Systematik, Geographie und Biologie der Tiere* **30**: 419–470.

Carlsson A. 1920. Über *Arctictis binturong. Acta Zoologica* **1**: 337–380.

Carlsson A. 1925. Über *Ailurus fulgens. Acta Zoologica* **6**: 269–308.

Carlon B, Hubbard C. 2012. Hip and thigh anatomy of the clouded leopard (*Neofelis nebulosa*) with comparisons to the domestic cat (*Felis catus*). *Anat. Rec.* **295**: 577–589.

Concha I, Adaro L, Borroni C, Altamirano C. 2004. Consideraciones anatómicas sobre la musculatura intrínseca del miembro torácio del puma (*Puma concolor*). *Int. J. Morphol.* **22**: 121–125.

Cuff AR, Sparkes EL, Randau M, Pierce SE, Kitchener AC, Goswami A, Hutchinson JR. 2016. The scaling of postcranial muscles in cats (Felidae) I: Forelimb, cervical, and thoracic muscles. *J. Anat.* **229**: 128–141.

Cuff AR, Sparkes EL, Randau M, Pierce SE, Kitchener AC, Goswami A, Hutchinson JR. 2016. The scaling of postcranial muscles in cats (Felidae) II: Hindlimb and lumbosacral muscles. *J. Anat.* **229**: 142–152.

Davis DD. 1949. The shoulder architecture of bears and other carnivores. *Fieldiana Zool.* **31**: 285–305.

Davis DD. 1964. The giant panda. A morphological study of evolutionary mechanisms. *Fieldiana (Zoology Memoirs)* **3**: 1–339.

Devis CW. 1868. Notes on the myology of *Viverra civetta. J. Anat. Physiol.* **2**: 207–217.

Diogo R, Pastor F, De Paz F, Potau JM, Bello-Hellegouarch G, Ferrero EM, Fisher RE. 2012. The head and neck muscles of the serval and tiger: Homologies, evolution, and proposal of a mammalian and a veterinary muscle ontology. *Anat. Rec. (Hoboken)* **295**: 2157–2178.

English AM. 1976. Functional anatomy of the hands of fur seals and sea lions. *Am. J. Anat.* **147**: 1–18.

English AW. 1977. Structural correlates of forelimb function in fur seals and sea lions. *J. Morphol.* **151**: 325–352.

Ercoli MD, Youlatos D. 2016. Integrating locomotion, postures and morphology: The case of the tayra, *Eira barbara* (Carnivora, Mustelidae). *Mamm. Biol.* **81**: 464–476.

Ercoli MD, Echarri S, Busker F, Alvarez A, Morales MM, Turazzini GF. 2013. The functional and phylogenetic implications of the myology of the lumbar region, tail, hind limbs of the lesser grison (*Galictis cuja*). *J. Mamm. Evol.* **20**: 309–336.

Ercoli MD, Alvarez A, Stefanini MI, Busker F, Morales MM. 2015. Muscular anatomy of the forelimbs of the lesser grison (*Galictis cuja*), and a functional and phylogenetic overview of Mustelidae and other Caniformia. *J. Mamm. Evol.* **22**: 57–91.

Evans HE. 1993. *Miller's Anatomy of the Dog.* W.B. Saunders: Philadelphia.

Evans HE, Quoc An N. 1980. Anatomy of the ferret. In *Biology and Diseases of the Ferret* (ed. Fox JG), pp. 16–69. Williams and Wilkins: Baltimore.

Fisher RE. 1942. *The Osteology and Myology of the California River Otter.* Stanford University Press: Stanford, CA.

Fisher RE, Adrian B, Barton M, Holmgren J, Tang SY. 2009. The phylogeny of the red panda (*Ailurus fulgens*): Evidence from the forelimb. *J. Anat.* **215**: 611–635.

Fisher RE, Adrian B, Elrod C, Hicks M. 2008. The phylogeny of the red panda (*Ailurus fulgens*): Evidence from the hind limb. *J. Anat.* **213**: 607–628.

Gambaryan PP, Karapetjan WS. 1961. Besonderheiten im Bau des Seelöwen (*Eumetopias californianus*), der Bajkalrobbe (*Phoca sibirica*) und des Seeotters (*Enhydra lutris*) in Anpassungen an die Fortbewegung im Wasser. *Zool. Jahrb.* **79**: 123–148.

Gilbert SG. 2002. *Pictorial Anatomy of the Cat.* University of Toronto Press: Toronto.

Gonyea WJ, Ericson GC. 1977. Morphological and histochemical organization of the flexor carpi radialis muscle in the cat. *Am. J. Anat.* **148**: 329–344.

Gowell RC. 1897. Myology of the hind limb of the raccoon. *Kan. Univ. Quar.* **6**: 121–126.

Hall ER. 1926. The muscular anatomy of three mustelid mammals, *Mephitis*, *Spilogale*, and *Martes*. *Univ. Calif. Publ. Zool.* **30**: 7–39.

Hall ER. 1927. The muscular anatomy of the American badger (*Taxidea taxus*). *Univ. Calif. Publ. Zool.* **30**: 205–219.

Haughton S. 1867a. On the muscular anatomy of the Irish terrier, as compared with that of the Australian dingo. *Proc. R. Irish Acad. Series 2* **IX**: 504–507.

Haughton S. 1867b. On the muscular anatomy of the otter (*Lutra vulgaris*). *Proc. R. Irish Acad. Series 2* **IX**: 511–515.

Haughton S. 1867c. On the muscular anatomy of the badger. *Proc. R. Irish Acad. Series 2* **IX**: 507–508.

Haughton S. 1867d. On the muscles of the Virginian bear. *Proc. R. Irish Acad. Series 2* **IX**: 508–511.

Haughton S. 1867e. On the muscular anatomy of the lion. Proc. R. Irish Acad. Sci. Series 2 **IX**: 85–93.

Howard LD. 1973. Muscular anatomy of the forelimb of the sea otter (*Enhydra lutris*). *Proc. Calif. Acad. Sci.* **39**: 411–500.

Howard LD. 1975. Muscular anatomy of the hind limb of the sea otter (*Enhydra lutris*). *Proc. Calif. Acad. Sci.* **40**: 335–416.

Hudson PE, Corr SA, Payne-Davis RC, Clancy SN, Lane E, Wilson AM. 2011. Functional anatomy of the cheetah (*Acinonyx jubatus*) forelimb. *J. Anat.* **218**: 375–385.

Hudson PE, Corr SA, Payne-Davis RC, Clancy SN, Lane E, Wilson AM. 2011. Functional anatomy of the cheetah (*Acinonyx jubatus*) hindlimb. *J. Anat.* **218**: 363–374.

Julik E, Zack S, Adrian B, Maredia S, Parsa A, Poole M, Starbuck A, Fisher RE. 2012. Functional anatomy of the forelimb muscles of the ocelot (*Leopardus pardalis*). *J. Mamm. Evol.* **19**: 277–304.

Julitz C. 1909. Osteologie und Myologie der Extremitäten und des Wickelschwanzes vom Wickelbären, *Cercoleptes caudivolvulus*: Mit besonderer Berücksichtigung der Anpassungserscheinungen an das Baumleben. *Arch. Naturgesch. Berlin* **75**: 143–188.

Junior PS, dos Santos LMRP, Nogueira DMP, Abidu-Figueiredo M, Santos ALQ. 2015. Occurrence and morphometrics of the brachioradialis muscle in wild carnivorans (Carnivora: Caniformia, Feliformia). *Zoologia* **32**: 23–32.

Kelley EA. 1888. Notes on the myology of *Ursus maritimus*. *Proc. Acad. Nat. Sci. Philadelphia* **40**: 141–154.

Leach D. 1977. The forelimb musculature of marten (*Martes americana* Turton) and fisher (*Martes pennanti* Erxleben). *Can. J. Zool.* **55**: 31–41.

Liu M, Zack SP, Lucas L, Allen D, Fisher RE. 2016. Hind limb myology of the ringtail (*Bassariscus astutus*) and the myology of hind foot reversal. *J. Mammal.* **97**: 211–233.

Macalister A. 1873a. On the anatomy of Aonyx. Proc. R. Irish Acad. Series 2 **I**: 539–547.

Macalister A. 1873b. The muscular anatomy of the civet and tayra. *Proc. R. Irish Acad. Series 2* **I**: 506–513.

Mackintosh HW. 1875. Notes on the myology of the coati-mondi (*Nasua narica* and *N. fusca*) and common marten (*Martes foina*). *Proc. R. Ir. Acad. Series 2* **II**: 48–55.

McClearn D. 1985. Anatomy of raccoon (*Procyon lotor*) and Coati (*Nasua narica* and *N. nasua*) forearm and leg muscles: Relations between fiber length, moment-arm length, and joint-angle excursion. *J. Morphol.* **183**: 87–115.

Moore AL, Budny JE, Russell AP, Butcher MT. 2013. Architectural specialization of the intrinsic thoracic limb musculature of the American badger (*Taxidea taxus*). *J. Morphol.* **274**: 35–48.

Morales MM, Moyano SR, Ortiz AM, Ercoli MD, Aguado LI, Cardozo SA, Giannini NP. 2018. Comparative myology of the ankle of *Leopardus wiedii* and *L. geoffroyi* (Carnivora: Felidae): Functional consistency with osteology, locomotor habits and hunting in captivity. Zoology **126**: 46–57.

Murie MD. 1871. Researches upon the anatomy of Pinnipedia. Part II. Descriptive anatomy of the sea-lion (Otaria jubata). *Trans. Zool. Soc. London* **7**: 527–506.

Murphy R, Beardsley A. 1974. Mechanical properties of the cat soleus muscle in situ. *Am. J. Physiol.* **227**: 1008–1013.

Ray, LJ. 1949. The myology of the inferior extremity of the Malay bear, *Ursus malayanus*. *J. Zool.* **119**: 121–132.

Reighard J, Jennings HS. 1902. *Anatomy of the Cat*. Henry Holt and Company: New York.

Ross FO. 1876. On the myology of the cheetah, or hunting leopard of India (Felis jubata). *Proc. R. Irish Acad.* **II**: 23–32.

Sacks RD, Roy RR. 1982. Architecture of the hind limb muscle of cats: Functional significance. *J. Morphol.* **173**: 185–195.

Scharlau B. 1925. Die Muskeln der oberen Extremität einer 18 jährigen Löwin. *Z. Anat. Entwicklungsges.* **77**: 187–211.

Shepherd FJ. 1883. Short notes on the myology of the American Black Bear (*Ursus americanus*). *J. Anat. Physiol.* **18**: 103–117.

Smith JL, Edgerton VR, Betts B, Collatos TC. 1977. EMG of slow and fast ankle extensors of cat during posture, locomotion, and jumping. *J. Neurophysiol.* **40**: 503–513.

Spoor CF, Badoux DM. 1986a. Nomenclatural review of long digital forelimb flexors in carnivores. *Anat. Rec.* **216**: 471–473.

Spoor CF, Badoux DM. 1986b. Descriptive and functional myology of the neck and forelimb of the striped hyena (*Hyaena hyaena*, L.1758). Anat. Anz. **161**: 375–387.

Taverne M, Fabre A-C, Herbin M, Herrel A, Lacroux C, Lowie A, Pagès F, Peigné S, Theil J-C, Böhmer C. 2018. Convergence in the functional properties of forelimb muscles in carnivorans: Adaptations to an arboreal lifestyle? *Biol. J. Linn. Soc.* **125**: 250–263.

Taylor ME. 1974. The functional anatomy of the forelimb of some African Viverridae (Carnivora). *J. Morphol.* **143**: 307–336.

Taylor WT, Weber RJ. 1951. *Functional Mammalian Anatomy: With Special Reference to the Cat*. Van Nostrand: New York.

Viranta S, Lommi H, Holmala K, Laakkonen J. 2016. Musculoskeletal anatomy of the Eurasian lynx, *Lynx lynx* (Carnivora: Felidae) forelimb: adaptations to capture large prey? *J. Morphol.* **277**: 753–765.

Walmsley B, Hodgson JA, Burke RE. 1978. Forces produced by medial gastrocnemius and soleus muscles during locomotion in freely moving cats. *J. Neurophysiol.* **41**: 1203–1216.

Watson M. (1882) On the muscular anatomy of *Proteles* as compared with that of *Hyaena* and *Viverra*. Proc. Zool. Soc. Lond. **50**: 579–586.

Watson M, Young AH. 1879. On the anatomy of *Hyaena crocuta* (*H. maculata*). Proc. Zool. Soc. Lond. **1879**: 79–107.

Williams RC. 1955. *The Osteology and Myology of the Ranch Mink (Mustela vison)*. Cornell University Press: Ithaca.

Williams SB, Daynes J, Peckham K, Payne RC. 2008a. Functional anatomy and muscle moment arms of the thoracic limb of an elite sprinting athlete: The racing greyhound (*Canis familiaris*). *J. Anat.* **213**: 373–382.

Williams SB, Wilson AM, Rhodes L, Andrews J, Payne RC. 2008b. Functional anatomy and muscle moment arms of the pelvic limb of an elite sprinting athlete: The racing greyhound (*Canis familiaris*). *J. Anat.* **213**: 361–372.

Windle BCA. 1888. Notes on the limb myology of *Procyon cancrivorus* and of the Ursidae. *J. Anat. Physiol.* **23**: 81–89.
Windle BCA, Parsons FG. 1897. On the myology of the terrestrial Carnivora. Part I. Muscles of the head, neck, and fore-limb. *Proc. Zool. Soc. London* **65**: 370–409.
Windle BCA, Parsons FG. 1898. The myology of the terrestrial Carnivora. Part II. *Proc. Zool. Soc. Lond.* **66**: 152–186.
Young AH. 1880. Myology of *Viverra civetta*. *J. Anat. Physiol.* **14**: 166–177.
Young AH, Robinson A. 1889. On the anatomy of *Hyaena striata*. Part II. J. Anat. Physiol. **23**: 187–200.
Young RP, Scott SH, Loeb GE. 1992. An intrinsic mechanism to stabilize posture-joint-angle-dependent moment arms of the feline ankle muscles. *Neurosci. Lett.* **45**: 137–140.
Young RP, Scott SH, Loeb GE. 1993. The distal hindlimb musculature of the cat: Multiaxis moment arms at the ankle joint. *Exp. Brain Res.* **96**: 141–151.

Chiroptera

Grassée PP. 1971. Muscles de la jambe et du pied - Les Chauves-souris. In *Traitée de Zoologie Tome 17* (ed. Grassé PP), pp. 386–391. Maison et Cie: Paris.
Humphry GM. 1869. The myology of the limbs of *Pteropus*. *J. Anat. Physiol.* **3**: 294–319.
Macalister A. 1872. The myology of the Cheiroptera. *Philos. Trans. R. Soc. London* **162**: 125–171.
Mori M. 1960. Muskulatur des *Pteropus edulis*. *Okajimas Folia Anat. Jpn.* **36**: 253–307.
Morra T. 1899. I muscoli cutanei della membrana alare dei chirotteri. *Bull. Mus. Zool. Ed Anat. Comp. R. Univ. Di Torino* **XIV**: 1–7.
Norberg UM. 1972. Functional osteology and myology of the wing of the Dog-faced bat *Rousettus aegyptiacus* (É. Geoffroy) (Mammalia, Chiroptera). Z. Morphol. Tiere **73**: 1–44.
Stanchak KE, Santana SE. 2018. Assessment of the hindlimb membrane musculature of bats: Implications for active control of the calcar. *Anat. Rec.* **301**: 267–278.
Tokita M, Abe T, Suzuki K. 2012. The developmental basis of bat wing muscle. *Nat. Comm.* **3**: 1302.
Vaughan TA. 1959. Functional morphology of three bats: *Eumops*, *Myotis*, *Macrotus*. *Univ. Kansas Publ. Mus. Nat. Hist.* **12**: 1–153.
Vaughan TA. 1966. Morphology and flight characteristics of Molossid bats. *J. Mamm.* **7**: 249–260.
Vaughan TA. 1970a. The muscular system. In *Biology of Bats* Vol. 1 (ed. Wimsatt WA), pp. 139–194. Academic Press: New York.

Insectivora

Campbell B. 1939. The shoulder anatomy of the moles. A study in phylogeny and adaptation. *Am. J. Anat.* **64**: 1–39.
Castiella MJ, Laville E, Renous S, Gasc J-P. 1992. Caractéristiques morphologiques du membre antérieur de la taube commune, *Talpa europaea* (Mammalia, Talpidae). *Mammalia* **56**: 265–285.
Endo H, Oishi M, Yonezawa T, Rakotondraparany F, Hasegawa M. 2007. The semifossorial function of the forelimb in the common rice tenrec (*Oryzorictes hova*) and the streaked tenrec (*Hemicentetes hemispinosus*). *Anat., Histol., Embryol.* **36**: 413–418.
Endo H, Yonezawa T, Rakotondraparany F, Sasaki M, Hasegawa M. 2006. The adaptational strategies of the hindlimb muscles in the Tenrecidae species including the aquatic web-footed tenrec (*Limnogale mergulus*). *Ann. Anat.* **188**: 383–390.
Gaughran GRL. 1954. A comparative study of the osteology and myology of the cranial and cervical regions of the shrew, *Blarina brevicauda*, and the mole, *Scalopus aquaticus*. *Misc. Publ. Mus. Zool.* **80**: 1–82.
Gupta BB. 1965. Anatomy of the thigh and posterior axial region of two genera of Indian hedgehogs. *J. Mamm.* **46**: 210–228.
Reed CA. 1951. Locomotion and appendicular anatomy in three Soricoid insectivores. *Am. Midl. Nat.* **45**: 513–671.
Suzuki A. 1990. Composition of myofiber types in limb muscles of the house shrew (*Suncus murinus*): Lack of type 1 myofibers. *Anat. Rec.* **228**: 23–30.
Whidden HP. 2000. Comparative myology of moles and the phylogeny of the Talpidae (Mammalia, Lipotyphla). *Am. Mus. Novit.* **3294**: 1–53.
Yalden DW. 1966. The anatomy of mole locomotion. *J. Zool.* **149**: 55–64.

Lagomorpha

Barone R, Pavaux C, Blin PC, Cuq P. 1973. *Atlas d'anatomie du lapin (Atlas of Rabbit Anatomy)*. Masson et Cie: Paris.
Crabb ED. 1890. *Principles of Functional Anatomy of the Rabbit*. P. Blakiston's Son & Co., Inc.: Philadelphia.
Krause W. 1884. *Anatomie des Kaninchens in topologischer und operativer Rücksicht*. Verlang Von Wilhelm Engelmann: Leipzig.
Williams SB, Payne RC, Wilson AM. 2007. Functional specialisation of the pelvic limb of the hare (*Lepus europeus*). *J. Anat.* **210**: 472–490.
Williams SB, Wilson AM, Payne RC. 2007. Functional specialisation of the thoracic limb of the hare (*Lepus europeus*). *J. Anat.* **210**: 491–505.

Marsupialia

Abdala V, Moro S, Flores DA. 2006. The flexor tendons in the didelphid manus. *Mastozool. Neotrop.* **13**: 193–204.
Argot C. 2001. Functional-adaptive anatomy of the forelimb in the Didelphidae, and the paleobiology of the Paleocene marsupials *Mayulestes ferox* and *Pucadelphys andinus*. *J. Morphol.* **247**: 51–79.
Argot C. 2002. Functional-adaptative analysis of the hind limb anatomy of extant marsupials and the paleobiology of the Paleocene marsupials *Mayulestes ferox* and *Pucadelphis andinus*. *J. Morphol.* **253**: 76–108.
Barbour RA. 1963. The musculature and limb plexus of *Trichosurus vulpecula*. *Aust. J. Zool.* **11**: 488–610.
Boardman W. 1941. On the anatomy and functional adaptation of the thorax and pectoral girdle in the wallaroo (*Macropus robustus*). *J. Linn. Soc. NSW* **66**: 349–387.
Carlsson A. 1914. Über *Dendrolagus dorianus*. *Zool. Jahrb.* **36**: 547–617.
Carlsson A. 1926. Über den Bau des *Dasyuroides byrnei* und seine Beziehungen zu den übrigen Dasyuridae. *Acta. Zool.* **7**: 249–275.
Cheng C. 1955. The development of the shoulder region of the opossum, *Didelphis virginiana*, with special reference to musculature. *J. Morphol.* **97**: 415–471.
Coues E. 1871. The osteology and myology of *Didelphis virginiana*. *Mem. Bost. Soc. Nat. Hist.* **2**: 41–149.
Cunningham DJ. 1878a. The intrinsic muscles of the hand of the thylacine (*Thylacinus cynocephalus*), cuscus (*Phalangista maculata*), and phascogale (*Phascogale calura*). *J. Anat. Physiol.* **12**: 434–444.
Cunningham DJ. 1878b. The nerves of the fore-limb of the thylacine (*Thylacinus cynocephalus* or *harrisii*) and cuscus (*Phalangista maculata*). *J. Anat. Physiol.* **12**: 427–433.
Cunningham DJ. 1881. The nerves of the hind-limb of the thylacine (*Thylacinus harrisii* or *cynocephalus*) and cuscus (*Phalangista maculata*). *J. Anat.* **15**: 265–277.
Cunningham DJ. 1882. Report on Some Points in the Anatomy of the thylacine (Thylacinus cynocephalus), cuscus (Phalangista maculata), and phascogale (Phascogale calura), collected during the Voyage of H. M. S. Challenger in the years 1873–1876; with an account of the Comparative Anatomy of the Intrinsic Muscles and Nerves of the Mammalian Pes. Voy H M S Challenger, Zool V, pt. XVI.
Harvey KJ, Warburton NM. 2010. Forelimb musculature of kangaroos with particular emphasis on the tammar wallaby *Macropus eugenii* (Desmarest, 1817). *Aust. Mammal.* **32**: 1–9.
Harvey KJ, Warburton N. 2010. Forelimb musculature of kangaroos with particular emphasis on the tammar wallaby *Macropus eugenii* (Desmares, 1817). *Aust. Mammal.* **32**: 1–9.
Hopwood PR. 1974. The intrinsic musculature of the pectoral limb of the Eastern Grey Kangaroo (*Macropus major* (Shaw) *Macropus giganteus* (Zimm)). *J. Anat.* **118**: 445–468.
Hopwood PR, Butterfield RM. 1976. The musculature of the proximal pelvic limb of the Eastern Grey Kangaroo *Macropus major* (Shaw) *Macropus giganteus* (Zimm). *J. Anat.* **121**: 259–277.
Horiguchi M. 1981. A comparative anatomical study of the pectoral muscle group in the brindled bandicoot (*Isoodon macrourus* Gould 1842) an Australian marsupial. *Acta Anat. Nippon* **56**: 375–399.
Jenkins Jr FA, Weijs WA. 1979. The functional anatomy of the shoulder in the Virginia opossum (*Didelphis virginiana*). *J. Zool.* **188**: 379–410.
MacAlister A. 1870. On the myology of the Wombat (*Phascolomys wombata*) and the Tasmanian devil (*Sarcophilus ursinus*). *Ann. Mag. Nat. Hist.* **5**: 153–173.
Macalister A. 1872. The muscular anatomy of the koala (*Phascolarctos cinereus*). *Ann. Mag. Nat. Hist.* **10**: 127–135.
Macalister A. 1872. Further observations on the myology of *Sarcophilus ursinus*. *Ann. Mag. Nat. Hist.* **10**: 17–20.
MacCormick A. 1886a. The myology of the limbs of *Dasyurus viverrinus*. A. Myology of the forelimb. *J. Anat. Physiol.* **21**: 103–137.
MacCormick A. 1886b. The myology of the limbs of *Dasyurus viverrinus*. B. Myology of the hindlimb. *J. Anat. Physiol.* **21**: 199–226.
MacCormick A. 1887. The myology of the limbs of *Dasyurus viverrinus*. A. Myology of the fore-limb. J. Anat. Physiol. **21**: 103–137.
MacCormick A. 1887. The myology of the limbs of *Dasyurus viverrinus*. B. Myology of the hind limb. *J. Anat. Physiol.* **21**: 199–226.
Parsons FG. 1903. Anatomy of the pig-footed bandicoot (*Chaeropus castanotis*). *J. Linn. Soc. Lond.* **29**: 64–80.

Sciote JJ, Rowlerson A. 1998. Skeletal fiber types and spindle distribution in limb and jaw muscles of adult and neonatal opossum (*Monodelphis domestica*). *Anat. Rec.* **251**: 548–562.

Shrivastava RK. 1962. The deltoid musculature of the Marsupialia. *Am. Midl. Nat.* **67**: 305–320.

Sonntag CF. 1922. On the myology and classification of the Wombat, Koala, and Phalangers. *Proc. Zool. Soc. London* **92**: 863–896.

Stein BR. 1981. Comparative limb myology of two opossums, *Didelphis* and *Chironectes*. *J. Morphol.* **169:** 113–140.

Warburton NM. 2006. Functional morphology of marsupial moles (Marsupialia, Notoryctidae). *Verh. Naturwiss. Ver. Hamburg* **42**: 39–149.

Warburton NM, Harvey KJ, Prideaux GJ, O'Shea JE. 2011. Functional morphology of the forelimb of living and extinct tree-kangaroos (Marsupialia: Macropodidae). *J. Morphol.* **272**: 1230–1244.

Warburton NM, Yakovleff M, Malric A. 2012. Anatomical adaptations of the hind limb musculature of tree-kangaroos for arboreal locomotion (Marsupialia: Macropodinae). *Aust. J. Zool.* **60**: 246–258.

Warburton NM, Bateman PW, Fleming PA. 2013. Sexual selection on forelimb muscles of western grey kangaroos (Skippy was clearly a female). *Biol. J. Linn. Soc.* **109**: 923–931.

Warburton NM, Gregoire L, Jacques S, Flandrin C. 2014. Adaptations for digging in the forelimb muscle anatomy of the southern brown bandicoot (*Isoodon obesulus*) and bilby (*Macrotis lagotis*). *Aust. J. Zool.* **61**: 402–419.

Warburton NM, Malric A, Yakovleff M, Leonard V, Cailleau C. 2015. Hind limb myology of the southern brown bandicoot (*Isoodon obesulus*) and greater bilby (*Macrotis lagotis*) (Marsupialia: Peramelemorphia). *Aust. J. Zool.* **63**: 147–162.

Warburton NM, Marchal C-R. 2017. Forelimb myology of carnivorous marsupials (Marsupialia: Dasyuridae): Implications fort he ancestral body plan oft he Australodelphia. *Anat. Rec.* **300**: 1589–1608.

Wilson JT. 1894. On the myology of *Notoryctes typhlops* with comparative notes. *Trans. R. Soc. S. Aust.* **18**: 3–74.

Young AH. 1879. The intrinsic muscles of the marsupial hand. *J. Anat. Physiol.* **14**: 149–167.

Young AH. 1882. The muscular anatomy of the Koala (*Phascolarctos cinereus*): with additional notes. *J. Anat. Physiol.* **16**: 217–242.

Monotremata

Gambaryan PP, Aristov AA, Dixon JM, Zubtsova GY. 2002. Peculiarities of the hind limb musculature in monotremes: An anatomical description and functional approach. *Russ. J. Theriol.* **1**: 1–36.

Gambaryan PP, Kuznetsov AN, Panyutina AA, Gerasimov SV. 2015. Shoulder and forelimb myology of extant Monotremata. *Russ. J. Theriol.* **14**: 1–56.

Howell AB. 1936. The musculature of antebrachium and manus in the platypus. *Am. J. Anat.* **59**: 425–432.

Low JW. 1929. Contributions to the development of the pelvic girdle. III. The pelvic girdle and its related musculature in Monotremes. *Proc. Zool. Soc. London* **99**: 245–265.

Mivart G. 1866. On some points in the anatomy of *Echidna hystrix*. *Trans. Linn. Soc. London* **25**: 379–403.

Pearson HS. 1926. Pelvic and thigh muscles of *Ornithorhynchus*. *J. Anat.* **60**: 152–163.

Walter LR. 1988. Appendicular musculature in the Echidna, *Tachyglossus aculeatus* (Monotremata: Tachyglossidae). *Aust. J. Zool.* **36**: 65–81.

Perissodactyla

Beddard FE, Treves F. 1889. On the anatomy of *Rhinoceros sumatrensis*. *Proc. Zool. Soc. Lond.* **1**: 7–25.

Budras K-D, Sack WO, Röck S. 2009. *Anatomy of the Horse*. Schlütersche: Hannover.

Sisson S. 1975. Equine myology. In *Sisson and Grossman's the Anatomy of the Domestic Animals* (ed. Getty R), pp. 376–453. W.B. Saunders Company: Philadelphia.

Windle BCA, Parsons FG. 1901. On the muscles of the Ungulata. Part I. Muscles of the head, neck, and fore-limb. *Proc. Zool. Soc. Lond.* **2**: 656–704.

Pholidota

Kawashima T, Thorington RW, Jr., Bohaska PW, Chen YJ, Sato F. 2015. Anatomy of shoulder girdle muscle modifications and walking adaptation in the scaly Chinese pangolin (*Manis pentadactyla pentadactyla*: Pholidota) compared with the partially osteoderm-clad armadillos (Dasypodidae). *Anat. Rec. (Hoboken)* **298**: 1217–1236.

Primates

Anapol F, Barry K. 1996. Fiber architecture of the extensors of the hindlimb in semiterrestrial and arboreal guenons. *Am. J. Phys. Anthropol.* **99**: 429–447.

Anemone RL. 1993. The functional anatomy of the hip and thigh in primates. In *Postcranial Adaptation in Nonhuman Primates* (ed. Gebo DL), pp. 150–74. Northern Illinois University Press: DeKalb, IL.

Ankel-Simons F. 2007. *An Introduction to Primate Anatomy.* Academic Press: San Diego.

Ashton EH, Oxnard CE. 1963. The musculature of the primate shoulder. *Trans. Zool. Soc. London* **29**: 553–650.

Aversi-Ferreira TA, Diogo R, Potau JM, Bello G, Pastor JF, Ashraf Aziz M. 2010. Comparative anatomical study of the forearm extensor muscles of *Cebus libidinosus* (Ryland's et al., 200 Primates, Cebidae), modern humans and other primates with comments on primate evolution, phylogeny and manipulatory behavior. Anat. Rec. **293**: 256–2070.

Aversi-Ferreira TA, Aversi-Ferreira RA, Bretas RV, Nishimaru H, Nishijo H. 2016. Comparative anatomy of the arm muscles of the Japanese monkey (*Macaca fuscata*) with some comments on locomotor mechanics and behavior. *J. Med. Primatol.* **45**: 165–179.

Carlson KJ. 2006. Muscle architecture of the common chimpanzee (*Pan troglodytes*): Perspectives for investigating chimpanzee behavior. *Primates* **47**: 218–229.

Channon AJ, Crompton RH, Gu MM. 2010. Muscle moment arms of the gibbon hind limb: Implications for hylobatid locomotion. *J. Anat.* **216**: 446–462.

Cheng EJ, Scott SH. 2000. Morphometry of *Macaca mulatta* Forelimb. I. Shoulder and elbow muscles and segment inertial parameters. J. Morphol. **245**: 206–224.

Diogo R, Wood B. 2011. Soft-tissue anatomy of the primates: Phylogenetic analyses based on the muscles of the head, neck, pectoral region and upper limb, with notes on the evolution of these muscles. *J. Anat.* **219**: 273–359.

Diogo R, Wood BA. 2012. *Comparative Anatomy and Phylogeny of Primate Muscles and Human Evolution.* CRC Press: Boca Raton.

Diogo R, Richmond BG, Wood B. 2012. Evolution and homologies of primate and modern human hand and forearm muscles, with notes on thumb movements and tool use. *J. Hum. Evol.* **63**: 64–78.

Diogo R, Potau JM, Pastor JF. 2013. *Photographic and Descriptive Musculoskeletal Atlas of Chimpanzees (Pan): With Notes on the Attachments, Variations, Innervation, Function and Synonymy and Weight of the Muscles.* Taylor & Francis: Boca Raton.

Diogo R, Pastor JF, Hartstone-Rose A, Muchlinski MN. 2014. *Baby Gorilla: Photographic and Descriptive Atlas of Skeleton, Muscles and Internal Organs Including CT Scans and Comparison with Adult Gorillas, Humans and Other Primates.* Taylor & Francis: Oxford, UK. p. 101.

Diogo R, Shearer B, Potau JM, Pastor JF, de Paz FJ, Arias-Martorell J, Turcotte C, Hammond A, Vereecke E, Vanhoof M, Nauwelaerts S, Wood B. 2017. *Photographic and Descriptive Musculoskeletal Atlas of Bonobos.* Springer: Cham.

Fitts RH, Bodine SC, Romatowski JG, Widrick JJ. 1998. Velocity, force, power, and Ca2+ sensitivity of fast and slow monkey skeletal muscle fibers. *J. Appl. Physiol.* **84**: 1776–1787.

Freeman RA. 1886. Anatomy of the shoulder and upper arm of the mole (*Talpa europoea*). *J. Anat. Physiol.* **20**: 201–219.

Forster A. 1934. La "pince palmaire" et la "pince plantaire" de Perodicticus potto. Archives d'anatomie, d'histologie et d'embryologie 17.

Gebo DL. 1987. Functional anatomy of the tarsier foot. *Am. J. Phys. Anthropol.* **73**: 9–31.

Gyambibi A, Lemelin P. 2013. Comparative and quantitative myology of the forearm and hand of prosimian primates. *Anat. Rec.* **296**: 1196–1206.

Jacobi U. 1966. Die Muskulatur des Unterarmes und der Hand bei *Macaca mulutta. Z. Morphol. Anthropol.* **58**: 48–73.

Jacobson MD, Raab R, Fazeli BM, Abrams RA. 1992. Architectural design of the human intrinsic hand muscles. *J. Hand Surg.* **17**: 804–809.

Jouffroy FK. 1975. Osteology and myology of the lemuriform postcranial skeleton. In *Lemur Biology* (ed. Tattersall I), pp. 149–192. Springer: New York.

Jouffroy FK. 1962. *La Musculature des membres chez les Lémuriens de Madagascar.* Mammalia: Paris.

Kanagasuntheram R, Jayawardene FLW. 1957. The intrinsic muscles of the hand in the slender loris. *Proc. Zool. Soc. Lond.* **128**: 301–312.

Kikuchi Y. 2009. Comparative analysis of muscle architecture in primate arm and forearm. *Anat. Histol. Embryol.* **39**: 93–106.

Kikuchi Y, Kuraoka A. 2014. Differences in muscle dimensional parameters between non-formalin-fixed (freeze-thawed) and formalin-fixed specimens in gorilla (*Gorilla gorilla*). *Mamm. Study* **39**: 65–72.

Langdon JH. 1990. Variations in cruropedal musculature. *Int. J. Primatol.* **11**: 575–606.

Larson SG. 1988. Subscapularis function in gibbons and chimpanzees: Implications for interpretation of humeral head torsion in hominoids. *Am. J. Phys. Anthropol.* **76**: 449–462.

Leischner C, Crouch M, Allen K, Marchi D, Pastor F, Hartstone-Rose A. 2018. Scaling of primate forearm muscle architecture as it relates to locomotion and posture. *Anat. Rec.* **301**: 484–495.

Lemelin P, Diogo R. 2016. Anatomy, function, and evolution of the primate hand musculature In *Evolution of the Primate Hand* (ed. Kivell TL, Lemelin P, Richmond BG, Schmitt D), pp. 155–193. Berlin: Springer Verlag.

Marchi D, Leischner CL, Pastor F, Hartstone-Rose A. 2018. Leg muscle architecture in primates and its correlation with locomotion patterns. *Anat. Rec.* **301**: 515–527.

Michilsens F, Vereecke EE, D'Aout K, Aerts P. 2009. Functional anatomy of the gibbon forelimb: Adaptations to a brachiating life-style. *J. Anat.* **215**: 335–354.

Miller RA. 1932. Evolution of the pectoral girdle and fore limb in the primates. *Am. J. Phys. Anthropol.* **17**: 1–56.

Miller RA. 1943. Functional and morphological adaptations in the forelimbs of the slow lemurs. *Am. J. Anat.* **73**: 153–183.

Miller RA. 1952. The musculature of *Pan paniscus*. *Dev. Dyn.* **91**: 183–232.

Molnar J, Esteve-Altava B, Rolian C, Diogo R. 2017. Comparison of musculoskeletal networks of the primate forelimb. *Sci. Rep.* **7**: 10520.

Morton DJ. 1924. The peroneus tertius muscle in gorillas. *Anat. Rec.* **27**: 323–328.

Murie J, Mivart STG. 1869. On the anatomy of the Lemuroidea. *Trans. Zool. Soc. Lond.* **7**: 1–113.

Myatt JP, Crompton RH, Thorpe SKS. 2011. Hindlimb muscle architecture in non-human great apes and a comparison of methods for analyzing inter-species variation. *J. Anat.* **219**: 150–166.

Oishi M, Ogihara N, Endo H, Ichihara N, Asari M. 2009. Dimensions of forelimb muscles in orangutans and chimpanzees. *J. Anat.* **215**: 373–382.

Payne RC, Crompton RH, Isler K, Savage R, Vereecke EE, Günther MM, Thorpe SKS, D'Août K. 2006. Morphological analysis of the hindlimb in apes and humans. I. Muscle architecture. *J. Anat.* **208**: 709–724.

Payne RC, Crompton RH, Isler K, Savage R, Vereecke EE, Günther MM, Thorpe SKS, D'Août K. 2006. Morphological analysis of the hindlimb in apes and humans. II. Moment arms. *J. Anat.* **208**: 725–742.

Stern JT. 1971. *Functional Myology of the Hip and Thigh of Cebid Monkeys and its Implications for the Evolution of Erect Posture*. S. Karger Basel: New York.

Stern JTJ, Wells JP, Vangor AK, Fleagle JG. 1977. Electromyography of some muscles of the upper limb in *Ateles* and *Lagothrix*. *Yrbk. Phys. Anthropol.* **20**: 498–507.

Swindler DR, Wood CD. 1982. *An Atlas of Primate Gross Anatomy: Baboon, Chimpanzee, and Man*. Krieger Publishing: New York, NY.

Thorpe SKS, Crompton RH, Günther MM, Ker RF, McNeill Alexander R. 1999. Dimensions and moment arms of the hind- and forelimb muscles of common chimpanzees (*Pan troglodytes*). *Am. J. Phys. Anthropol.* **110**: 179–199.

Thorpe SKS, Crompton RH, Wang WJ. 2004. Stresses exerted in the hindlimb muscles of common chimpanzees (*Pan troglodytes*) during bipedal locomotion. *Folia Primatol.* **75**: 253–265.

Tuttle R. 1972. Relative mass of cheridial muscles in catarrhine primates. In *Functional and Evolutionary Biology of Primates* (ed. Tuttle R), pp. 262–291. Wenner-Gren: Chicago.

Tuttle RH. 1969. Quantitative and functional studies on the hands of the anthropoidea. I. The hominoidea. *J. Morphol.* **128**: 309–363.

Van Campen F, Van der Hoeven J. 1859. Ontleedkundig onderzoek van den potto van Bosman. Koninklijke Akademie van Wetenschappen.

Youlatos D. 2000. Functional anatomy of forelimb muscles in Guianan Atelines (Platyrrhini: Primates). *Ann. Sci. Nat.* **21**: 137–151.

Rodentia

Beddard FE. 1891. Notes on the anatomy of *Dolichotis patagonica*. *Proc. Zool. Soc. Lond.* **1891**: 236–244.

Bezuidenhout AJ, Evans HE. 2005. *Anatomy of the Woodchuck (Marmota monax)*. Allen Press: Lawrence. p. 180.

Candela AM, Picasso MBJ. 2008. Functional anatomy of the limbs of Erethizontidae (Rodentia, Caviomorpha): Indicators of locomotor behavior in Miocene porcupines. *J. Morphol.* **269**: 552–593.

Carrizo LV, Tulli MJ, Dos Santos DA, Abdala V. 2014. Interplay between postcranial morphology and locomotor types in Neotropical sigmodontine rodents. *J. Anat.* **224**: 469–481.

Charles JP, Cappellari O, Spence AJ, Hutchinson JR, Wells DJ. 2016. Musculoskeletal geometry, muscle architecture and functional specialisations of the mouse hindlimb. *PLoS One* **11**: e0147669.

Fry JF. 1961. Musculature and innervation of the pelvis and hind limb of the mountain beaver. *J. Morphol.* **109**: 173–197.

Gambaryan PP, Gasc J-P. 1993. Adaptive properties of the musculoskeletal system in the mole-rat *Myospalax myospalax* (Mammalia, Rodentia), cinefluographical, anatomical, and biomechanical analyses of burrowing. *Zool. Jb. Anat.* **123**: 363–401.

Gambaryan PP, Zherebtsova OV. 2014. Short muscles of the hand and foot in *Laonastes aenigmamus* (Rodentia: Diatomyidae) and some other rock-dwellers. *Russ. J. Theriol.* **13**: 83–95.

Gambaryan PP, Zherebtsova OV, Perepelova AA. 2013. Comparative analysis of forelimb musculature in *Laonastes aenigmamus* (Rodentia: Diatomyidae). *Proc. Zool. Inst. RAS* **317**: 226–245.

García-Esponda CM, Candela AM. 2010. Anatomy of the hind limb musculature in the cursorial caviomorph *Dasyprocta azarae* Lichtenstein, 1823 (Rodentia, Dasyproctidae): Functional and evolutionary significance. *Mammalia* **74**: 407–422.

Garcia-Esponda CM, Candela AM. 2015. The hip adductor muscle group in caviomorph rodents: Anatomy and homology. *Zoology* **118**: 203–212.

Garcia-Esponda CM, Candela AM. 2016. Hindlimb musculature of the largest living rodent *Hydrochoerus hydrochaeris* (Caviomorpha): Adaptations to semiaquatic and terrestrial styles of life. *J. Morphol.* **277**: 286–305.

Hill JE. 1937. Morphology of the pocket gopher mammalian genus *Thomomys*. *Univ. Calif. Publ. Zool.* **42**: 81–172.

Howell AB. 1926. Anatomy of the wood rat. Comparative anatomy of the subgenera of the American wood rat (Genus Neotoma). *Monogr. Amer. Soc. Mammal.* **1**: 1–225.

Kawashima T, Thorington RW, Jr, Bohaska PW, Sato F. 2017. Evolutionary transformation of the palmaris longus muscle in flying squirrels (Pteromyini: Sciuridae): An anatomic consideration of the origin of the uniquely specialized styliform cartilage. *Anat. Rec.* **300**: 340–352.

Klingener D. 1964. The comparative myology of four Dipodoid rodents (Genera *Zapus*, *Napaeozapus*, *Sicista*, and *Jaculus*). *Misc. Publ. Mus. Zool.* **124**: 1–100.

Lehmann WH. 1963. The forelimb architecture of some fossorial rodents. *J. Morphol.* **113**: 59–76.

Luff AR. 1981. Dynamic properties of the inferior rectus, extensor digitorum longus, diaphragm and soleus muscles of the mouse. *J. Physiol.* **313**: 161–171.

McEvoy JS. 1982. Comparative myology of the pectoral and pelvic appendages of the North American porcupine (*Erethizon dorsatum*) and the prehensile-tailed porcupine (*Coendou prehensilis*). *Bull. Am. Mus. Nat. Hist.* **173**: 337–421.

Mivart G, Murie J. 1866. On the anatomy of the crested agouti (*Dasyprocta cristata*, Desm.). Proc. Zool. Soc. Lond. **1866**: 383–417.

Parsons FG. 1894. On the myology of the sciuromorphine and histricomorphine rodents. *Proc. Zool. Soc. Lond.* **1894**: 251–296.

Peterka HE. 1936. A study of the myology and osteology of tree sciurids with regard to adaptation to arboreal, glissant and fossorial habits. *Trans. Kans. Acad. Sci.* **39**: 313–332.

Powell PL, Roy RR, Kanim P, Bello MA, Edgerton VR. 1984. Predictability of skeletal muscle tension from architectural determinations in guinea pig hindlimbs. *J. Appl. Physiol.* **57**: 1715–1721.

Rinker GC. 1954. The comparative myology of the mammalian genera *Sigmodon*, *Oryzomys*, *Neotoma*, and *Peromyscus* (Cricetinae), with remarks on their intergeneric relationships. *Misc. Publ. Mus. Zool.* **83**: 1–124.

Rocha-Barbosa O, Loguercio MFC, Renous S, Gasc J-P. 2007. Comparative study on the forefoot and hindfoot intrinsic muscles of some cavioidea rodents (Mammalia, Rodentia). *Zoology* **110**: 58–65.

Rupert JE, Joll JE, Elkhatib WY, Organ JM. 2018. Mouse hind limb skeletal muscle functional adaptation in a simulated fine branch arboreal habitat. *Anat. Rec.* **301**: 434–440.

Ryan JM. 1989. Comparative myology and phylogenetic systematics of the Heteromyidae (Mammalia, Rodentia). *Misc. Publ. Mus. Zool. Univ. Mich.* **176**: 1–103.

Stalheim-Smith A. 1984. Comparative study of the forelimbs of the semifossorial prairie dog, *Cynomys gunnisoni*, and the scansorial fox squirrel, *Sciurus niger*. *J. Morphol.* **180**: 55–68.

Stein BR. 1988. Morphology and allometry in several genera of semiaquatic rodents (*Ondatra*, *Nectomys*, and *Oryzomys*). *J. Mammal.* **69**: 500–511.

Stein BR. 1986. Comparative limb myology of four Arvicolid rodent genera (Mammalia, Rodentia). *J. Morphol.* **187**: 321–342.

Stein BR. 1990. Limb myology and phylogenetic relationships in the superfamily Dipodoidea (birch mice, jumping mice, and jerboas). *Z. Zool. Syst. Evolutionsforsch.* **28**: 299–314.

Thorington RW, Jr., Darrow K, Betts AD. 1997. Comparative myology of the forelimb of squirrels (Sciuridae). *J. Morphol.* **234**: 155–182.

Windle BCA. 1897. On the myology of *Dolichotis patagonica* and *Dasyprocta isthmica*. *J. Anat.* **31**: 343–353.

Wood AE, White RR. 1950. The myology of the chinchilla. *J. Morphol.* **86**: 547–597.

Woods CA. 1972. Comparative myology of jaw, hyoid, and pectoral appendicular regions of new and old world hystricomorph rodents. *Bull. Am. Mus. Nat. Hist.* **147**: 117–198.

Scandentia

Carlsson A. 1922. Über die Tupaiidae und ihre Beziehungen zu den Insectivora und den Prosimiae. *Acta Zool.* **3**: 227–270.

George RM. 1977. The limb musculature of the Tupaiidae. *Primates* **18**: 1–34.

Le Gros Clark WE. 1924. The myology of the tree-shrew (*Tupaia minor*). *Proc. Zool. Soc.* **31**: 497–497.

Sargis EJ. 2002. Functional morphology of the hindlimb of tupaiids (Mammalia, Scandentia) and its phylogenetic implications. *J. Morphol.* **254**: 149–185.

Xenarthra

Galton JC. 1868. The myology of the upper and lower extremities of Orycteropus capensis. *Trans. Linn. Soc. London* **26**: 567–608.

Mendel FC. 1981. Foot of two-toed sloths: Its anatomy and potential uses relative to size of support. *J. Morphol.* **170**: 357–372.

Mendel FC. 1981. The hand of two-toed sloths (*Choloepus*): Its anatomy and potential uses relative to size of support. *J. Morphol.* **169**: 1–19.

Miller RA. 1935. Functional adaptations in the forelimb of the sloths. *J. Mammal.* **16**: 38–51.

Nyakatura JA, Fischer MS. 2011. Functional morphology of the muscular sling at the pectoral girdle in tree sloths: convergent morphological solutions to new functional demands? *J. Anat.* **219**: 360–374.

Olson RA, Glenn ZD, Cliffe RN, Butcher MT. 2017. Architectural properties of sloth forelimb muscles (Pilosa: Bradypodidae). *J. Mamm. Evol.* **25**: 573–588.

Olson RA, Womble MD, Thomas DR, Glenn ZD, Butcher MT. 2016. Functional morphology of the forelimb of the Nine-Banded armadillo (*Dasypus novemcinctus*): Comparative perspectives on the myology of Dasypodidae. *J. Mamm. Evol.* **23**: 49–69.

Taylor BK. 1978. The anatomy of the forelimb in the anteater (*Tamandua*) and its functional implications. *J. Morphol.* **157**: 347–367.

Taylor BK. 1985. Functional anatomy of the forelimb in vermilinguas (anteaters). In *The Evolution and Ecology of Armadillos, Sloths, and Vermilinguas* (ed. Montgomery GG), pp. 151–171. Smithsonian Institution Press: Washington.

Toledo N, Bargo MS, Vizcaino SF. 2013. Muscular reconstruction and functional morphology of the forelimb of early Miocene sloths (Xenarthra, Folivora) of Patagonia. *Anat. Rec. (Hoboken)* **296**: 305–325.

Toledo N, Bargo MS, Vizcaino SF. 2015. Muscular reconstruction and functional morphology of the hind limb of santacrucian (Early Miocene) sloths (Xenarthra, Folivora) of Patagonia. *Anat. Rec. (Hoboken)* **298**: 842–864.

Vassal PA, Jouffroy FK, Lessertisseur J. 1962. Musculature de la main et du pied du paresseux aï (*Bradypus tridactylus* L.). *Folia Clin. Biol.* **31**: 142–153.

Index

E

F

G

H

Taylor & Francis Group
an **informa** business

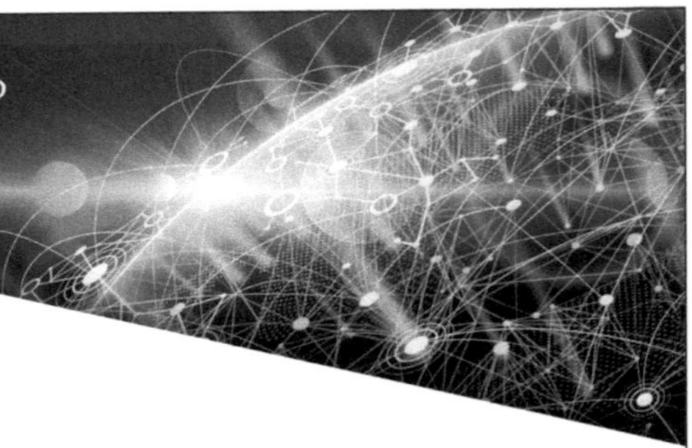

Taylor & Francis eBooks

www.taylorfrancis.com

A single destination for eBooks from Taylor & Francis with increased functionality and an improved user experience to meet the needs of our customers.

90,000+ eBooks of award-winning academic content in Humanities, Social Science, Science, Technology, Engineering, and Medical written by a global network of editors and authors.

TAYLOR & FRANCIS EBOOKS OFFERS:

- A streamlined experience for our library customers
- A single point of discovery for all of our eBook content
- Improved search and discovery of content at both book and chapter level

REQUEST A FREE TRIAL

support@taylorfrancis.com

Routledge
Taylor & Francis Group

CRC Press
Taylor & Francis Group